A Handbook of Piping

For Plumbing, Irrigation, Heating Systems, Steam Power and other uses

by Carl L. Svenson

with an introduction by Roger Chambers

This work contains material that was originally published in 1918.

This publication is within the Public Domain.

*This edition is reprinted for educational purposes
and in accordance with all applicable Federal Laws.*

Introduction Copyright 2017 by Roger Chambers

Self Reliance Books

Get more historic titles on animal and stock breeding, gardening and old fashioned skills by visiting us at:

http://selfreliancebooks.blogspot.com/

Introduction

I am pleased to present yet another title on Homesteading and Farm Life.

The work is in the Public Domain and is re-printed here in accordance with Federal Laws.

As with all reprinted books of this age that are intended to perfectly reproduce the original edition, considerable pains and effort had to be undertaken to correct fading and sometimes outright damage to existing proofs of this title. At times, this task is quite monumental, requiring an almost total "rebuilding" of some pages from digital proofs of multiple copies. Despite this, imperfections still sometimes exist in the final proof and may detract from the visual appearance of the text.

I hope you enjoy reading this book as much as I enjoyed making it available to readers again.

Roger Chambers

PREFACE

There are many things which every engineer is assumed to know about piping, but the sources of such information are not always so readily available as to justify this assumption. In designing some pieces of work requiring the use of piping, the designer has often been under the necessity of searching through collections of catalogs, handbooks, and even fittings themselves, perhaps without finding the details desired. The inconvenience and loss of time resulting from the lack of a ready source of information regarding the use of pipe and its accessories would seem to justify the publication of a book devoted to it.

This work is thus offered for the purpose of supplying in convenient form information and data regarding piping, fittings, pipe joints, valves, piping drawings, and pipe lines and their accessories. It is hoped that the variety and extent of the tables, illustrations and formulae will be sufficient to make it of value to both engineers and students. The tables have been prepared with care, and are all uniform in arrangement, to facilitate their use. In the case of tables of sizes the names of the different companies have been given, which it is believed will add to their value. The illustrations have all been especially drawn for the book. A list of books and references is given in Chapter XIX with a view to extending the usefulness of this work.

Various authorities have been consulted, and no claim for orginality can be made for the substance of the information thus obtained, but it is hoped that the form of presentation will commend itself.

The author wishes to express his appreciation of the complete and valuable responses with which his inquiries were met by the

PREFACE

companies and individuals mentioned in the text, and in particular the services of Prof. Thomas E. French and Mr. W. J. Norris.

Suggestions and criticisms will be welcomed by both publishers and author.

CARL L. SVENSEN

COLUMBUS, OHIO.
April 8, 1917

CONTENTS

	PAGE
PREFACE	iii

CHAPTER I

PIPE 1
    Historical — Wrought Iron and Steel — Briggs Standard — Outside Diameter Pipe — Manufacture of Steel Pipe — Cast Iron — Copper — Brass — Lead — Riveted Pipe — Strength of Materials.

CHAPTER II

DIMENSIONS AND STRENGTH OF PIPE 11
    General Formula — Formulae for Cast Iron — Cast Iron Cylinder Tests — Cast Iron Hub and Spigot Pipe — Plain Cast Iron Pipe — Briggs Standard Dimensions — Bursting Pressures of Pipe — — Mill Tests — English Pipe — Riveted Pipe — Copper and Brass Pipe — Lead Pipe — Wooden Stave Pipe.

CHAPTER III

PIPE THREADS 35
    American Pipe Threads — Standard Pipe Thread Gages — Pipe Threading — Pipe Tools — English Pipe Threads — Foreign Pipe Threads.

CHAPTER IV

PIPE FITTINGS 44
    Screw Fittings — Couplings — Elbows — Tees, Crosses, Bushings, Caps, Plugs — Nipples — Cast Iron Fittings — Screwed Reducing Fittings — Brass Fittings — Malleable Iron Fittings — Extra Heavy Cast Steel Screwed Fittings — Strength of Fittings — Flanged Fittings — Reducing Fittings — Cast Steel Fittings — Ammonia Fittings — British Standard Pipe Flanges and Fittings.

CHAPTER V

PIPE JOINTS 76
    Welded Joints — Screw Unions — Flange Unions — Bolt Circles and Drillings — Flange Facing — Flange Joints for Steel Pipe —

CONTENTS

Pipe Flange Tables — Special Connections — Converse Joints — Matheson Joints — Flanges for Copper Pipe — Lead Pipe Joints — Joints for Riveted Pipe — Joints for Cast Iron Pipe.

CHAPTER VI

STANDARD VALVES 98
Valves — Materials — Globe and Gate Valves — Valve Seats — Gate Valves — By-Pass Valves — Valve Stem Arrangements — Strength of Gate Valves — Standard Pressures and Dimensions — Check Valves — Operation of Valves — Location.

CHAPTER VII

SPECIAL VALVES 114
Butterfly Valves — Blow-off Valves — Plug Valves — Boiler Stop Valves — Foster Automatic Valve — Emergency Stop Valves — Crane-Erwood Automatic Valve — Reducing Valves — Reducing Valve Sizes — Pump Governors — Back Pressure Valves — Automatic Exhaust Relief Valves — Safety Valves — Installation of Pop Safety Valves — Extracts from Report of American Society of Mechanical Engineers' Boiler Code Committee.

CHAPTER VIII

STEAM PIPING 137
General Considerations — Header System — Direct System with Cross-over Header — Ring System — Duplicate System — Steam Velocity — Size of Pipe — Equalization of Pipes — Superheated Steam — Effect of High Temperature on Metals and Alloys — Live Steam Header — Connections between Boiler and Header — Pipe Lines from Main Header — Auxiliary and Small Steam Lines for Engines, Pumps, etc. — Steam Loop — Injector Piping — Live Steam Feed Water Purifier — Method of Piping Purifier — Water Column Piping — The Placing of Thermometers in Pipes — Steam Gages.

CHAPTER IX

DRIP AND BLOW-OFF PIPING 161
Drainage — Separators — Drip Pockets — Steam Traps — Drips from Steam Cylinders — Drainage Fittings — Automatic Pump and Receiver — Blow-Off Piping.

CHAPTER X

EXHAUST PIPING AND CONDENSERS 172
Exhaust Piping — Exhaust from Small Engines, Pumps, etc. — Exhaust Heads — Vacuum Exhaust Pipes — Classes of Condensers — Surface Condensers — Piping for Surface Condenser — Jet Con-

CONTENTS

densers — Jet Condenser Piping — Barometric Condenser — Piping for Barometric Condenser — Multi-jet Educator Condenser.

CHAPTER XI

FEED WATER HEATERS 188
Uses and Types of Heaters — Closed Feed Water Heaters — Closed Heater Piping — Open Feed Water Heaters — Open Heater Piping.

CHAPTER XII

PIPING FOR HEATING SYSTEMS 201
Piping for Heating Systems — Steam Heating Piping Systems — Steam Radiator Pipe Connections — Sizes of Steam Heating Pipes — Hot Water Heating Systems — Expansion Tanks — Hot Water Radiator Pipe Connections — Sizes of Hot Water Pipes — Exhaust Steam Heating — The Webster Vacuum System of Steam Heating — Radiator Pipe Connections — Typical Arrangement Webster Systems — Atmospheric System of Steam Heating — Central Station Heating — Underground Steam Mains — Underdrainage — Installation in Wood Casings — Expansion and Contraction — Interior Piping for Central Station Heat.

CHAPTER XIII

WATER AND HYDRAULIC PIPING 226
Water Piping — Gravity Pipe Lines — Flow of Water in Pipes — Pump Suction Piping — Pump Discharge Piping — Boiler Feed Piping — Interior Water Piping — Hydraulic Pipe and Fittings — Hydraulic Valves.

CHAPTER XIV

COMPRESSED AIR, GAS AND OIL PIPING 237
Compressed Air Piping — Compressed Air Transmission — The Air Lift Pumping System — Gas Fitting — Materials — Location of Piping — Sizes of Pipes — Testing — Gas Meters — Gas Piping Specifications — Pressure Test — Obstructions and Jointing — Slope of Piping — Protection of Piping — Outlets — Gas Engine Connection — Explanation of Piping Schedule — Use of Piping Schedule — Plan of Piping — Stems — Arms — General — Oil Piping — Oil Piping for Lubrication — Richardson Individual Oiling System — Phenix Individual Oiling System — Oil Pipe Fittings — Oil Piping Drawing — Sight Feed Lubricator Connections — Oil Fuel Piping.

CHAPTER XV

ERECTION — WORKMANSHIP — MISCELLANEOUS 269
Handling Pipe — Putting Up Pipe — Pipe Dopes — Gaskets —

CONTENTS

Valves — Vibration and Support — Expansion — Pipe Bends — Bending Pipe — Nozzles — Pipe Saddles — Supporting Large Thin Pipe — Flexible Metal Hose — Aluminum Piping and Tubing — Brass and Copper Tubing — Boiler Tubes — Color System to Designate Piping.

CHAPTER XVI

PIPING INSULATION 289
    Pipe Coverings — Tests on Pipe Coverings — Low Pressure Steam, Hot and Cold Water Pipes — Cold Pipes — Forms of Pipe Coverings — Underground Piping — Out-of-Doors Piping.

CHAPTER XVII

PIPING DRAWINGS 306
    Classification of Piping Drawings — Erection Drawings — Conventional Representation — Dimensioning — Flanges — Coils — Sketching — Developed or Single Plane Drawings — Isometric Drawing — Oblique Drawings.

CHAPTER XVIII

SPECIFICATIONS 329
    Standard Piping Schedule — Standard Specifications (Stone & Webster) — Model Specifications (Walworth).

CHAPTER XIX

LIST OF BOOKS AND REFERENCES 347

INDEX 353

APPENDIX

    Plate 1 — Main Steam Lines — Plan.
    Plate 2 — Main Steam Lines — Elevations.
    Plate 3 — Auxiliary Exhaust Lines — Plan.
    Plate 4 — Auxiliary Exhaust Lines — Elevations.
    Plate 5 — Boiler Feed Lines — Plan.
    Plate 6 — Boiler Feed Lines — Elevation.
    Plate 7 — Boiler Blow-Off Lines.
    Plate 8 — Heater Suction and City Water Lines.

A HANDBOOK ON PIPING

CHAPTER I

PIPE

Historical. — All branches of engineering involve the conveying of fluids — gas, air, water, etc. For this purpose pipes made of various materials are used. Wood was probably one of the first piping materials, and a piece of early wood piping is shown in Fig. 1. Pipes made of hollow hemlock logs were used with the first waterworks constructed in America, at Boston, Massachusetts, in 1652.

In tropical countries bamboo tubes are used for conveying water short distances and it is likely that the practice dates from

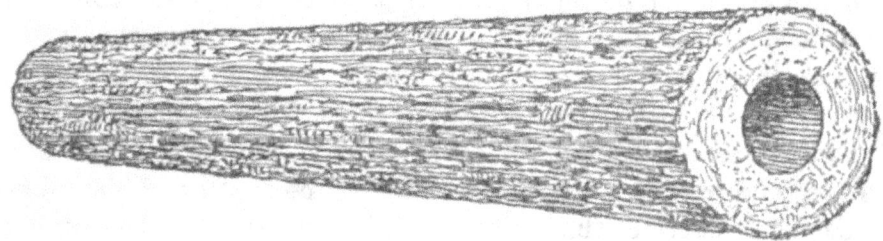

Fig. 1. A Piece of Wood Piping.

ancient times. Tubes made of pottery have been found in prehistoric ruins and lead pipes were in use as early as the first century A.D. Wrought iron tubes were first made for gun barrels. The method employed was to bend an iron plate to form a skelp. A smith then welded the edges of the red hot metal piecemeal by hammering over a rod. Machinery for welding tubes was patented in 1812 by an Englishman named Osborn. For conveying gas for lighting purposes old gun barrels were screwed together to form the first continous pipes. In 1824 James Russell filed a "specification for an improvement in the manufacture of tubes for gas and other purposes," by which the weld could be formed either with or without a mandrel, and the edges butted against each other. The basis of the present process was invented by Cornelius Whitehouse in 1825. Between 1830 and 1834 the first butt-welding furnace in the United States was built by Morris, Tasker and Morris in Philadelphia. In 1849 Walworth & Nason built the Wanalancet Iron & Tube Works at Malden, Mass., of which Robert Briggs was construction engineer. Other early pipe mills were those of Griffith Brothers, Allison & Company, and Girard Tube Company, Philadelphia, and Seyfert, McManus & Company, Reading, Pa.

Materials ordinarily used for pipe are clay, cement, cast iron, wrought iron or steel, steel plate, brass, copper, lead, lead lined and tin lined iron or steel.

Wrought Iron and Steel. — Wrought iron or steel piping is most generally used for conveying steam, gas, air, and water. Wrought iron pipe because of its expense has been largely displaced by steel pipe. Through custom the term "Wrought Iron Pipe" is often taken to refer to the Briggs Standard sizes rather than to the material of which the pipe is made, and so it is necessary to specify exactly what is wanted. "Steel," "wrought steel," and "wrought pipe," are terms sometimes used and refer to welded pipe made of steel. If real wrought iron pipe made from puddled iron is required the terms "genuine wrought iron," "guaranteed wrought iron," or the manufacturer's brand or name should be used. There are differences of opinion as to the superiority of one over the other, especially in the matter of corrosion. Some people consider that the cinder which remains in the wrought iron breaks up the continuity of the metal and tends to impede corrosion. Many authorities hold that there is little or no difference in the rust-resisting qualities of the two materials. Steel pipe has a higher tensile strength than wrought iron. In 1915 approximately 90 per cent. of the wrought pipe was made of steel, a reversal of conditions of twenty years ago when wrought iron was mostly used.

Briggs Standard. — Both wrought iron and steel pipe are made to the same standard of sizes. Standard pipe is known by its nominal inside diameter. This nominal diameter differs from the actual diameter by varying amounts as an inspection of Table 4 in Chapter II will show. It is necessary to guard against underweight pipe known as "merchant weight," of which the reputable companies have given up the manufacture. This varies from standard full weight pipe and is usually 5 to 10 per cent. thinner. It should be carefully avoided in work of any importance as the extra cost of maintenance will soon overbalance the small difference in first cost. Besides standard weight there is made extra strong and double extra strong pipe. The outside diameter remains the same, but the thickness is increased by decreasing the inside diameter. Fig. 2 shows sections of the three weights of pipe of the same nominal inside diameter. Above 125 pounds per square inch extra strong pipe should be used. Standard weight is sometimes used for pressures up to 200

Fig. 2. Sections of $\frac{1}{2}''$, Standard, Extra Heavy, and Double Extra Heavy Wrought Pipe.

pounds per square inch, but this is not advisable. Double extra strong pipe is used for hydraulic work.

Outside Diameter Pipe. — Above 12 inches in diameter pipe is known as O. D. or outside diameter pipe. It is then specified by its outside diameter. The thickness varies with the diameter and the use for which it is required. For large sizes it is always advisable to specify the outside diameter and the thickness of the metal. Especially is this true if the pipe is to be threaded, as sufficient thickness must be allowed to maintain the strength of the pipe after cutting the threads. The thickness should not be less than $\frac{5}{16}$ inches.

When used for water wrought iron or steel pipe may be galvanized, or otherwise treated to prevent corrosion and pitting.

Manufacture of Steel Pipe. — The manufacture of steel pipe by the National Tube Company is described in one of their books, from which the following is abstracted:

"Welded tubes and pipe are made either by the lap or butt-weld process.

" The lap-weld process consists of two operations, bending and welding. The plate, rolled to the necessary width and gage for the size of pipe intended, is brought to a red heat in a suitable furnace, and then passed through a set of rolls which bevel the edges, so that when overlapped and welded the seam will be neat and smooth. It now passes immediately to the bending machine where it takes roughly the cylindrical shape of a pipe with the two edges overlapping. In this form it is again heated in another furnace, Fig. 3, and when brought to the welding temperature the bent skelp is pushed out of the furnace into the welding rolls,

Fig. 3. Lap-Weld Furnace — Bent Plate ready to Charge.

Fig. 4. Each of these rolls has a semi-circular groove forming a circular pass, corresponding to the size of pipe being made. A cast iron ball, or mandrel, held in position between the welding rolls by a stout bar, serves to support the inside of the pipe as it is carried through. This 'ball' or mandrel is shaped like a projectile and the pipe slides over it on being drawn through the rolls. Thus every portion of the lapped edge is subjected to a compression between the ball on the inside and the rolls on the outside, which reduces the lap to the same thickness as the rest of the pipe, and welds the overlapping portions solidly together.

"The pipe then enters similarly shaped rolls called the sizing rolls, which correct any irregularities in shape and give the exact outside diameter required. Any variation in gage makes a proportional variation in the internal diameter. Finally the tube is passed through the straightening or cross rolls, consisting of two rolls set with their axes askew. The surfaces of these rolls are so curved that the tube is in contact with each for the whole

Fig. 4. Welding Rolls for Lap-Weld, Mandrel in Position.

length of the roll, and is passed forward and rapidly rotated when the rolls are revolved. The tube is made practically straight by the cross rolls, and is also given a clean finish with a thin, firmly adhering scale.

" After this last operation the tube is rolled up an inclined cooling table, so that the metal will cool off slowly and uniformly without internal strain. When cool enough the rough ends are removed by cold saws or in a cutting-off machine, after which the tube is ready for inspection and testing. In the case of threaded pipe the ends are threaded before testing.

"In the case of some sizes of double-extra-strong pipe (3-inch to 8-inch) made by the lap-weld process, two pipes are first made to such sizes as will telescope one within the other, the respective welds being placed opposite each other; these are then returned to the furnace, brought to the proper welding heat, and given a pass through the welding rolls. While a pipe made in this way is, in respect to its resistance to internal pressure, as strong or stronger than when made from one piece of skelp, it is not necessarily welded at all points between the two tubular surfaces;

Fig. 5. Drawing Butt-Weld Pipe.

however, each piece is first thoroughly welded at the seam before telescoping.

"Skelp used in making butt-welded pipe comes from the rolling department of the steel mills with a specified length, width, and gage, according to the size pipe for which it is ordered. The edges are slightly beveled with the face of the skelp, so that the surface of the plate which is to become the inside of the pipe is not quite as wide as that which forms the outside; thus when the edges are brought together they meet squarely.

"The skelp for all butt-welded pipe is heated uniformly to the welding temperature. The strips of steel when properly heated are seized by their ends with tongs and drawn from the furnaces through bell-shaped dies or 'bells,' as they are called, Fig. 5. The inside of these bells is so curved that the plate is gradually formed in the shape of a tube, the edges being forced squarely together and welded. For some sizes the pipe is drawn through two bells consecutively at one heat, one bell being just behind the other, the second one being of a slightly smaller diameter than the first.

"The pipe is then run through sizing and cross rolls similar to those used in the lap-weld process, to secure the correct outside diameter and finish.

"The pull required to draw double-extra strong (hydraulic) pipe by this process is so great, on account of the thickness of the skelp, that it is found necessary to weld a strong bar on the end of the skelp, thereby more evenly distributing the strain. With this bar the skelp is drawn through several bells of decreas-

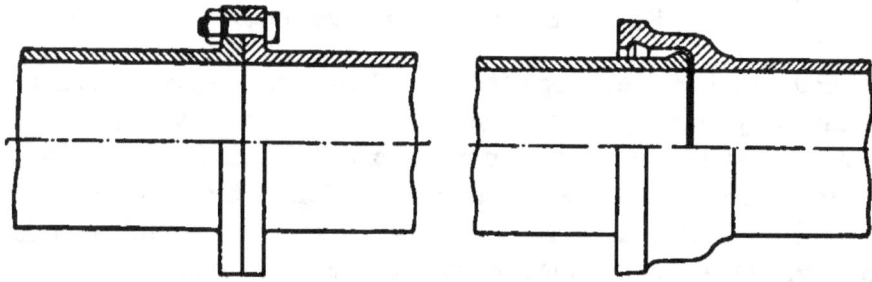

Fig. 6. Cast Iron Pipe — Flanged.   Fig. 7. Cast Iron Pipe — Bell and Spigot.

ing size, and is reheated between draws until the seam is thoroughly welded. It is evident that the skelp is put to a severe test in this operation, and, unless the metal is sound and homogeneous, the ends are likely to be pulled off."

Cast Iron. — Cast iron is commonly used for underground water pipes, gas mains, and sanitation piping, and it may be used for any low pressure work. Because of its uncertain nature it should not be used for high pressures. Cast iron does not corrode as readily as wrought iron or steel pipe. It is cheap and easily shaped. Cast iron pipe must be well supported because of its great weight. Supports should be placed from ten to twelve feet apart. Cast iron pipe is made with either flanged ends, Fig. 6, or bell and spigot ends, Fig. 7. For sanitation piping and underground work the bell and spigot end pipe is used. There is a certain amount of flexibility with this form of joint which adapts it to variations in level. The joint is leaded and calked.

Flanged pipe is bolted together with gaskets between the flanges. This is the usual form when the pipe is above ground.

Copper. — Copper pipe is expensive, and is used only where its flexibility makes it superior to other materials, such as on shipboard or for expansion bends, for small oil piping, and for stills and chemical work. At high temperature it becomes brittle. Copper pipe is sometimes wound with steel or copper wire under tension to increase its strength. The same result is

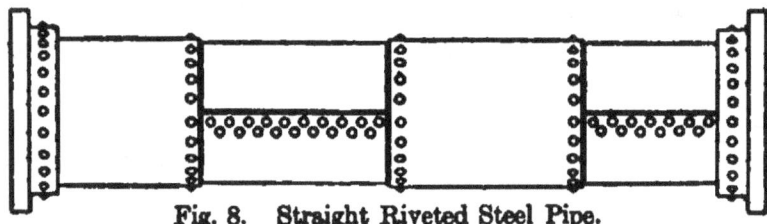

Fig. 8. Straight Riveted Steel Pipe.

secured by using steel hoops at frequent spaces. Copper pipe may be made by brazing plates together (in which cases the joint is a source of weakness) or they may be solid drawn in iron pipe sizes.

Brass. — Brass pipe is safe and strong but is too expensive for general use. It is used for hot water where iron would corrode rapidly, generally in small sizes. Up to four inches diameter seamless drawn brass tubes come in twelve foot lengths in iron pipe sizes. Such pipe is called iron pipe size to distinguish it from thin brass tubing and plumbers' brass pipe.

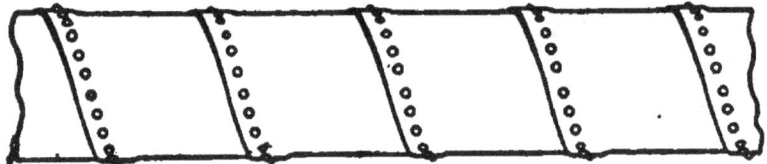

Fig. 9. Spiral Riveted Steel Pipe.

Lead. — Lead piping is used for water and waste pipes and for acid and various chemical solutions which would rapidly corrode iron pipe. It is made in sizes up to twelve inches diameter and of the thicknesses and qualities of lead as given in Table 14 in Chapter II.

Pipe is also made of tin, of lead lined with tin, and of steel lined with lead for special purposes or conditions.

Riveted Pipe. — Large pipes may be made up of steel plates forming riveted steel pipe. They may be either straight riveted,

Fig. 8, or spiral riveted, Fig. 9. They may be joined by flanges of cast or pressed steel. These flanges are riveted to the ends of the pipe. The riveted ends are calked, and then the pipe is generally galvanized. Such pipe is largely used for low pressure work, as exhaust mains, drains, etc. The spiral riveted pipe has only one seam and consequently is stronger than the straight riveted pipe for the same diameter and thickness of plate.

Strength of Materials. — Some average values for the properties of various materials used for piping, valves, and fittings are given in the following tabulations. These values will be found to vary somewhat with different manufacturers, but the ultimate strength should not be much more than five per cent. lower.

Material	Ultimate Tensile Strength	Elastic Limit	Elongation	Reduction in area
Cast Iron	23,000			
Semi Steel	33,400			
Malleable Iron	37,000			
Brass	18,000 to 30,000	15,000		
Hard Metal Composition	33,000	25,000	15% in 2 inches	10%
Cast Steel	65,000	35,000	25% in 2 inches	35%
Monel Metal	75,000	36,000	32% in 2 inches	40%
Crucible Steel	80,000			
Wrought Iron	40,000			
Soft Steel	50,000	30,000	18% in 2 inches	50%
Lead Pipe	1,650			

The following is the composition of two casting alloys of the U. S. Navy Bureau of Steam Engineering.

	Copper	Tin	Zinc	Iron (Max.)	Lead (Max.)
Gun Bronze	87–89%	9–11%	1–3%	.06%	.2%
Screw Pipe Fittings, Brass	77–80%	4% (Min.)	13–19%	.1%	3.0%

The gun bronze is suitable for all composition valves four inches in diameter and above; expansion joints, flanged pipe fittings, gear wheels, bolts and nuts, miscellaneous brass castings, all parts where strength is required of brass castings, or where subjected to salt water, and for all purposes where no other alloy is specified. Composition valves; safety and relief, feed, check and stop, surface blow, drain, air and water cocks, main stop, throttle reducing, sea, safety sluice, and manifolds at pumps. This gun bronze has an ultimate tensile strength (minimum) of 30,000 pounds, yield point (minimum) of 15,000 pounds, and elongation in two inches (minimum) of 15 per cent. The brass listed is suitable for composition screwed fittings. This brass has an ultimate tensile strength (minimum) of about 40,000 pounds, yield point (minimum) of about 20,000 pounds and elongation in two inches (minimum) of about 20 per cent. No physical tests are specified however.

CHAPTER II

DIMENSIONS AND STRENGTH OF PIPE

All kinds of pipe are now manufactured in standard sizes and thicknesses, so that it is not often necessary to figure them. Various formulae are here given for use where it is desirable to check sizes, to have pipe made to specifications, or for any other reason. The properties of materials are given in the tabulation at the end of Chapter I.

General Formula. — The general formula for cylinders subject to internal pressure is obtained as follows:

In Fig. 10, let d = inside diameter in inches.
t = thickness of cylinder wall in inches.
l = length of cylinder wall in inches.
p = internal fluid pressure in lbs. per sq. in.
f = stress induced in material in lbs. per sq. in.

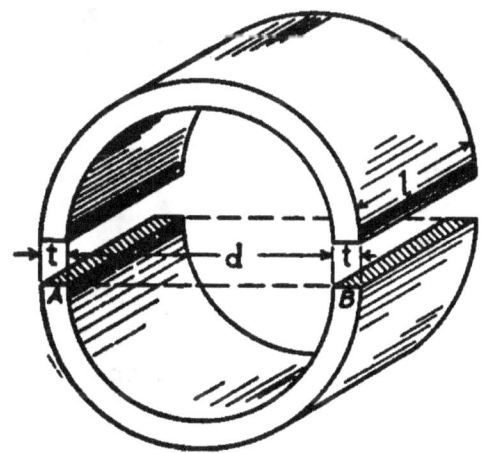
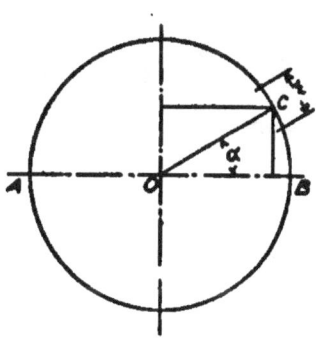

Fig. 10. General Formula for Pipe.   Fig. 11.

The pressure will be exerted at right angles to the surface. Considering a very small portion of the circumference, w, Fig. 11, the arc may be assumed equal to the chord, and the area about point C will be wl square inches. The pressure at C will then be pwl.

Let α = angle $C\,O\,B$

Let $O\,C$ = pressure at C = $p\,w\,l$

The vertical component of $O\,C$ will then be $plw \sin \alpha$

Each point may be treated in the same manner, and the algebraic sum of the upward pressures will equal the algebraic sum of the downward pressures. This will be a measure of the tendency to separate the cylinder at A and B and is equal to

$\Sigma\,plw \sin \alpha = pld$

Resisting this pressure is the metal at A and B, the strength of which is $2ltf$.

Equating this to the pressure gives

$$pld = 2ltf$$

$$p = \frac{2tf}{d} \dotfill (1)$$

or $\quad t = \frac{pd}{2f} \dotfill (2)$

This formula may be used for wrought iron or steel, assuming a proper factor of safety. For cast iron it does not give practical thicknesses, and a constant is generally added.

Formulae for Cast Iron. — Several formulae are here given for cast iron pipe. The formula for pressures above 100 pounds per square inch is

$$t = \frac{pd}{4000} + \frac{1''}{2} \dotfill (3)$$

Another common formula is

$$t = \frac{pd}{2000} \dotfill (4)$$

The American Society of Mechanical Engineers' formula for cast iron pipe is

$$t = \left[\frac{p+100}{4f}d + 0.333\left(1 - \frac{d}{100}\right)\right]1.2 \dotfill (5)$$

in which $f = 1800$

Fanning's formula for cast iron water pipe is

$$t = 0.00006\,(h + 230)\,d + 0.333 - 0.0033d \dotfill (6)$$

in which h = head in feet

DIMENSIONS AND STRENGTH OF PIPE

Francis' formula for cast iron water pipe is
$$t = 0.000058\,hd + 0.0152d + 0.312 \quad\quad\quad\quad (7)$$

Cast Iron Cylinder Tests. — In the A. S. M. E. Trans. Vol. 19, page 597, Prof. C. H. Benjamin gives the results of some tests of cast iron cylinders made at Case School of Applied Science. The cylinders were 10 1/8 inches in diameter, 20 inches long, 3/4 inches thick and had covers bolted on the ends. Water pressure was used.

Cylinder	1	2	3	4	Average
Bursting pressure	1350	1400	1350	1200	1350
Unit stress $f = \dfrac{pd}{2t}$	9040	10200	9735	9080	9500

Cast Iron Hub and Spigot Pipe. — The dimensions of hub and spigot pipe given in Tables 1 and 2 are from the "Standard

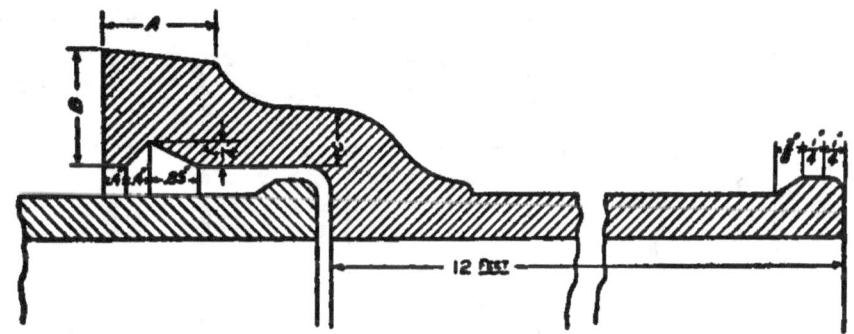

Fig. 12. Cast Iron Bell and Spigot Pipe.

Specifications for Cast Iron Pipe" of the American Society for Testing Materials, which give complete information as to materials, allowable variation in weight, methods of inspection, testing, etc. The hydrostatic tests for various classes of pipe are given as follows:

	20-Inch Diameter and Larger Pounds per Sq. In.	Less than 20-Inch Diameter Pounds per Sq. In.
Class A Pipe	150	300
Class B Pipe	200	300
Class C Pipe	250	300
Class D Pipe	300	300

TABLE 1 (Fig. 12)

Cast Iron Hub and Spigot Pipe

Weights are for Twelve Feet Laying Lengths and Standard Sockets

Nominal Inside Diam. Inches	Class A 100 Ft. Head 43 Lbs. Pressure			Class B 200 Ft. Head 86 Lbs. Pressure			Class C 300 Ft. Head 130 Lbs. Pressure			Class D 400 Ft. Head 173 Lbs. Pressure		
	Thickness Ins.	Weight per Foot	Weight per Length	Thickness Ins.	Weight per Foot	Weight per Length	Thickness Ins.	Weight per Foot	Weight per Length	Thickness Ins.	Weight per Foot	Weight per Length
4	0.42	20.0	240	0.45	21.7	260	0.48	23.3	280	0.52	25.0	300
6	0.44	30.8	370	0.48	33.3	400	0.51	35.8	430	0.55	38.3	460
8	0.46	42.9	515	0.51	47.5	570	0.56	52.1	625	0.60	55.8	670
10	0.50	57.1	685	0.57	63.8	765	0.62	70.8	850	0.68	76.7	920
12	0.54	72.5	870	0.62	82.1	985	0.68	91.7	1100	0.75	100.0	1200
14	0.57	89.6	1075	0.66	102.5	1230	0.74	116.7	1400	0.82	129.2	1550
16	0.60	108.3	1300	0.70	125.0	1500	0.80	143.8	1725	0.89	158.3	1900
18	0.64	129.2	1550	0.75	150.0	1800	0.87	175.0	2100	0.96	191.7	2300
20	0.67	150.0	1800	0.80	175.0	2100	0.92	208.3	2500	1.03	229.2	2750
24	0.76	204.2	2450	0.89	233.3	2800	1.04	279.2	3350	1.16	306.7	3680
30	0.88	291.7	3500	1.03	333.3	4000	1.20	400.0	4800	1.37	450.0	5400
36	0.99	391.7	4700	1.15	454.2	5450	1.36	545.8	6550	1.58	625.0	7500
42	1.10	512.5	6150	1.28	591.7	7100	1.54	716.7	8600	1.78	825.0	9900
48	1.26	666.7	8000	1.42	750.0	9000	1.71	908.3	10900	1.96	1050.0	12600
54	1.35	800.0	9600	1.55	933.3	11200	1.90	1141.7	13700	2.23	1341.7	16100
60	1.39	916.7	11000	1.67	1104.2	13250	2.00	1341.7	16100	2.38	1583.3	19000

TABLE 2 (Fig. 12)
Cast Iron Hub and Spigot Pipe

Nominal Diam. Inches	Classes	Actual Outside Diam. Inches	Diam. of Sockets		Depth of Sockets		A	B	C
			Pipe Inches	Special Castings Inches	Pipe Inches	Special Castings Inches			
4	A — B	4.80	5.60	5.70	3.50	4.00	1.5	1.30	.65
4	C — D	5.00	5.80	5.70	3.50	4.00	1.5	1.30	.65
6	A — B	6.90	7.70	7.80	3.50	4.00	1.5	1.40	.70
6	C — D	7.10	7.90	7.80	3.50	4.00	1.5	1.40	.70
8		9.05	9.85	10.00	4.00	4.00	1.5	1.50	.75
8	C — D	9.30	10.10	10.00	4.00	4.00	1.5	1.50	.75
10	A — B	11.10	11.90	12.10	4.00	4.00	1.5	1.50	.75
10	C — D	11.40	12.20	12.10	4.00	4.00	1.5	1.60	.80
12	A — B	13.20	14.00	14.20	4.00	4.00	1.5	1.60	.80
12	C — D	13.50	14.30	14.20	4.00	4.00	1.5	1.70	.85
14	A — B	15.30	16.10	16.10	4.00	4.00	1.5	1.70	.85
14	C — D	15.65	16.45	16.45	4.00	4.00	1.5	1.80	.90
16	A — B	17.40	18.40	18.40	4.00	4.00	1.75	1.80	.90
16	C —	17.80	18.80	18.80	4.00	4.00	1.75	1.90	1.00
18	A — B	19.50	20.50	20.50	4.00	4.00	1.75	1.90	.95
18	C — D	19.92	20.92	20.92	4.00	4.00	1.75	2.10	1.05
20	A — B	21.60	22.60	22.60	4.00	4.00	1.75	2.00	1.00
20	C — D	22.06	23.06	23.06	4.00	4.00	1.75	2.30	1.15
24	A — B	25.80	26.80	26.80	4.00	4.00	2.00	2.10	1.05
24	C — D	26.32	27.32	27.32	4.00	4.00	2.00	2.50	1.25
30	A	31.74	32.74	32.74	4.50	4.50	2.00	2.50	1.15
30	B	32.00	33.00	33.00	4.50	4.50	2.00	2.30	1.15
30	C	32.40	33.40	33.40	4.50	4.50	2.00	2.60	1.32
30	D	32.74	33.74	33.74	4.50	4.50	2.00	3.00	1.50
36	A	37.96	38.96	38.96	4.50	4.50	2.00	2.50	1.25
36	B	38.30	39.30	39.30	4.50	4.50	2.00	2.80	1.40
36	C	38.70	39.70	39.70	4.50	4.50	2.00	3.10	1.60
36	D	39.16	40.16	40.16	4.50	4.50	2.00	3.40	1.80
42	A	44.20	45.20	45.20	5.00	5.00	2.00	2.80	1.40
42	B	44.50	45.50	45.50	5.00	5.00	2.00	3.00	1.50
42	C	45.10	46.10	46.10	5.00	5.00	2.00	3.40	1.75
42	D	45.58	46.58	46.58	5.00	5.00	2.00	3.80	1.95
48	A	50.50	51.50	51.50	5.00	5.00	2.00	3.00	1.50
48	B	50.80	51.80	51.80	5.00	5.00	2.00	3.30	1.65
48	C	51.40	52.40	52.40	5.00	5.00	2.00	3.80	1.95
48	D	51.98	52.98	52.98	5.00	5.00	2.00	4.20	2.20
54	A	56.66	57.66	57.66	5.50	5.50	2.25	3.20	1.60
54	B	57.10	58.10	58.10	5.50	5.50	2.25	3.60	1.80
54	C	57.80	58.80	58.80	5.50	5.50	2.25	4.00	2.15
54	D	58.40	59.40	59.40	5.50	5.50	2.25	4.40	2.45
60	A	62.80	63.80	63.80	5.50	5.50	2.25	3.40	1.70
60	B	63.40	64.40	64.40	5.50	5.50	2.25	3.70	1.90
60	C	64.20	65.20	65.20	5.50	5.50	2.25	4.20	2.25
60	D	64.82	65.82	65.82	5.50	5.50	2.25	4.70	2.60

Plain Cast Iron Pipe. — For flanged cast iron pipe the weight of the flanges must be added to the weight of the plain pipe. The weight of two flanges is equal to the weight of one foot of pipe. Table 3 gives the approximate weight per foot of length for cast iron pipe of various thicknesses.

TABLE 3

WEIGHT IN POUNDS PER FOOT OF PLAIN CAST IRON PIPE

Diam. Ins.	Thickness of Metal in Inches								
	¼	⅜	½	⅝	¾	⅞	1	1⅛	1¼
	lbs.	lbs.	lbs.	lbs.	lbs.	lbs.	lbs.	lbs.	lbs.
2	5.52	8.74	12.27	16.11	20.25	24.70	29.45	34.52	39.88
2½	6.75	10.58	14.73	19.18	23.95	28.99	34.36	40.04	46.02
3	7.93	12.43	17.18	22.24	27.61	32.29	39.27	45.56	52.16
3½	9.20	14.27	19.64	25.31	31.29	37.58	44.18	51.08	58.29
4	10.43	16.11	22.09	28.38	34.98	41.88	49.09	56.60	64.43
4½	11.66	17.95	24.54	31.45	38.66	46.18	54.00	62.13	70.56
5	12.89	19.79	27.00	34.52	42.34	50.47	58.91	67.65	76.70
5½	14.11	21.63	29.45	37.58	46.02	54.76	63.81	73.17	82.84
6	15.34	23.47	31.91	40.65	49.70	59.06	68.72	78.69	88.97
7	17.79	27.15	36.82	46.79	57.06	67.65	78.54	89.74	101.24
8	20.25	30.83	41.72	52.92	64.43	76.24	88.36	100.78	113.52
9	22.70	34.52	46.63	59.06	71.79	84.83	98.18	111.83	125.79
10	25.16	38.20	51.54	65.19	79.15	93.42	107.99	122.87	138.06
11	27.61	41.88	56.45	71.33	86.52	102.01	117.81	133.92	150.33
12	30.07	46.56	61.36	77.47	93.88	110.60	127.63	144.96	162.60
13	32.52	49.24	66.27	83.60	101.24	119.19	137.45	156.01	174.87
14	34.98	52.92	71.18	89.74	108.61	127.78	147.26	167.05	187.15
15	....	56.60	76.09	95.87	115.97	136.37	157.08	178.10	199.42
16	....	60.29	80.99	102.01	123.33	144.96	166.90	189.14	211.69
18	....	67.65	90.81	114.28	138.06	162.14	186.53	211.23	236.23
20	....	....	100.63	126.55	152.79	179.32	206.17	233.32	260.78
22	....	....	110.45	138.83	167.51	196.50	225.80	255.41	285.32
24	....	....	120.26	151.10	182.24	213.68	245.44	277.50	309.87

Briggs Standard Dimensions. — Wrought iron and steel pipe as used for steam, gas, air, and water is known as the Briggs Standard, the dimensions of which have been established as noted in Chapter I. The sizes and information are given in Tables 4, 5, and 6 for standard weight, extra strong, and double extra strong pipe.

DIMENSIONS AND STRENGTH OF PIPE

TABLE 4
STANDARD WROUGHT IRON OR STEEL PIPE

Nominal Size	Internal Diameter Inches	External Diameter Inches	Thickness Inches	Weight per Foot		Area of Internal Cross Section Sq. Inches	Length of Pipe in Feet per Square Foot of	
				Plain Pounds	Threaded and Coupled Pounds		Internal Surface	External Surface
⅛	.269	.405	.088	.244	.245	.057	14.199	9.431
¼	.364	.540	.088	.424	.425	.104	10.493	7.073
⅜	.493	.675	.091	.567	.568	.191	7.747	5.658
½	.622	.840	.109	.850	.852	.304	6.141	4.547
¾	.824	1.050	.113	1.130	1.134	.533	4.635	3.637
1	1.049	1.315	.133	1.678	1.684	.864	3.641	2.904
1¼	1.380	1.660	.140	2.272	2.281	1.495	2.767	2.301
1½	1.610	1.900	.145	2.717	2.731	2.036	2.372	2.010
2	2.067	2.375	.154	3.652	3.678	3.355	1.847	1.608
2½	2.469	2.875	.203	5.793	5.819	4.788	1.547	1.328
3	3.068	3.500	.216	7.575	7.616	7.393	1.245	1.091
3½	3.548	4.000	.226	9.109	9.202	9.886	1.076	.954
4	4.026	4.500	.237	10.790	10.889	12.730	.948	.848
4½	4.506	5.000	.247	12.538	12.642	15.947	.847	.763
5	5.047	5.563	.258	14.617	14.810	20.006	.756	.686
6	6.065	6.625	.280	18.974	19.185	28.891	.629	.576
7	7.023	7.625	.301	23.544	23.769	38.738	.543	.500
8	8.071	8.625	.277	24.696	25.000	51.161	.473	.442
9	8.941	9.625	.342	33.907	34.188	62.786	.427	.396
10	10.192	10.750	.279	31.201	32.000	81.585	.374	.355
10	10.136	10.750	.307	34.240	35.000	80.691	.376	.355
10	10.020	10.750	.365	40.483	41.132	78.855	.381	.355
11	11.000	11.750	.375	45.557	46.247	95.033	.347	.325
12	12.090	12.750	.330	43.773	45.000	114.800	.315	.299
12	12.000	12.750	.375	49.562	50.706	113.097	.318	.299

TABLE 5

Extra Strong Wrought Pipe

Nominal Size	External Diameter Inches	Internal Diameter Inches	Thickness Inches	Weight per Foot Plain Ends Pounds	Internal Area Sq. Inches	Length of Pipe per Square Foot of	
						External Surface Feet	Internal Surface Feet
⅛	.405	.215	.095	.314	.036	9.431	17.766
¼	.540	.302	.119	.535	.072	7.073	12.648
⅜	.675	.423	.126	.738	.141	5.658	9.030
½	.840	.546	.147	1.087	.234	4.547	6.995
¾	1.050	.742	.154	1.473	.433	3.637	5.147
1	1.315	.957	.179	2.171	.719	2.904	3.991
1¼	1.660	1.278	.191	2.996	1.283	2.301	2.988
1½	1.900	1.500	.200	3.631	1.767	2.010	2.546
2	2.375	1.939	.218	5.022	2.953	1.608	1.969
2½	2.875	2.323	.276	7.661	4.238	1.328	1.644
3	3.500	2.900	.300	10.252	6.605	1.091	1.317
3½	4.000	3.364	.318	12.505	8.888	.954	1.135
4	4.500	3.826	.337	14.983	11.497	.848	.998
4½	5.000	4.290	.355	17.611	14.455	.763	.890
5	5.563	4.813	.375	20.778	18.194	.686	.793
6	6.625	5.761	.432	28.573	26.067	.576	.663
7	7.625	6.625	.500	38.048	34.472	.500	.576
8	8.625	7.625	.500	43.388	45.663	.442	.500
9	9.625	8.625	.500	48.728	58.426	.396	.442
10	10.750	9.750	.500	54.735	74.662	.355	.391
11	11.750	10.750	.500	60.075	90.763	.325	.355
12	12.750	11.750	.500	65.415	108.434	.299	.325

As stated in Chapter I, pipe above 12 inches is called O. D. pipe because it is known by its actual outside diameter. The inside diameter changes with the variation in thickness. Information concerning O. D. pipe is given in Table 7.

Pipe is sold in random lengths which are 18 to 21 feet for standard, and 12 to 22 feet for extra strong and for double extra strong pipe. These lengths have recently been doubled and pipe is now made in lengths from 35 to 42 feet. Ordinarily standard pipe is threaded and supplied with couplings, while extra and double extra pipe have plain ends.

Bursting Pressures of Pipe. — From an investigation and comparison of five formulae, Reid T. Stewart in a paper in A. S. M. E. Trans. Vol. 34 concludes that for all ordinary calculations pertaining to the bursting strength of commercial tubes, pipes, and cylinders, Barlow's formula is to be preferred. This formula assumes that because of the elasticity of the material, the different circumferential fibres will have their diameters increased in such

TABLE 6
DOUBLE EXTRA STRONG WROUGHT PIPE

Nominal Size	External Diameter Inches	Approximate Internal Diameter Inches	Thickness Inches	Weight per Foot Plain Ends Pounds	Internal Area Sq. Inches	Length of Pipe per Square Foot of	
						External Surface Feet	Internal Surface Feet
½	.840	.252	.294	1.714	.050	4.547	15.157
¾	1.050	.434	.308	2.440	.148	3.637	8.801
1	1.315	.599	.358	3.659	.282	2.904	6.376
1¼	1.660	.896	.382	5.214	.630	2.301	4.263
1½	1.900	1.100	.400	6.408	.950	2.010	3.472
2	2.375	1.503	.436	9.029	1.774	1.608	2.541
2½	2.875	1.771	.552	13.695	2.464	1.328	2.156
3	3.500	2.300	.600	18.583	4.155	1.091	1.660
3½	4.000	2.728	.636	22.850	5.845	.954	1.400
4	4.500	3.152	.674	27.541	7.803	.848	1.211
4½	5.000	3.580	.710	32.530	10.066	.763	1.066
5	5.563	4.063	.750	38.552	12.966	.686	.940
6	6.625	4.897	.864	53.160	18.835	.576	.780
7	7.625	5.875	.875	63.079	27.109	.500	.650
8	8.625	6.875	.875	72.424	37.122	.442	.555

TABLE 7
WEIGHT OF OUTSIDE DIAMETER WROUGHT PIPE

Outside Diameter of Pipe Inches	Weight in Pounds per Foot							
	¼ Inch Thick	$\frac{5}{16}$ Inch Thick	⅜ Inch Thick	$\frac{7}{16}$ Inch Thick	½ Inch Thick	$\frac{9}{16}$ Inch Thick	⅝ Inch Thick	¾ Inch Thick	1 Inch Thick
14	36.71	45.68	54.57	63.37	72.10	80.73	89.28	106.13	138.84
15	39.38	49.02	58.57	68.04	77.43	86.73	95.95	114.14	149.52
16	42.05	52.36	62.58	72.72	82.77	92.74	102.63	122.15	160.20
17	44.72	55.69	66.58	77.39	88.11	98.75	109.30	130.16	170.88
18	47.39	59.03	70.59	82.06	93.45	104.76	115.98	138.17	181.56
20	57.00	65.71	78.60	91.41	104.13	116.77	129.33	154.19	202.92
21	59.20	69.04	82.60	96.08	109.47	122.78	136.00	162.20	
22	62.60	72.38	86.61	100.75	114.81	128.79	142.68	170.21	
24	68.00	85.00	94.62	110.10	125.49	140.80	156.03	186.23	
26	74.00	93.00	102.63	119.44	136.17	152.82	169.38	202.25	
28	80.00	100.00	120.00	128.79	146.85	164.83	182.73	218.27	
30	85.00	107.00	128.00	138.13	157.53	176.85	196.08	234.30	

a manner as to keep the area of cross section constant; and that the length of the tube is unaltered by the internal fluid pressure. As neither of these assumptions is theoretically correct the result is approximate. Barlow's formula is

$$\frac{p}{f} = 2\frac{t}{D};\ p = 2f\frac{t}{D};\ t = \tfrac{1}{2}D\frac{p}{f};\ f = \tfrac{1}{2}D\frac{p}{t} \ldots\ldots (8)$$

D = outside diameter in inches.
t = nominal or average thickness of wall in inches.
p = internal fluid pressure in pounds per square inch.
f = fibre stress in pounds per square inch.
n = safety factor based on ultimate strength.

$$f = \frac{40000}{n}\text{ for butt-welded steel pipe}$$

$$f = \frac{50000}{n}\text{ for lap-welded steel pipe}$$

$$f = \frac{60000}{n}\text{ for seamless steel tubes}$$

$$f = \frac{28000}{n}\text{ for wrought iron pipe}$$

The average values of f are based on a large number of tests on commercial tubes and pipes made at one of the mills of the National Tube Company, which gave the following values:

Butt-welded steel pipe	41686
Butt-welded wrought iron pipe	29168
Lap-welded steel pipe	52225
Lap-welded wrought iron pipe	30792

The average bursting pressures for a number of the tests referred to above are shown graphically in Fig. 13.

It is understood that recent improvements in the manufacture of butt-welded pipe *i.e.* 3 inches and smaller, have resulted in such strengthening of the weld that the bursting strength is approximately equal to that of lap-welded pipe.

Mill Tests. — The various pipe mills have their own standard of test pressures which are applied to wrought pipe. National Tube Company test pressures are as follows:

DIMENSIONS AND STRENGTH OF PIPE

STANDARD PIPE

Nominal Size	Method of Manufacture	Test Pressure
⅛ inch to 2 inches (inclusive)	Butt-weld	700 pounds
2½ inches and 3 inches	" "	800 "
Up to 8 inches	Lap-weld	1000 "
9 and 10 inches	" "	900 "
11 and 12 inches	" "	800 "
13 and 14 inches	" "	700 "
15 inch	" "	600 "

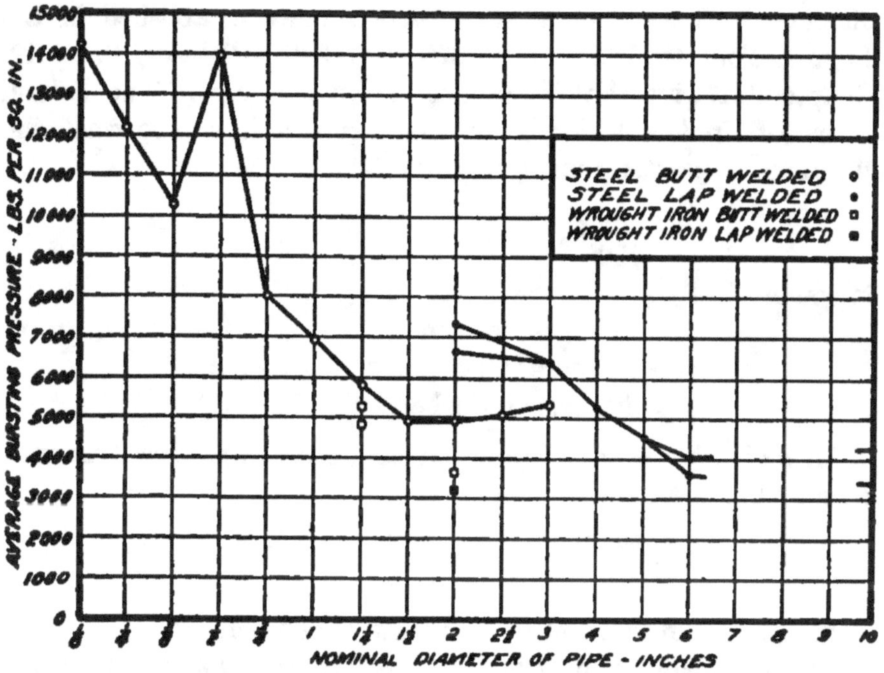

Fig. 13. Diagram Showing Bursting Strength of Wrought Iron and Steel Pipe.

EXTRA STRONG PIPE

Nominal Size	Method of Manufacture	Test Pressure
⅛ inch to 1 inch (inclusive)	Butt-weld	700 pounds
1¼ inches to 3 inches (inclusive)	" "	1500 "
1½ and 2 inches	Lap-weld	2500 "
2½ to 4 inches (inclusive)	" "	2000 "
4½ to 6 inches (inclusive)	" "	1800 "
7 inches to 9 inches (inclusive)	" "	1500 "
10 inches	" "	1200 "
11 and 12 inches	" "	1100 "
13 inches to 15 inches (inclusive)	" "	1000 "

Double Extra Strong Pipe

Nominal Size	Method of Manufacture	Test Pressure
½ inch to 1 inch (inclusive)	Butt-weld	700 pounds
1¼ inches to 2½ inches (inclusive)	" "	2200 "
1½ inches to 3 inches (inclusive)	Lap-weld	3000 "
3½ inches and 4 inches	" "	2500 "
4½ inches to 8 inches (inclusive)	" "	2000 "

English Pipe. — English standard wrought pipe differs slightly from the Briggs Standard. The ruling dimension is the external diameter, but the sizes are designated by the nominal internal diameter. These nominal sizes were mainly established in the English Tube trade between 1820 and 1840. Tables 18 and 19, Chapter III, give the dimensions of English pipe. The British Board of Trade rule for lap welded wrought iron pipe when the thickness is greater than ¼ inch is

$$t = \frac{pd}{6000} \quad \quad (9)$$

in which

t = thickness in inches.
p = pressure in pounds per square inch.
d = diameter in inches.

Riveted Pipe. — For spiral riveted steel pipe the following formula may be used.

$$t = \frac{pd}{2fe} \quad \quad (10)$$

in which e = efficiency of riveted joint in per cent.

The dimensions and weight of Root spiral riveted pipe, made by Abendroth & Root, as given in Table 8, are for piping to be used for conveying water, oil, gas, exhaust steam, compressed air, etc. Spiral riveted pipe is two-thirds stronger and is more rigid than straight seam pipe of equal weight. This great rigidity is due to the absence of seams having a tendency to weaken the pipe, there being but one continuous helical seam from one end to the other, and this forms a stiffening rib. When spiral riveted pipe has been tested to destruction, fracture has always occurred toward the center of the strip rather than at the seam.

For underground water work systems and exposed work where the temperature is less than 100 degrees F., asphalted pipe is advised. It is made in lengths up to 30 feet. For conveying exhaust steam, paper pulp, and all hot liquids, especially such as are acid or alkaline, galvanized pipe is advised. It may be single or double galvanized and is made in lengths up to 20 feet.

TABLE 8

ABENDROTH AND ROOT BLACK SPIRAL RIVETED PIPE

Diameter in Inches	Thickness B. W. Gauge	Approximate Bursting Pressure in Lbs. per Sq. Inch	Weight in Lbs. per 100 Feet		
			Plain End Pipe	With A & R Flanges, Bolts and Gaskets	With Root Bolted Joints
3	22	1060	115	139	153
	20	1325	147	171	185
	18	1860	205	229	243
4	20	1000	195	227	247
	18	1390	273	305	325
	16	1845	360	392	412
5	20	795	242	282	304
	18	1100	340	380	402
	16	1480	448	488	510
6	18	930	385	433	475
	16	1220	505	555	595
	14	1580	653	701	743
	12	2060	858	906	948
7	18	790	446	510	540
	16	1060	588	652	682
	14	1340	755	819	849
	12	1780	992	1056	1086
8	18	690	507	587	604
	16	945	669	749	766
	14	1180	860	940	957
	12	1540	1130	1210	1227
9	16	820	753	873	863
	14	1040	967	1087	1077
	12	1380	1271	1391	1381
10	16	740	835	963	1025
	14	945	1071	1199	1261
	12	1024	1408	1536	1598
11	16	670	916	1060	1122
	14	860	1176	1320	1382
	12	1120	1546	1690	1752
12	16	615	1003	1163	1215
	14	790	1287	1447	1499
	12	1025	1692	1852	1904
	10	1265	2080	2240	2292

TABLE 8 (Continued)

Diameter in Inches	Thickness B. W. Gauge	Approximate Bursting Pressure in Pounds per Square Inch	Weight in Pounds per 100 Feet		
			Plain End Pipe	With A and R Flanges, Bolts and Gaskets	With Root Bolted Joints
13	16	570	1106	1274	1346
	14	730	1420	1588	1660
	12	950	1866	2034	2106
	10	1165	2294	2462	2534
14	16	530	1199	1399	1465
	14	675	1539	1739	1805
	12	890	2022	2222	2288
	10	1090	2486	2686	2752
15	14	630	1649	1889	1973
	12	825	2167	2407	2491
	10	1015	2664	2904	2988
16	14	590	1771	2051	2149
	12	770	2327	2607	2705
	10	950	2861	3141	3239
18	14	525	1974	2334	2394
	12	690	2593	2953	3013
	10	850	3188	3548	3608
20	14	470	2180	2556	2606
	12	620	2863	3239	3291
	10	760	3521	3897	3949
22	14	430	2390	2830	2830
	12	565	3140	3580	3580
	10	695	3860	4300	4300
24	14	395	2604	3108	3084
	12	515	3421	3925	3901
	10	635	4216	4720	4696
26	12	475	3558	4718	4090
	10	580	4380	5540	4912
28	12	440	3894	5274	4478
	10	545	4790	6100	5304
30	12	410	4115	5531	4755
	10	510	5063	6479	5703

DIMENSIONS AND STRENGTH OF PIPE

The following information and Tables 9, 10, and 11 are based upon the American Spiral Pipe Works publications. In manufacturing Taylor's spiral riveted pipe, a strip of sheet metal is wound into helical shape with one edge overlapping the other for riveting the seam. The sheet is drawn and formed in such a manner as to obtain metal to metal contact, in order that the pipe may be more nearly smooth inside. The riveting is done cold by compression or squeezing under enormous pressure, thus insuring complete filling of the rivet holes with slight countersink. The pipe comes from the machines in a continuous piece, and is cut to any desired length. American Spiral pipe is made of various thicknesses, in sizes from 3 inches to 40 inches diameter, and is furnished in any length desired up to 30 feet for asphalt coated pipe and 20 feet for galvanized pipe.

TABLE 9

Taylor's Spiral Riveted Pipe

Standard Thickness

Diameter in Inches	Thickness U. S. Standard Gauge	Approximate Weight per Foot Asphalted Pounds	Approximate Bursting Pressure in Pounds per Square Inch	Diameter in Inches	Thickness U. S. Standard Gauge	Approximate Weight per Foot Asphalted Pounds	Approximate Bursting Pressure in Pounds per Square Inch
3	20	1.9	1500	16	14	18.1	585
4	18	3.0	1500	18	14	19.9	520
5	18	3.7	1200	20	14	22.1	470
6	16	5.3	1250	22	12	33.7	595
7	16	6.2	1070	24	12	36.5	540
8	16	7.1	935	26	12	39.5	505
9	16	8.0	835	28	10	51.7	605
10	16	8.8	750	30	10	56.8	560
11	16	9.7	680	32	10	61.6	525
12	16	10.6	625	34	10	65.4	490
13	16	11.4	575	36	10	69.1	470
14	14	15.9	670	40	10	76.7	420
15	14	17.0	625				

Above weights are for plain ends without connections.

Working pressure should not be more than 25 per cent. of the ultimate strength or bursting pressure.

TABLE 10

Taylor's Spiral Riveted Pipe

Extra Heavy Thickness

Diameter in Inches	Thickness U. S. Standard Gauge	Approximate Weight per Foot Asphalted Pounds	Approximate Bursting Pressure in Pounds per Square Inch	Diameter in Inches	Thickness U. S. Standard Gauge	Approximate Weight per Foot Asphalted Pounds	Approximate Bursting Pressure in Pounds per Square Inch
3	18	2.3	2000	16	12	25.2	820
4	16	3.7	1875	18	12	27.6	730
5	16	4.5	1500	20	12	30.6	660
6	14	6.6	1560	22	10	42.2	765
7	14	7.7	1340	24	10	45.7	705
8	14	8.8	1170	26	10	49.5	650
9	14	9.9	1045	28	8	63.6	735
10	14	11.0	935	30	8	68.7	685
11	14	12.0	850	32	8	74.3	645
12	14	13.0	780	34	8	78.8	600
13	14	14.1	720	36	8	83.4	570
14	12	22.2	940	40	8	92.4	515
15	12	23.7	875				

TABLE 11

Taylor's Spiral Riveted Pipe

Double Extra Heavy Thickness

Diameter in Inches	Thickness U. S. Standard Gauge	Approximate Weight per Foot Asphalted Pounds	Approximate Bursting Pressure in Pounds per Square Inch	Diameter in Inches	Thickness U. S. Standard Gauge	Approximate Weight per Foot Asphalted Pounds	Approximate Bursting Pressure in Pounds per Square Inch
6	12	9.2	2170	18	10	34.5	940
7	12	10.7	1860	20	10	38.3	840
8	12	12.3	1640	22	8	50.8	940
9	12	13.9	1460	24	8	55.2	820
10	12	15.3	1310	26	8	59.8	795
11	12	16.6	1200	28	6	76.6	870
12	12	18.2	1080	30	6	80.5	810
13	12	19.7	1010	32	6	87.1	760
14	10	27.6	1210	34	6	93.6	715
15	10	29.6	1125	36	6	97.8	680
16	10	31.5	1050	40	6	108.5	610

Some of the advantages claimed for riveted pipe as compared with cast iron pipe in large sizes are given in a pamphlet by Edwin Burhorn, M.E. These are uniformity in thickness and materials, absence of blow holes, no shrinkage strains, lessened freight and haulage charges (straight riveted pipe can be shipped "knocked down" and nested, the sheets being properly curved, punched, fitted, and marked ready for erection), cheapened erection and handling costs as its weight is only about one third that of corresponding cast iron pipe, lessened resistance to flow of contents, safety against damage due to hidden defects. The pamphlet also describes and illustrates straight riveted pipe which has been built and which is advocated for high pressure steam mains, exhaust steam systems, vacuum exhausts for engines and turbines, discharge pipe from hydraulic dredges, water power distribution, pneumatic power and air supply, gas power and pipe lines, etc.

The thickness of material and character of the joint on riveted pipe depend entirely upon the service for which the pipe is required. The lap and butt joint may be single, double, or triple riveted, designed for the special conditions, and flanges may be either single or double riveted to the pipe.

When pipe is straight riveted the computation becomes the same as for a steel tank or boiler shell. Information with regard to straight riveted pipe is given in Table 12.

TABLE 12

Straight Seam Riveted Pipe

Inside Diameter Inches	Thickness of Material		Theoretical Safe Working Head, Feet	Approximate Weight per Lineal Foot Pounds
	U. S. Standard Gauge	Inches		
16	16	.062	190	13.00
16	14	.078	237	16.00
16	12	.109	332	22.25
16	11	.125	379	24.50
16	10	.140	425	28.50
18	16	.062	168	14.75
18	14	.078	210	18.50
18	12	.109	295	25.25
18	11	.125	337	29.00
18	10	.140	378	32.50
18	8	.171	460	40.00
20	16	.062	151	16.00
20	14	.078	189	19.75

TABLE 12 (*Continued*)

Straight Seam Riveted Pipe

Inside Diameter Inches	Thickness of Material		Theoretical Safe Working Head, Feet	Approximate Weight per Lineal Foot Pounds
	U. S. Standard Gauge	Inches		
20	12	.109	265	27.50
20	11	.125	304	31.50
20	10	.140	340	35.00
20	8	.171	415	45.50
22	16	.062	138	17.75
22	14	.078	172	22.00
22	12	.109	240	30.50
22	11	.125	276	34.50
22	10	.140	309	39.00
22	8	.171	376	50.00
24	14	.078	158	23.75
24	12	.109	220	32.00
24	11	.125	253	37.50
24	10	.140	283	42.00
24	8	.171	346	50.00
24	6	.200	405	59.00
26	14	.078	145	25.50
26	12	.109	203	35.50
26	11	.125	233	39.50
26	10	.140	261	44.25
26	8	.171	319	54.00
26	6	.200	373	64.00
28	14	.078	135	27.25
28	12	.109	188	38.00
28	11	.125	216	42.25
28	10	.140	242	47.50
28	8	.171	295	58.00
28	6	.200	346	69.00
30	12	.109	176	39.50
30	11	.125	202	45.00
30	10	.140	226	50.50
30	8	.171	276	61.75
30	6	.200	323	73.00
30	¼	.250	404	90.00
36	11	.125	168	54.00
36	10	.140	189	60.50
36	3/16	.187	252	81.00
36	¼	.250	337	109.00
36	5/16	.312	420	135.00
40	3/16	.187	226	90.00
40	¼	.250	303	120.00
40	5/16	.312	378	150.00
40	3/8	.375	455	180.00

DIMENSIONS AND STRENGTH OF PIPE

The safe working heads given in the Table are theoretical and are based on ordinary working conditions, so judgment should be used in deciding the safe heads for a particular case. The values given in the Table are for double-riveted longitudinal seams and single-riveted circumferential seams. Proper allowances should be made for possible water hammer, settling, expansion and contraction of the pipe, and causes which would tend to collapse the pipe.

Copper and Brass Pipe. — Copper pipe may be figured by the British Board of Trade rule which for well made pipe with brazed joints is

$$t = \frac{pd}{6000} + \frac{1''}{16} \quad\quad\quad\quad (11)$$

and for solid drawn pipe of 8 inches diameter or less

$$t = \frac{pd}{6000} + \frac{1''}{32} \quad\quad\quad\quad (12)$$

t = thickness in inches.
p = pressure in pounds per square inch.
d = diameter in inches.

Table 13 gives dimensions and weights of brass and copper pipe.

TABLE 13
Seamless Drawn Brass and Copper Pipe

	Standard Weight					Extra Heavy		
Nominal Diameter	Inside Diameter Inches	Outside Diameter Inches	Approximate Weight per Lineal Foot		Nominal Diameter	Inside Diameter Inches	Approximate Weight per Lineal Foot	
			Brass Pounds	Copper Pounds			Brass Pounds	Copper Pounds
⅛	.281	.405	.25	.26	⅛	.205	.370	.388
¼	.375	.540	.43	.45	¼	.294	.625	.650
⅜	.494	.675	.62	.65	⅜	.421	.830	.870
½	.625	.840	.90	.95	½	.542	1.200	1.33
¾	.822	1.05	1.25	1.31	¾	.736	1.660	1.75
1	1.062	1.315	1.70	1.79	1	.951	2.360	2.478
1¼	1.368	1.66	2.50	2.63	1¼	1.272	3.300	3.465
1½	1.600	1.90	3.00	3.15	1½	1.494	4.250	4.462
2	2.062	2.375	4.00	4.20	2	1.933	5.460	5.733

TABLE 13 (Continued)

Seamless Drawn Brass and Copper Pipe

Standard Weight — *Extra Heavy*

Nominal Diameter	Inside Diameter Inches	Outside Diameter Inches	Approximate Weight per Lineal Foot		Nominal Diameter	Inside Diameter Inches	Approximate Weight per Lineal Foot	
			Brass Pounds	Copper Pounds			Brass Pounds	Copper Pounds
2½	2.500	2.875	5.75	6.04	2½	2.315	8.300	8.715
3	3.062	3.50	8.30	8.72	3	2.892	11.200	11.760
3½	3.500	4.00	10.90	11.45	3½	3.358	13.700	14.385
4	4.000	4.50	12.70	13.33	4	3.818	16.500	17.325
4½	4.500	5.00	13.90	14.60	5	4.813	22.800	23.940
5	5.062	5.563	15.75	16.54	6	5.750	32.00	33.60
6	6.125	6.625	18.31	19.23				
7	7.062	7.625	26.28	27.60				
8	7.982	8.625	29.88	31.37				

Lead Pipe. — As mentioned in Chapter I, lead pipe was in use in very early times. It was made by the Romans by bending sheets of lead and soldering the seams. Lead pipe is now made by extrusion, using the hydraulic press to produce continuous pieces of almost any length. For lead pipe the Chadwick-Boston Company give the following formulae and Tables 14 and 15:

$$t = \frac{pd}{2f} \quad \quad (13)$$

$$t = \frac{hd}{750} \quad \quad (14)$$

t = thickness in inches.
d = diameter in inches.
f = fibre stress in pounds per square inch.
p = internal fluid pressure in pounds per square inch.
h = head in feet.

DIMENSIONS AND STRENGTH OF PIPE

TABLE 14
Sizes and Weights of Lead Pipe

Calibre Inches		A = Outside Diameter, Inches B = Weight per Foot, Pounds, Ounces								
$1/8$	A	$1/4$								
	B	0–2½								
$1/4$	A	$3/8$	$7/16$	$31/64$						
	B	0–5	0–8	0–11						
$3/8$	A	$31/64$	$34/64$	$9/16$	$7/12$	$39/64$	$41/64$	$35/50$	$3/4$	$25/32$	$41/48$
	B	0–6	0–8	0–10	0–12	0–14	1–0	1–4	1–8	1–12	2–0
$1/2$	A	$39/64$	$21/32$	$43/64$	$7/10$	$34/48$	$49/64$	$13/16$	$41/48$	$7/8$	$46/48$
	B	0–8	0–10	0–12	0–14	1–0	1–4	1–8	1–12	2–0	2–8
	A	$1^{1}/_{48}$	$1^{1}/_{8}$								
	B	3–0	4–0								
$5/8$	A	$3/4$	$25/32$	$13/16$	$27/32$	$57/64$	$59/64$	$49/50$	1	$1^{1}/_{24}$	$1^{1}/_{20}$
	B	0–13	0–14	1–0	1–4	1–8	1–12	2–0	2–4	2–8	2–12
	A	$1^{5}/_{64}$	$1^{1}/_{8}$	$1^{5}/_{20}$	$1^{1}/_{8}$	$1^{17}/_{64}$					
	B	3–0	3–4	3–8	4–0	4–8					
$3/4$	A	$7/8$	$57/64$	$58/64$	$59/64$	$43/50$	$63/64$	$1^{1}/_{64}$	$1^{1}/_{16}$	$1^{5}/_{64}$	$1^{1}/_{8}$
	B	0–12	0–14	1–0	1–2	1–4	1–8	1–12	2–0	2–4	2–8
	A	$1^{3}/_{20}$	$1^{1}/_{6}$	$1^{14}/_{64}$	$1^{17}/_{64}$	$1^{15}/_{48}$	$1^{17}/_{48}$				
	B	2–12	3–0	3–8	4–0	4–8	5–0				
1	A	$1^{9}/_{50}$	$1^{3}/_{16}$	$1^{14}/_{64}$	$1^{1}/_{4}$	$1^{17}/_{64}$	$1^{7}/_{24}$	$1^{11}/_{32}$	$1^{3}/_{8}$	$1^{5}/_{12}$	$1^{17}/_{32}$
	B	1–4	1–8	1–12	2–0	2–4	2–8	3–0	3–8	4–0	5–0
	A	$1^{29}/_{48}$	$1^{21}/_{32}$	$1^{3}/_{4}$							
	B	6–0	7–0	8–0							
$1^{1}/_{4}$	A	$1^{5}/_{12}$	$1^{9}/_{20}$	$1^{15}/_{32}$	$1^{31}/_{64}$	$1^{17}/_{32}$	$1^{37}/_{64}$	$1^{5}/_{8}$	$1^{13}/_{20}$	$1^{7}/_{10}$	$1^{3}/_{4}$
	B	1–12	2–0	2–4	2–8	3–0	3–8	4–0	4–8	5–0	6–0
	A	$1^{13}/_{16}$	$1^{57}/_{64}$	$1^{31}/_{32}$							
	B	7–0	8–0	9–0							
$1^{1}/_{2}$	A	$1^{9}/_{12}$	$1^{7}/_{10}$	$1^{3}/_{4}$	$1^{25}/_{32}$	$1^{13}/_{16}$	$1^{27}/_{32}$	$1^{57}/_{64}$	$1^{61}/_{64}$	$2^{1}/_{32}$	
	B	2–0	2–8	3–0	3–8	4–0	4–8	5–0	6–0	7–0	
	A	$2^{3}/_{32}$	$2^{3}/_{16}$	$2^{5}/_{16}$							
	B	8–0	10–0	12–0							
$1^{3}/_{4}$	A	$1^{31}/_{32}$	$2^{1}/_{32}$	$2^{3}/_{32}$	$2^{5}/_{32}$	$2^{17}/_{64}$	$2^{5}/_{12}$	$2^{1}/_{2}$			
	B	3–0	4–0	5–0	6–0	8–0	10–0	12–0			
2	A	$2^{3}/_{16}$	$2^{1}/_{4}$	$2^{5}/_{16}$	$2^{3}/_{8}$	$2^{5}/_{12}$	$2^{15}/_{32}$	$2^{17}/_{32}$	$2^{7}/_{12}$	$2^{11}/_{16}$	
	B	3–0	4–0	5–0	6–0	7–0	8–0	9–0	10–0	12–0	

TABLE 14 (*Continued*)

Sizes and Weights of Lead Pipe

Calibre Inches		A = Outside Diameter, Inches B = Weight per Foot, Pounds, Ounces								
$2^1/_2$	A	$2^{11}/_{16}$	$2^{25}/_{32}$	$2^{28}/_{32}$	$2^{11}/_{12}$	3	$3^1/_8$	$3^1/_3$		
	B	3-8	5-0	7-0	8-0	11-0	14-0	18-0		
3	A	$3^1/_8$	$3^7/_{32}$	$3^9/_{32}$	$3^1/_3$	$3^{13}/_{32}$	$3^1/_2$	$3^{41}/_{64}$	$3^{11}/_{16}$	$3^3/_4$
	B	4-0	5-0	6-0	8-0	10-0	13-0	16-0	17-0	19-0
$3^1/_2$	A	$3^{11}/_{16}$	$3^{23}/_{32}$	$3^7/_8$	4	$4^1/_{12}$				
	B	4-8	6-0	10-0	15-0	19-0				
4	A	$4^5/_{32}$	$4^1/_5$	$4^1/_4$	$4^{19}/_{64}$	$4^3/_8$	$4^1/_2$	$4^{15}/_{16}$		
	B	5-0	6-0	8-0	10-0	12-0	18-0	21-0		
$4^1/_2$	A	$4^{11}/_{16}$	$4^{29}/_{32}$	$4^{57}/_{64}$	5					
	B	7-0	8-0	14-0	20-0					
5	A	$5^{13}/_{64}$	$5^{15}/_{64}$	$5^3/_8$	$5^1/_2$					
	B	8-0	9-0	15-0	22-0					
6	A	$6^1/_8$	$6^3/_{16}$	$6^1/_2$	$6^3/_4$					
	B	10-0	12-0	26-0	33-0					

TABLE 15

Weight of Lead Pipe for Various Pressures

Calibre Inches	Pressure in Pounds per Square Inch						
	15	20	25	38	50	75	100
	lbs. oz.	lbs. oz.	lbs. oz.	lbs. oz.	lbs. oz.	lbs. oz.	lbs. oz.
$^3/_8$	0-8	0-10 0-12	0-12	1-0	1-4	1-4 1-8	1-8
$^1/_2$	0-12	0-14 1-0	1-4	1-8 1-12	2-0	2-8	3-0
$^5/_8$	1-4 1-8	1-12	1-12 2-0	2-4 2-8	2-12 3-0	3-4 3-8	4-0
$^3/_4$	1-4 1-8	1-12 2-0	2-4 2-8	3-0 3-8	4-0	4-8	5-0
1	1-12 2-0	2-8	3-0	4-0	5-0	6-0	7-0
$1^1/_4$	2-8	3-0	4-0	4-8 5-0	7-0	9-0	12-0
$1^1/_2$	3-8	4-0	4-8 5-0	6-0	10-0	12-0	15-0

DIMENSIONS AND STRENGTH OF PIPE

Wooden Stave Pipe. — Continuous wooden stave pipe is used for conveying water long distances and especially where the expense of cast iron or steel pipe would be prohibitive. Sizes ordinarily range from two to ten feet in diameter. The staves are generally made of redwood or fir, and of thicknesses ranging from $1^5/_8$ inches net thickness for sizes up to 44 inches diameter, 2 inches up to 60 inches, and $2^1/_2$ inches up to 8 feet diameter.

The bands for wooden stave pipe should be of soft steel with an ultimate tensile strength of about 60,000 pounds per square inch, and an elongation of at least 25 per cent. in 8 inches. The ends of the bands should have either rolled threads or be upset

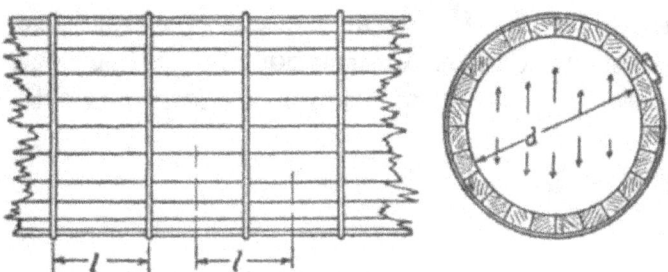

Fig. 14. Wood Stave Pipe.

so as to have the same strength as the unthreaded portion. The usual sizes of bands vary from $3/_8$ inches for pipe 2 feet outside diameter to $3/_4$ inches for pipe $4^1/_2$ feet outside diameter. The spacing may be figured from formula 15.

In Fig. 14 let

A = area of section of hoop in square inches.
f = unit stress of material of hoop in pounds per square inch.
d = diameter of pipe in inches.
l = spacing of hoops in inches.
p = pressure in pounds per square inch.

Then

pdl = force tending to separate pipe,
$2Af$ = force resisting separation of pipe,

equating

$$pdl = 2Af.$$

Introducing a coefficient C to allow for the stress due to swelling of the wood including a factor of safety of four or five for the bands, this equation becomes

$$l = \frac{2Af}{pdC} \quad \quad \quad (15)$$

or

$$A = \frac{pdlC}{2f} \quad \quad \quad (16)$$

It is not considered desirable to have the band spacing exceed 10 inches, and good practice often indicates even closer spacing, regardless of pressure requirements.

Bulletin 155, of the U. S. Department of Agriculture, by S. O. Jayne, gives considerable information on this subject, and has been referred to in the preparation of the foregoing article.

CHAPTER III

PIPE THREADS

Screw threads form a part of many types of joints and fittings used for piping. The kinds used for such purposes will be described in this chapter.

American Pipe Threads. — The thread used on piping in the United States is known as the Briggs Standard. This standard

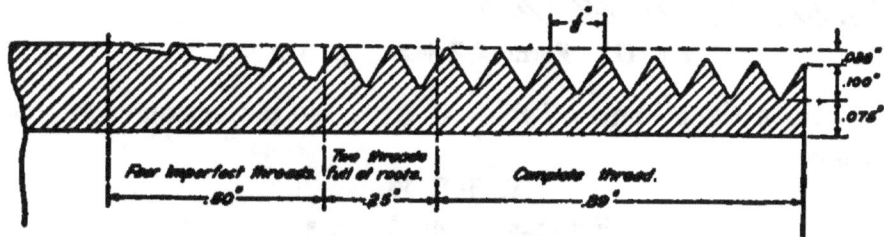

Fig. 15. Enlarged Section of 2½" Pipe Thread.

is due to Robert Briggs, C. E., who prepared a paper on "American Practice in Warming Buildings by Steam," for the Institution of Civil Engineers of Great Britain. This paper was presented and read after his death. An enlarged longitudinal section of a nominal $2^1/_2$-inch pipe is shown in Fig. 15. The end of the pipe has a taper of 1 in 16 or $^3/_4$ inch per foot, Fig. 16. The thread has an angle of 60 degrees and is rounded at the top and bottom, so that the depth of the thread is .8 of the pitch. Fig. 17 shows this form. The length of perfect thread, which is the distance the pipe should enter, is given by the formula

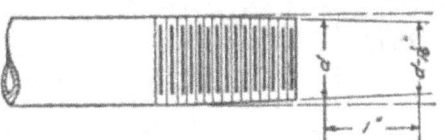

Fig. 16. Taper of Threaded Pipe End.

$$L = (4.8 + .8D)\frac{1}{N} \quad \ldots \ldots \ldots \ldots \ldots \ldots (17)$$

D = actual external diameter of pipe.
N = number of threads per inch.

Preceding the perfect threads are two threads perfect at the bottom and imperfect at the top. Preceding these are four threads imperfect at both top and bottom. The number of threads per inch is arbitrary, and comes from usage along with the nominal size of the pipe. They are finer in pitch than ordinary bolt threads because of the thinness of the metal and to

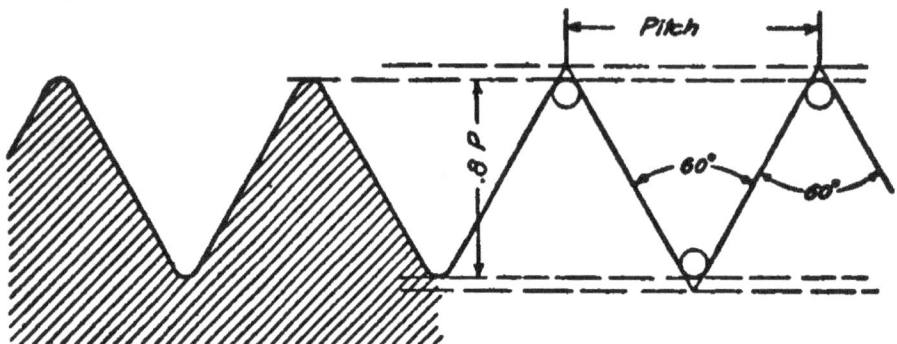

Fig. 17. Form of Briggs' Pipe Thread.

maintain a tight joint. Table 16 gives the dimensions for pipe threads.

TABLE 16
Standard Pipe Threads

Size Inches	Threads per Inch	Diameter of Tap Drill Inches	Outside Diameter of Threads at End of Pipe Inches	Depth of Threads Inches	Number of Perfect Threads	Length of Perfect Threads Inches
1/8	27	21/64	.393	.029	5.13	.19
1/4	18	27/64	.522	.044	5.22	.29
3/8	18	9/16	.656	.044	5.4	.30
1/2	14	11/16	.815	.057	5.46	.39
3/4	14	29/32	1.025	.057	5.6	.40
1	11 1/2	1 1/8	1.283	.069	5.87	.51
1 1/4	11 1/2	1 15/32	1.626	.069	6.21	.54
1 1/2	11 1/2	1 23/32	1.866	.069	6.33	.55
2	11 1/2	2 3/16	2.339	.069	6.67	.58
2 1/2	8	2 9/16	2.819	.100	7.12	.89
3	8	3 3/16	3.441	.100	7.6	.95
3 1/2	8	3 11/16	3.938	.100	8.0	1.00
4	8	4 3/16	4.434	.100	8.4	1.05
4 1/2	8	4 3/4	4.931	.100	8.8	1.10
5	8	5 5/16	5.490	.100	9.28	1.16
6	8	6 5/16	6.546	.100	10.08	1.26
7	8	....	7.540	.100	10.88	1.36
8	8	....	8.534	.100	11.68	1.46
9	8	....	9.527	.100	12.56	1.57
10	8	....	10.645	.100	13.44	1.68

PIPE THREADS

Standard Pipe Thread Gages. — In order to avoid variation in the number of threads which pipe will screw into fittings tapped at different shops it is necessary to have a definite standard for the proper depth of thread. The following is from the report of the committee on Standardization of Pipe Threads of the American Society of Mechanical Engineers.

"The gages shall consist of one plug and one ring gage of each size.

"The plug gage shall be the Briggs standard pipe thread as adopted by the manufacturers of pipe fittings and valves, and recommended by The American Society of Mechanical Engineers in 1886. The plug is to have a flat or notch indicating the distance that the plug shall enter the ring by hand.

"The ring gage is to be known as the American Briggs standard adopted by the Manufacturers' Standardization Committee in 1913, and recommended by The American Society of Mechanical Engineers, the Committee on International Standard for Pipe Threads, and the Pratt & Whitney Company, manufacturers of gages. The thickness of the ring is given in Table 17. It shall be flush with the small end of the plug. This will locate the flat notch on the plug flush with the large side of the ring.

TABLE 17 (FIG. 18)

STANDARD PIPE THREAD GAGES

Pipe Size Inches	Ring Gage Thickness Inches	Pipe Size Inches	Ring Gage Thickness Inches
1/8	.180	5	.937
1/4	.200	6	.958
3/8	.240	7	1.000
1/2	.320	8	1.063
3/4	.339	9	1.130
1	.400	10	1.210
1 1/4	.420	12	1.360
1 1/2	.420	14	1.562
2	.436	15	1.687
2 1/2	.682	16	1.812
3	.766	18	2.000
3 1/2	.821	20	2.125
4	.844	22	2.250
4 1/2	.875	24	2.375

"The Table indicates the dimensions of the ring gage, *A*, shown in Fig. 18, which are the figures adopted by the Manufacturers' Standardization Committee.

"In the use of the plug gage shown in Fig. 18, the notch on the plug is to gage, and one thread large or one thread small must be the inspection limits.

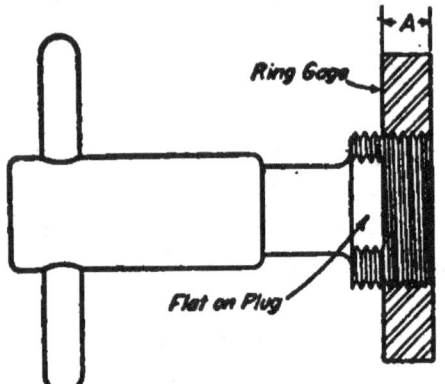

Fig. 18. Standard Plug and Ring Gage.

"In the use of the ring gage, male threads are to gage when flush with small end of ring, and one thread large or one thread small must be inspection limits."

Pipe Threading. — Pipe threads may be cut either by hand or in a machine. When cut by hand a pipe tap or die is used, shown in Fig. 19. For machine threading a lathe may be used, setting a properly shaped tool at right angles to the axis of the pipe, not perpendicular to the taper. A Saunders' pipe threading machine is shown in Fig. 20. A good threaded joint requires clean, smoothly

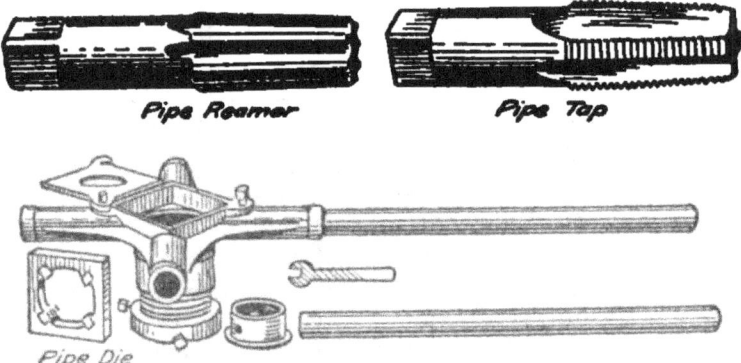

Fig. 19. Pipe Reamer, Hand Tap, Die and Die Stock.

cut threads. To make sure of such threads the die must be made with proper consideration as to lip, chip space, clearance, lead, and sufficient number of chasers. Valuable information along the following lines is given in National Tube Company's Bulletin No. 6.

PIPE THREADS 39

The lip is the inclination of the cutting edge of the chaser to the surface of the pipe, as shown in Fig. 21. This lip angle should

Fig. 20. Pipe Threading Machine.

be from 15 degrees to 25 degrees, depending upon conditions, and may be obtained by milling the cutting face of the chaser as shown by the full lines, or inclining the chaser as in the dotted

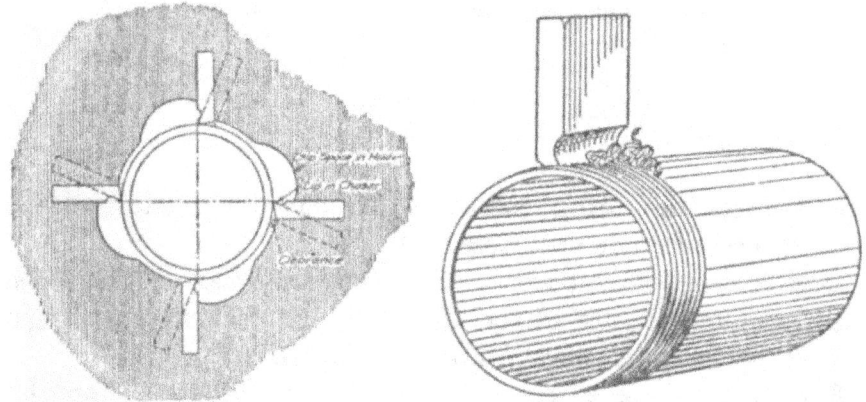

Fig. 21. Thread Cutting.   Fig. 22. Thread Cutting.

lines. Chip space should be provided as shown in the figure, as otherwise the chips will clog and tear the threads. Fig. 22 shows the working of a properly made chaser.

Clearance is the angle between the threads of the chasers and those of the pipe. Lead is the angle which is ground or machined on the front of each chaser to enable the die to start on the pipe and to distribute the work of cutting. The proper amount of

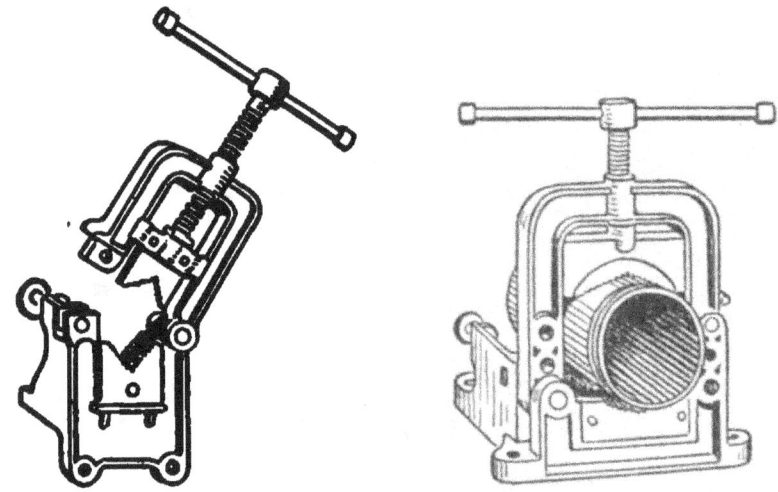

Fig. 23. Pipe Vises.

lead is about three threads. The number of chasers to obtain good results in threading at one cut is as follows:

1¼"	to	4"	should have approximately		6	chasers
4"	"	7"	"	"	8	"
7"	"	10"	"	"	10	"
10"	"	12"	"	"	12	"
12"	"	14"	"	"	14	"
14"	"	18"	"	"	16	"
18"	"	20"	"	"	18	"

In all cases the cutting tools should be kept well lubricated with good lard oil or crude cottonseed oil.

Pipe Tools. — Examples of various forms of vises, cutting tools, wrenches, etc., for use in the threading and making up of pipe are illustrated and named in Figs. 23 and 24.

English Pipe Threads. — "British Standard Pipe Threads," as given in the report of the Engineering Standards Committee,

PIPE THREADS 41

are shown in Fig. 25. This is the Whitworth form of thread. The tops and bottoms are rounded so that the depth is about .64 of

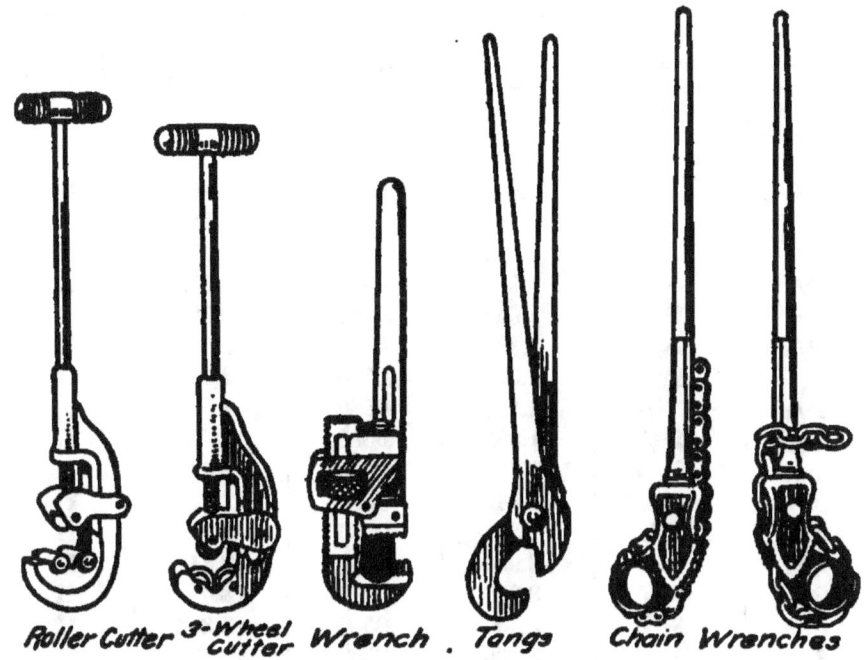

Fig. 24. Pipe Cutters, Tongs, and Wrenches.

the pitch, and the angle is 55 degrees. Ordinary pipe ends or "short screws" taper $3/4$ inch to the foot or $1/16$ inch per inch of length measured on the diameter, as in the Briggs system. Long screws are made straight. Table 18 gives information on

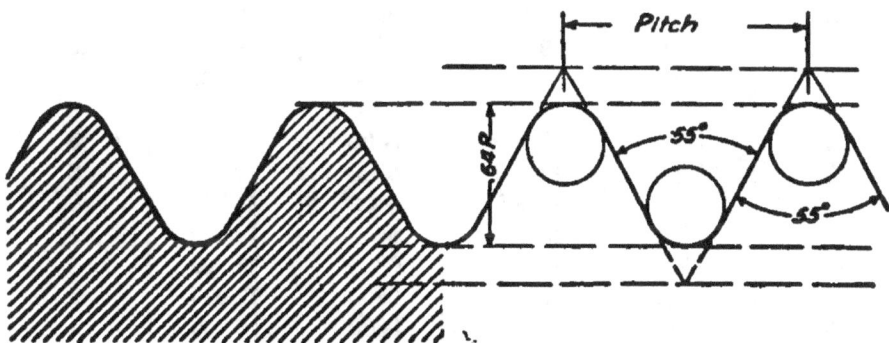

Fig. 25. Form of Whitworth Pipe Thread.

British pipe threads as approved by the above committee, for sizes up to 18 inches diameter.

TABLE 18

British Standard Pipe Threads

Nominal Inside Diameter Inches	Approximate Outside Diameter Inches	Gage Diameter Top of Thread Inches	Depth of Thread Inches	Core Diameter Inches	Number of Threads per Inch
1/8	11/32	.383	.0230	.337	28
1/4	17/32	.518	.0335	.451	19
3/8	11/16	.656	.0335	.589	19
1/2	27/32	.825	.0455	.734	14
5/8	15/16	.902	.0455	.811	14
3/4	1 1/16	1.041	.0455	.950	14
7/8	1 7/32	1.189	.0455	1.098	14
1	1 11/32	1.309	.0580	1.193	11
1 1/4	1 11/16	1.650	.0580	1.534	11
1 1/2	1 29/32	1.882	.0580	1.766	11
1 3/4	2 5/32	2.116	.0580	2.000	11
2	2 3/8	2.347	.0580	2.231	11
2 1/4	2 5/8	2.587	.0580	2.471	11
2 1/2	3	2.960	.0580	2.844	11
2 3/4	3 1/4	3.210	.0580	3.094	11
3	3 1/2	3.460	.0580	3.344	11
3 1/4	3 3/4	3.700	.0580	3.584	11
3 1/2	4	3.950	.0580	3.834	11
3 3/4	4 1/4	4.200	.0580	4.084	11
4	4 1/2	4.450	.0580	4.334	11
4 1/2	5	4.950	.0580	4.834	11
5	5 1/2	5.450	.0580	5.334	11
5 1/2	6	5.950	.0580	5.834	11
6	6 1/2	6.450	.0580	6.334	11
7	7 1/2	7.450	.0640	7.322	10
8	8 1/2	8.450	.0640	8.322	10
9	9 1/2	9.450	.0640	9.322	10
10	10 1/2	10.450	.0640	10.322	10
11	11 1/2	11.450	.0800	11.290	8
12	12 1/2	12.450	.0800	12.290	8
13	13 3/4	13.680	.0800	13.520	8
14	14 3/4	14.680	.0800	14.520	8
15	15 3/4	15.680	.0800	15.520	8
16	16 3/4	16.680	.0800	16.520	8
17	17 3/4	17.680	.0800	17.520	8
18	18 3/4	18.680	.0800	18.520	8

The Whitworth Standard Threads are given in Table 19 for sizes up to 4 inches in diameter.

TABLE 19

WHITWORTH STANDARD PIPE THREADS

Nominal Size	Actual Outside Diameter Inches	Diameter at Bottom of Thread Inches	No. of Threads per Inch	Diameter of Tap Drill Inches	Nominal Size	Actual Outside Diameter Inches	Diameter at Bottom of Thread Inches	No. of Threads per Inch	Diameter of Tap Drill Inches
1/8	.3825	.3367	28	5/16	1 7/8	2.245	2.1285	11	
1/4	.518	.4506	19	27/64	2	2.347	2.2305	11	1 5/32
3/8	.6563	.5889	19	9/16	2 1/8	2.467	2.3505	11	
1/2	.8257	.7342	14	11/16	2 1/4	2.5875	2.4710	11	2 13/32
5/8	.9022	.8107	14	25/32	2 3/8	2.794	2.6775	11	
3/4	1.041	.9495	14	29/32	2 1/2	3.0013	2.8848	11	2 23/32
7/8	1.189	1.0975	14	1 1/16	2 5/8	3.124	3.0075	11	
1	1.309	1.1925	11	1 1/8	2 3/4	3.247	3.1305	11	3 1/32
1 1/8	1.492	1.3755	11		2 7/8	3.367	3.2505	11	
1 1/4	1.650	1.5335	11	1 15/32	3	3.485	3.3685	11	3 9/32
1 3/8	1.745	1.6285	11		3 1/4	3.6985	3.5820	11	3 1/2
1 1/2	1.8825	1.7660	11	1 22/32	3 1/2	3.912	3.7955	11	3 3/4
1 5/8	2.021	1.9045	11		3 3/4	4.1255	4.0090	11	4
1 3/4	2.047	1.9305	11	1 15/16	4	4.339	4.2225	11	4 1/4

Foreign Pipe Threads. — The author is indebted to Mr. William J. Baldwin for notes on foreign practice. In the practice of Germany and France (comparing the German and French systems with the Briggs system), Germany uses straight threads nearly altogether. The pitch and form of thread is about the same as the English except that the thread as a whole is not tapered. France is more irregular in practice, the Navy following one method and private shops other methods. The French Navy, however, leans toward tapered threads.

South American countries have no fixed standards, but import from the United States and England and use the method of the country from which they import. Canada uses the Briggs standard. In Mexico a great deal of American pipe and fittings is used, but Mexico and the South and Central American countries use the methods of those from whom they buy, as a general rule.

CHAPTER IV

PIPE FITTINGS

Screw Fittings. — Since there is a practical limit to the length of pieces of pipe as well as the necessity for connections and convenient changes in direction, pipe fittings have been devised. There are two general classes of fittings, namely: screwed fittings and flanged fittings. As a rule the screwed fittings are used with

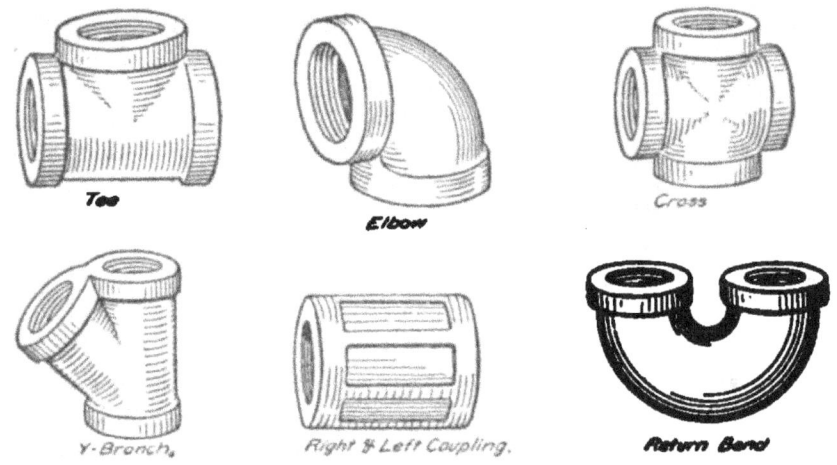

Fig. 26. Pipe Fittings.

the smaller sizes of pipe and for low pressure work. The flanged fittings are used for higher pressures and for larger sizes of pipe. Fig. 26 shows a variety of screwed fittings for "making up" standard pipe.

Couplings. — For joining two lengths of pipe, couplings are used. These may have right hand threads at both ends or may have right hand threads at one end and left hand threads at the other for convenience in connecting and disconnecting. Right and left couplings generally have bars running lengthwise to distinguish them from couplings with right hand threads. Sometimes reducing couplings are used where a change in size of pipe is desired. Couplings are made of cast iron, wrought iron, steel,

PIPE FITTINGS

malleable iron, and brass. A coupling is included on one end of each full length of standard pipe. Forms of couplings are shown

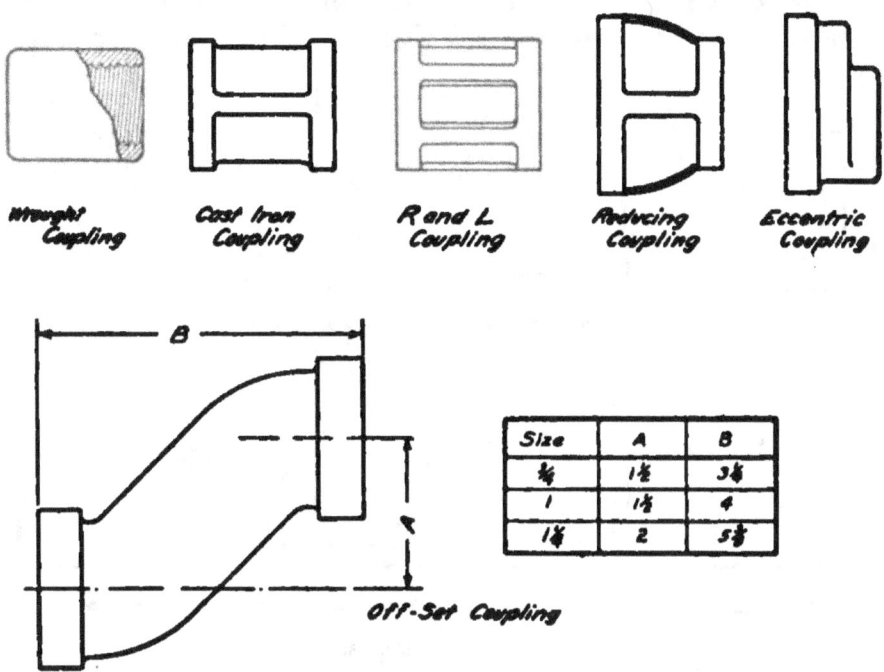

Fig. 27. Couplings.

in Fig. 27 and Table 20 gives the dimensions of standard wrought iron couplings.

TABLE 20

STANDARD WROUGHT IRON COUPLINGS

Size of Pipe Inches	Outside Diameter Inches	Length Inches	Average Weight Pounds	Size of Pipe Inches	Outside Diameter Inches	Length Inches	Average Weight Pounds
$1/8$	$19/32$	$15/16$	.03	$3 1/2$	$4 7/16$	$3 7/16$	3.40
$1/4$	$3/4$	$1 1/32$	.07	4	$4 15/16$	$3 7/16$	3.50
$3/8$	$29/32$	$1 3/32$	.11	$4 1/2$	$5 17/32$	$3 5/8$	4.70
$1/2$	$1 3/32$	$1 5/16$	.15	5	$6 1/4$	$4 1/8$	8.50
$3/4$	$1 11/32$	$1 9/16$	.25	6	$7 9/32$	$4 1/8$	9.70
1	$1 5/8$	$1 13/16$	.42	7	$8 9/32$	$4 1/8$	11.10
$1 1/4$	$1 31/32$	$2 1/16$	.60	8	$9 1/4$	$4 5/8$	13.60
$1 1/2$	$2 15/64$	$2 5/16$	.81	9	$10 5/16$	$5 1/8$	17.40
2	$2 23/32$	$2 9/16$	1.18	10	$11 5/8$	$6 1/8$	31.10
$2 1/2$	$3 5/16$	$2 7/8$	1.70	12	$13 7/8$	$6 1/8$	44.20
3	$3 15/16$	$3 1/16$	2.45				

46 A HANDBOOK ON PIPING

Elbows. — For turning corners elbows or ells are used, Fig. 28. Reducing ells are used to change the size of pipe at a corner. Sometimes ells are provided with an opening at the side in which case they are called side outlet elbows. Elbows are also made

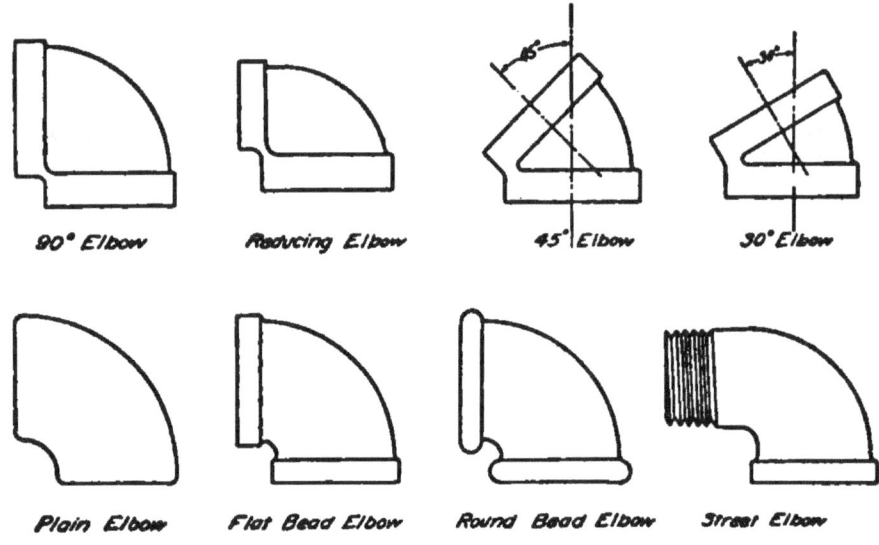

Fig. 28. Elbows.

for angles other than 90 degrees and are then specified by the angle, as 45 degree ell, 30 degree ell, or 60 degree ell, etc. It will be noticed that the angle is the one made with the axis or run of the pipe.

Tees, Crosses, Bushings, Caps, Plugs. — For a branch at right angles to the pipe tees are used. When the three openings are

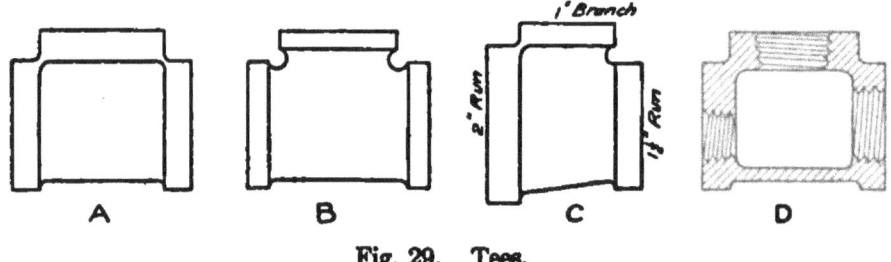

Fig. 29. Tees.

the same, the fitting is specified by the size of the pipe, as a 1-inch tee, or 2-inch tee, etc. When the branch is of a different size the size of run is given first and then the outlet, as $2'' \times 1\frac{1}{4}''$ tee. When the three openings are different they are all specified, giving the sizes of the run first, as $2'' \times 1\frac{1}{2}'' \times 1''$ tee, as shown at C in Fig. 29. Side outlet tees are also made. For a branch at other angles a Y, Fig. 30, may be used. Note that the angle is

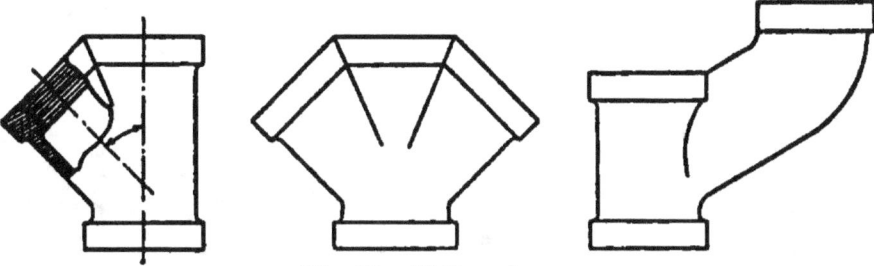

Fig. 30. Y-Branches.

the smaller of the two made with the run of the pipe. The use of a cross (+) is evident from the figure as well as the notation. Figs. 31 and 32 show other fittings. A bushing is used to bush or reduce the size of an opening so that a smaller pipe may be

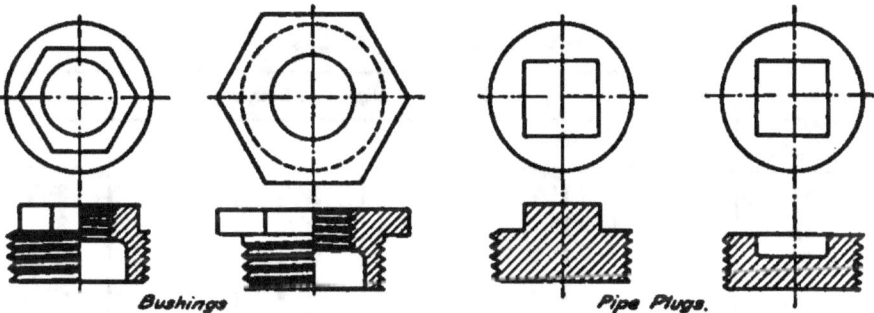

Fig. 31. Bushings and Plugs.

used. For closing an opening a pipe plug is used. For closing the end of a pipe a cap is used. A pipe nut is sometimes used as a locknut when a pipe is screwed into sheet metal. There are many special forms of fittings and the catalogs of standard manufacturers should be consulted. Tables 20 to 35 in this chapter give dimensions of screwed fittings sufficiently close for most purposes.

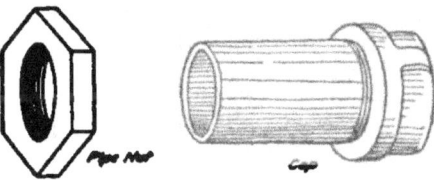

Fig. 32. Pipe Nut and Cap.

Nipples. — Short pieces of pipe used to join fittings which are near together are called nipples, and may be purchased with threads cut ready for use. They are known as close nipples when the threads run the entire length (*A*, Fig. 33), short or shoulder nipples when there is a small amount of unthreaded pipe (*B*, Fig. 33). Long nipples and extra long nipples have various lengths. Extra long nipples range from sizes given up to twelve

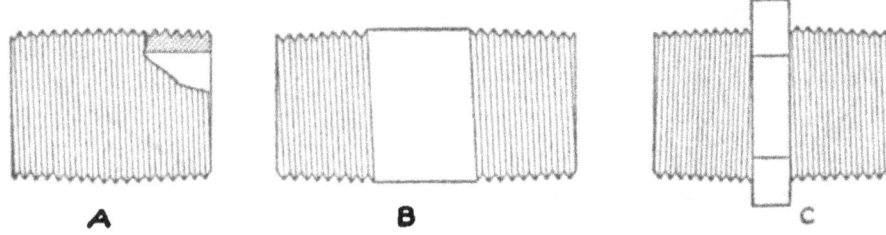

Fig. 33. Nipples.

inches in length. Table 21 gives the ordinary sizes of wrought iron nipples when both ends have right hand threads.

TABLE 21

Wrought Iron Nipples

Size of Pipe Inches	Length of Nipple in Inches					
	Close	Short	Long			
1/8	3/4	1 1/2	2	2 1/2	3	3 1/2
1/4	7/8	1 1/2	2	2 1/2	3	3 1/2
3/8	1	1 1/2	2	2 1/2	3	3 1/2
1/2	1 1/8	1 1/2	2	2 1/2	3	3 1/2
3/4	1 3/8	2	2 1/2	3	3 1/2	4
1	1 1/2	2	2 1/2	3	3 1/2	4
1 1/4	1 5/8	2 1/2	3	3 1/2	4	4 1/2
1 1/2	1 3/4	2 1/2	3	3 1/2	4	4 1/2
2	2	2 1/2	3	3 1/2	4	4 1/2
2 1/2	2 1/2	3	3 1/2	4	4 1/2	5
3	2 1/2	3	3 1/2	4	4 1/2	5
3 1/2	2 3/4	4	4 1/2	5	5 1/2	6
4	3	4	4 1/2	5	5 1/2	6
4 1/2	3	4	4 1/2	5	5 1/2	6
5	3 1/4	4 1/2	5	5 1/2	6	6 1/2
6	3 1/4	4 1/2	5	5 1/2	6	6 1/2
7	3 1/2	5				
8	3 1/2	5				
9	4	5				
10	4	5				
12	4	5				

PIPE FITTINGS

The sizes when threaded one end right hand and the other end left hand are the same for sizes up to four inches diameter. A right and left nipple of malleable iron with a hexagon center is shown at C in Fig. 33. These are made in sizes ranging from ¼ inch to 4 inches. Variations will be found as there are no standard dimensions.

Cast Iron Fittings. — Pipe fittings for screwed pipe are made of various materials and in various designs to suit the requirements of pressure and medium to be conveyed. For steam, water, etc., under pressure less than 125 pounds per square inch

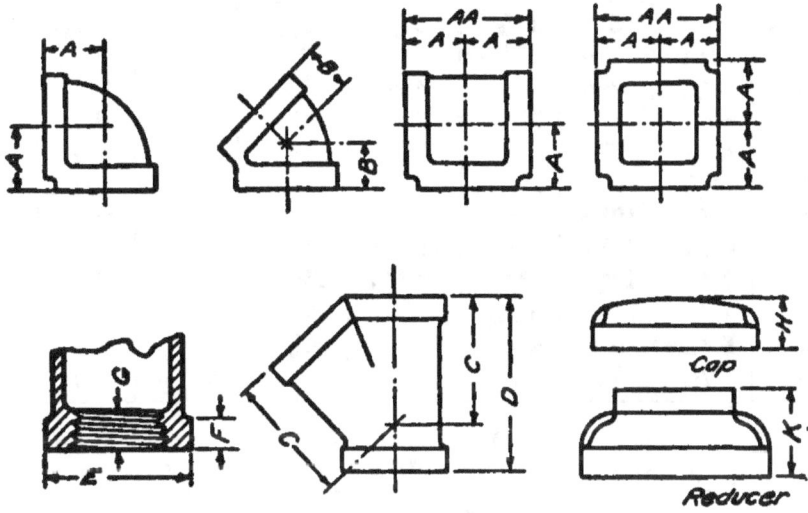

Fig. 34. Screwed Fittings.

standard weight fittings of cast iron are generally used. The question of strength involves much more than the pressure from within the pipe which induces a comparatively low stress in the material. The greater strains come from expansion, support, and "making up." For severe service or pressures from 125 to 250 pounds per square inch extra heavy cast iron fittings may be used. The dimensions of cast iron screwed fittings are not standardized and a variation will be found in the products of different manufacturers. For this reason the dimensions for standard weight, extra heavy, and long sweep cast iron fittings, and malleable iron fittings, as made by a number of companies have been given in Tables 22 to 29 inclusive. These will be found to give sufficient information for most purposes.

A HANDBOOK ON PIPING

TABLE 22 (Fig. 34)
Walworth Co. Standard Cast Iron Fittings

Size of Pipe Inches	A Inches	A-A Inches	B Inches	C Inches	D Inches	E Inches	F Inches	G Inches
1/4	3/4	1 1/2	7/16	...	...	1	1/4	3/8
3/8	7/8	1 3/4	9/16	1 7/16	2 1/16	1 1/8	5/16	7/16
1/2	1 1/16	2 1/8	11/16	1 7/8	2 9/16	1 7/16	3/8	1/2
3/4	1 5/16	2 5/8	13/16	2 1/16	2 3/4	1 3/4	7/16	9/16
1	1 1/2	3	15/16	2 1/2	3 1/4	2 1/16	1/2	5/8
1 1/4	1 13/16	3 5/8	1 1/16	3	3 3/4	2 1/2	9/16	11/16
1 1/2	2	4	1 3/16	3 1/4	4 3/4	2 3/4	5/8	13/16
2	2 3/8	4 3/4	1 3/8	4	5 1/2	3 3/8	11/16	7/8
2 1/2	2 7/8	5 3/4	1 5/8	5	6 13/16	4 1/8	13/16	1
3	3 5/16	6 5/8	1 7/8	5 5/8	7 5/8	4 3/4	15/16	1
3 1/2	3 11/16	7 3/8	2 1/16	6 3/8	8 3/4	5 1/4	1	1 1/16
4	4	8	2 1/4	7 1/8	9 3/4	6	1 1/16	1 1/8
4 1/2	4 7/16	8 7/8	2 7/16	7 7/8	10 1/2	6 9/16	1 1/8	1 1/4
5	4 11/16	9 3/8	2 9/16	8 1/2	11 5/16	7 1/16	1 1/8	1 1/4
6	5 5/16	10 5/8	2 13/16	9 15/16	13 1/8	8 3/8	1 1/8	1 3/8
7	6 1/16	12 1/8	3 1/8	11 1/4	14 5/8	9 3/4	1 3/16	1 1/2
8	6 13/16	13 5/8	3 9/16	12 15/16	16 13/16	10 7/8	1 3/8	1 5/8
9	7 1/2	15	3 7/8	14 1/2	19	12 1/8	1 7/16	1 3/4
10	8 1/4	16 1/2	4 5/16	16	20 7/8	13 1/4	1 5/8	1 3/4
12	9 9/16	19 1/8	4 7/8	...	...	15 5/8	1 3/4	1 7/8

TABLE 23 (Fig. 34)
Walworth Co. Extra Heavy Cast Iron Fittings

Size of Pipe Inches	A Inches	A-A Inches	B Inches	E Inches	F Inches	G Inches
1/2	1 5/32	2 5/16	3/4	1 21/32	7/16	9/16
3/4	1 3/8	2 3/4	7/8	1 29/32	1/2	5/8
1	1 19/32	3 3/16	1	2 5/16	9/16	11/16
1 1/4	1 15/16	3 7/8	1 3/16	2 5/8	11/16	13/16
1 1/2	2 1/16	4 1/8	1 1/4	3 1/16	3/4	7/8
2	2 1/2	5	1 1/2	3 3/4	7/8	1
2 1/2	3	6	1 3/4	4 9/16	1	1 1/8
3	3 11/16	7 3/8	2 1/4	5 3/8	1 1/4	1 3/8
3 1/2	4 1/32	8 1/16	2 7/16	6	1 5/16	1 7/16
4	4 15/32	8 15/16	2 11/16	6 13/16	1 7/16	1 9/16
4 1/2	4 27/32	9 11/16	2 7/8	7 3/8	1 9/16	1 11/16
5	5 7/32	10 7/16	3 1/8	7 15/16	1 11/16	1 13/16
6	5 13/16	11 5/8	3 5/16	9 5/16	1 3/4	1 7/8
8	7 3/16	14 3/8	3 15/16	11 9/16	1 7/8	1 15/16

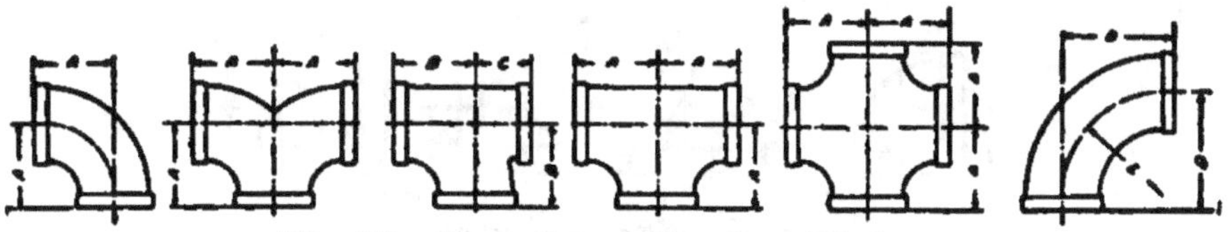

Fig. 35. Long Sweep Cast Iron Fittings.

TABLE 24 (Fig. 35)

WALWORTH CO. LONG SWEEP CAST IRON FITTINGS

Size of Pipe Inches	A Inches	B Inches	C Inches	D Inches	E Inches
1	2 1/4	2 5/8	1 1/2		
1 1/4	2 5/8	3 1/4	1 7/8		
1 1/2	3	3 3/8	2		
2	3 5/8	3 7/8	2 5/8		
2 1/2	4 3/4	4 1/2	3 1/4	6 1/8	5 1/4
3	5 1/2	5 1/4	3 1/2	7 1/4	5 7/8
3 1/2	5 3/4	5 3/4	4	8 1/4	6 3/4
4	6 1/8	6 3/8	4 3/8	10 5/8	9 1/8
4 1/2	6 1/4	6 7/8	4 5/8	10 7/8	9 1/4
5	7	7 3/4	5 1/8	11 1/8	9 1/2
6	7 1/2	9	6 1/4	13	11 3/8
7	8 7/8	10 3/8	6 5/8	14 1/2	12 3/4
8	9 1/2	11 7/8	7 3/8	18 1/4	16 1/2
9	10 3/4	.....	....	21 1/2	19 3/4
10	11 1/2	11 5/8	11 5/8	24 3/4	22 7/8
12	12 3/4	....	....	31	28 3/4

TABLE 25 (Fig 36)

NATIONAL TUBE COMPANY STANDARD CAST IRON FITTINGS

Size of Pipe Inches	A Inches	A–A Inches	B Inches	Size of Pipe Inches	A Inches	A–A Inches	B Inches
1/4	11/16	1 3/8	...	3 1/2	3 1/2	7	2 5/16
3/8	13/16	1 5/8	5/8	4	3 13/16	7 5/8	2 3/8
1/2	1 1/16	2 1/8	23/32	4 1/2	4 3/16	8 5/16	2 11/16
3/4	1 1/4	2 1/2	15/16	5	4 1/2	9	2 3/4
1	1 1/2	3	1 1/8	6	5 3/16	10 5/16	3 3/8
1 1/4	1 5/8	3 1/4	1 1/4	7	5 13/16	11 5/8	3 7/16
1 1/2	1 13/16	3 5/8	1 3/8	8	6 1/2	12 15/16	3 11/16
2	2 3/16	4 3/8	1 7/16	9	7 7/16	14 7/8	4 1/4
2 1/2	2 5/8	5 1/4	1 13/16	10	8 1/4	16 1/2	4 5/8
3	3 1/16	6 1/8	2 1/16	12	9 9/16	19 1/2	5 1/2

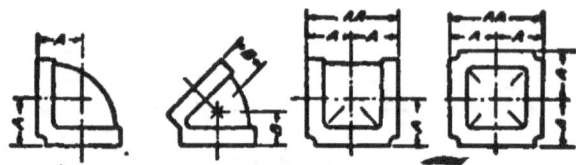

Fig. 36. Screwed Fittings.

TABLE 26 (FIG. 36)

NATIONAL TUBE COMPANY, EXTRA HEAVY CAST IRON FITTINGS

Size of Pipe Inches	A Inches	A-A Inches	Size of Pipe Inches	A Inches	A-A Inches
1/4	7/8	1 3/4	3 1/2	3 3/4	7 1/2
3/8	1	2	4	4 1/8	8 1/4
1/2	1 3/16	2 3/8	4 1/2	4 9/16	9 1/8
3/4	1 3/8	2 3/4	5	4 7/8	9 3/4
1	1 9/16	3 1/8	6	5 5/8	11 1/4
1 1/4	1 7/8	3 3/4	7	6 7/16	12 7/8
1 1/2	2 1/8	4 1/4	8	6 13/16	13 5/8
2	2 1/2	5	9	7 11/16	15 3/8
2 1/2	2 15/16	5 7/8	10	8 1/2	17
3	3 3/8	6 3/4	12	9 13/16	19 5/8

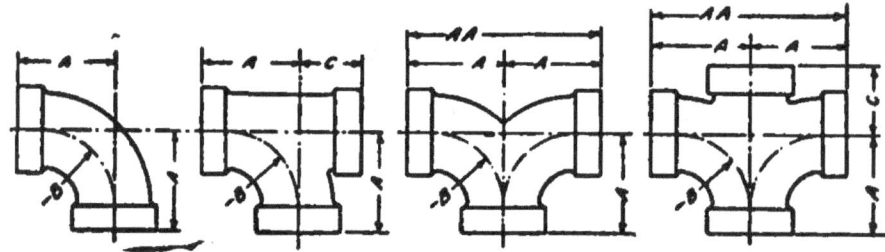

Fig. 37. Long Sweep Cast Iron Fittings.

TABLE 27 (FIG. 37)

NATIONAL TUBE COMPANY, LONG SWEEP CAST IRON FITTINGS

Size of Pipe Inches	A Inches	A-A Inches	B Inches	C Inches	Size of Pipe Inches	A Inches	A-A Inches	B Inches	C Inches
1	2 1/2	5	2	1 5/16	4 1/2	7 3/16	14 3/8	5 15/16	3 15/16
1 1/4	2 3/4	5 1/2	2 1/8	1 1/2	5	7 3/4	15 1/2	6 7/16	4 5/16
1 1/2	2 15/16	5 7/8	2 1/4	1 3/4	6	9 3/16	18 3/8	7 13/16	4 7/16
2	3 7/16	6 7/8	2 3/4	2 1/2	7	10 1/4	20 1/2	8 13/16	5 3/8
2 1/2	4 1/8	8 1/4	3 1/8	2 1/2	8	11 1/2	23	9 15/16	6
3	5 1/32	10 1/16	4 15/16	2 15/16	10	14 1/2	29	12 11/16	6 7/16
3 1/2	5 11/16	11 3/8	4 1/2	3 1/4	12	16	32	14	8
4	6 3/8	12 3/4	5 3/16	3 5/8					

PIPE FITTINGS

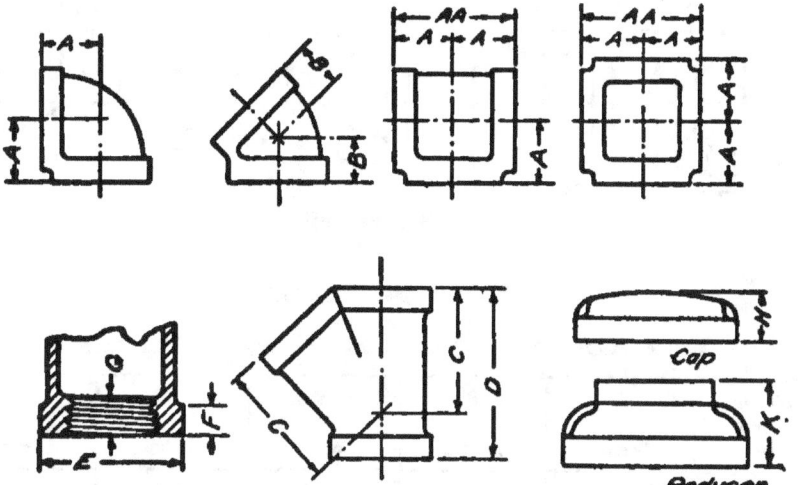

Fig. 34. Screwed Fittings.

TABLE 28 (FIG. 34)
CRANE COMPANY, STANDARD CAST IRON FITTINGS

Size of Pipe Inches	A Inches	B Inches	C Inches	D Inches	K Inches	H Inches
1/4	13/16	3/4				
3/8	15/16	13/16				
1/2	1 1/8	7/8	1 7/8	2 1/2		
3/4	1 5/16	1	2 1/4	3		
1	1 7/16	1 1/8	2 3/4	3 1/2		
1 1/4	1 3/4	1 5/16	3 1/4	4 1/4	2 1/8	
1 1/2	1 15/16	1 7/16	3 13/16	4 7/8	2 1/4	
2	2 1/4	1 11/16	4 1/2	5 3/4	2 7/16	
2 1/2	2 11/16	1 15/16	5 3/16	6 1/4	2 11/16	
3	3 1/8	2 3/16	6 1/8	7 7/8	2 15/16	
3 1/2	3 7/16	2 3/8	6 7/8	8 3/8	3 1/8	
4	3 3/4	2 5/8	7 5/8	9 3/4	3 3/8	2 1/16
4 1/2	4 1/16	2 13/16	9 1/4	11 3/8	3 5/8	2 3/16
5	4 7/16	3 1/16	9 1/4	11 3/8	3 7/8	2 3/8
6	5 1/8	3 7/16	10 3/8	13 7/16	4 3/8	2 5/8
7	5 13/16	3 7/8	12 1/4	15 1/4	4 13/16	2 7/8
8	6 1/2	4 1/4	13 5/8	16 15/16	5 1/4	3 1/8
9	7 3/16	4 11/16	16 3/8	20 11/16	5 11/16	3 3/8
10	7 7/8	5 3/16	16 3/4	20 11/16	6 3/16	3 5/8
12	9 1/4	6	19 5/8	24 1/8	7 1/8	4 1/4

A HANDBOOK ON PIPING

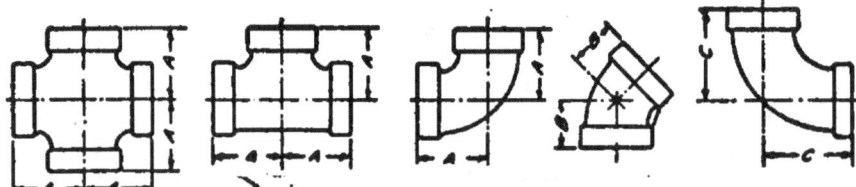

Fig. 38. Screwed Fittings.

TABLE 29 (Fig. 38)

CRANE COMPANY, EXTRA HEAVY CAST IRON FITTINGS

Size of Pipe Inches	A Inches	B Inches	Size of Pipe Inches	A Inches	B Inches
1	2	$1^3/8$	$4^1/2$	$5^1/2$	3
$1^1/4$	$2^1/4$	$1^1/2$	5	$6^1/8$	$3^5/16$
$1^1/2$	$2^9/16$	$1^5/8$	6	$7^1/4$	$3^3/4$
2	3	$1^{15}/16$	7	$8^1/8$	4
$2^1/2$	$3^1/2$	$2^1/4$	8	$9^1/8$	$4^3/4$
3	$4^1/8$	$2^1/2$	10	$11^3/8$	$4^7/8$
$3^1/2$	$4^{11}/16$	$2^9/16$	12	$13^3/8$	$5^1/2$
4	$5^1/8$	$2^3/4$			

Screwed Reducing Fittings. — The centre to face and face to face dimensions for Walworth Standard Weight cast iron screwed reducing tees and crosses, Fig. 39, are determined as follows:

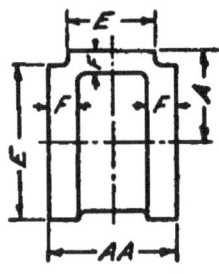

Fig. 39. Reducing Tees and Crosses.

For AA face to face, add to the outside diameter E of outlet bead, twice the width F of the run-bead. For A centre to face, add to the width F of outlet bead, one half the diameter E of the run-bead. Thus for a $2'' \times 3/4''$ tee the dimensions are

$$AA = 1^3/4 + {}^{11}/16 + {}^{11}/16 = 3^1/8''.$$
$$A = {}^7/16 + 1^{11}/16 = 2^1/8''.$$

See Table 22 for necessary dimensions.

Brass Fittings. — Brass fittings are made in both standard and extra heavy weights. They are used for feed water pipes where bad water makes steel pipes undesirable. Brass fittings may be had in iron pipe sizes. The dimensions as made by the Lunkenheimer Company are given in Table 30 for pressures up to 175 pounds, and in Table 31 for pressures up to 300 pounds.

PIPE FITTINGS

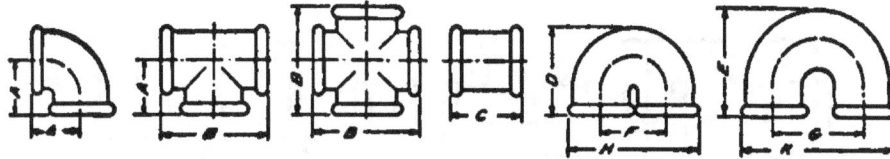

Fig. 40. Brass Fittings.

TABLE 30 (Fig. 40)
LUNKENHEIMER BRONZE FITTINGS, MEDIUM PATTERN

Size of Pipe Inches	A Inches	B Inches	C Inches	D Inches	E Inches	F Inches	G Inches	H Inches	K Inches
1/8	9/16	1 1/8	7/8	7/8	1 1/16	5/8	1	1 1/4	1 5/8
1/4	3/4	1 1/2	1	1 3/16	1 13/32	13/16	1 5/16	1 11/16	2 1/8
3/8	7/8	1 3/4	1 1/16	1 3/8	1 9/16	15/16	1 7/16	1 15/16	2 3/8
1/2	1	2	1 1/4	1 5/8	1 27/32	1 3/16	1 11/16	2 7/16	2 13/16
3/4	1 3/16	2 3/8	1 7/16	2	2 3/8	1 9/16	2 1/8	3	3 5/8
1	1 7/16	2 7/8	1 5/8	2 5/16	2 3/4	1 3/4	2 5/8	3 1/2	4 5/16
1 1/4	1 11/16	3 1/4	1 13/16	2 13/16	3 5/16	2 3/8	3 5/16	4 7/16	5 3/8
1 1/2	1 7/8	3 11/16	2	3	3 3/4	2 3/8	3 13/16	4 11/16	6 1/8
2	2 3/16	4 1/2	2 5/16	3 13/16	4 5/8	2 3/8	4 3/4	5 15/16	7 5/8
2 1/2	2 5/8	5 1/4	2 5/8	....	....	3 1/16	5 3/4		
3	3 1/16	6 1/16	2 13/16						

TABLE 31 (Fig. 40)
LUNKENHEIMER BRONZE FITTINGS, EXTRA HEAVY PATTERN

Size of Pipe Inches	A Inches	B Inches	C Inches	Size of Pipe Inches	A Inches	B Inches	C Inches
1/8	11/16	1 3/8	15/16	1 1/4	1 13/16	3 9/16	
1/4	7/8	1 11/16	1 1/16	1 1/2	2	4	
3/8	1	1 15/16	1 3/16	2	2 3/8	4 11/16	
1/2	1 1/8	2 1/4	1 3/8	2 1/2	2 3/4	5 7/16	
3/4	1 5/16	2 5/8	...	3	3 3/16	6 5/16	
1	1 1/2	3					

Malleable Iron Fittings. — Malleable iron fittings are made plain and beaded and for various pressures. Plain fittings are for low pressure work only. Standard beaded fittings may be used up to 150 pounds; extra heavy beaded fittings up to 250 pounds, and double extra heavy fittings for hydraulic work up to 800 pounds. The principle dimensions of malleable fittings as made by the National Tube Company are given in Tables 32 and 33. Extra heavy malleable iron fittings for pressures up to 250 pounds as made by Crane Company are dimensioned in Table 34.

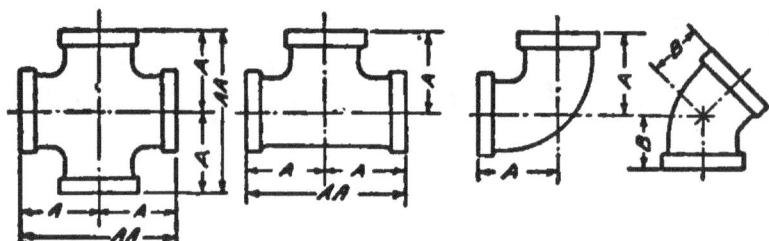

Fig. 41. Standard Malleable Fittings.

TABLE 32 (FIG. 41)

NATIONAL TUBE COMPANY STANDARD FLAT BEAD MALLEABLE FITTINGS

Size of Pipe Inches	A Inches	A-A Inches	B Inches	Size of Pipe Inches	A Inches	A-A Inches	B Inches
1/8	25/32	1 9/16	17/32	2 1/2	2 7/8	5 3/4	1 15/16
1/4	7/8	1 3/4	5/8	3	3 13/32	6 13/16	2 5/16
3/8	1 1/32	2 1/16	11/16	3 1/2	3 27/32	7 11/16	2 19/32
1/2	1 5/32	2 5/16	25/32	4	4 9/32	8 9/16	2 7/8
3/4	1 11/32	2 11/16	29/32	4 1/2	4 11/16	9 3/8	3 1/8
1	1 9/16	3 1/8	1 1/16	5	5 3/16	10 3/8	3 1/2
1 1/4	1 7/8	3 3/4	1 1/4	6	6 3/32	12 3/16	4 1/16
1 1/2	2 1/16	4 1/8	1 13/32	7	6 15/16	13 7/8	4 5/8
2	2 15/32	4 15/16	1 21/32	8	7 25/32	15 9/16	5 1/4

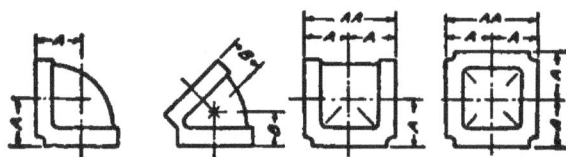

Fig. 36. Screwed Fittings.

TABLE 33 (FIG. 36)

NATIONAL TUBE COMPANY, EXTRA HEAVY FLAT BEAD MALLEABLE FITTINGS

Size of Pipe Inches	A Inches	A-A Inches	B Inches	Size of Pipe Inches	A Inches	A-A Inches	B Inches
1/4	11/16	1 3/8	....	3 1/2	3 1/2	7	2 5/16
3/8	13/16	1 5/8	5/8	4	3 13/16	7 5/8	2 3/8
1/2	1 1/16	2 1/8	23/32	4 1/2	4 3/16	8 5/16	2 11/16
3/4	1 1/4	2 1/2	15/16	5	4 1/2	9	2 3/4
1	1 1/2	3	1 1/8	6	5 3/16	10 5/16	3 3/8
1 1/4	1 5/8	3 1/4	1 1/4	7	5 13/16	11 5/8	3 7/16
1 1/2	1 13/16	3 5/8	1 3/8	8	6 1/2	12 15/16	3 11/16
2	2 3/16	4 3/8	1 7/16	9	7 7/16	14 7/8	4 1/4
2 1/2	2 5/8	5 1/4	1 13/16	10	8 1/4	16 1/2	4 5/8
3	3 1/16	6 1/8	2 1/16	12	9 9/16	19 1/8	5 1/2

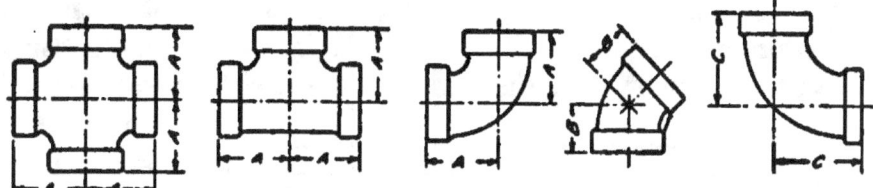

Fig. 38. Screwed Fittings.

TABLE 34 (FIG. 38)

CRANE COMPANY, EXTRA HEAVY MALLEABLE FITTINGS

Size of Pipe Inches	A Inches	B Inches	C Inches	Size of Pipe Inches	A Inches	B Inches	C Inches
1/4	1 1/16	3/4	....	2 1/2	3 1/2	2 1/4	4 3/4
3/8	1 1/4	7/8	....	3	4 1/8	2 1/2	5 1/2
1/2	1 1/2	1	....	3 1/2	4 5/8	2 5/8	6 1/4
3/4	1 3/4	1 1/8	....	4	5 1/8	2 13/16	7
1	2	1 5/16	2 1/2	4 1/2	5 5/8	....	7 3/4
1 1/4	2 1/4	1 1/2	3	5	6 1/4	....	8 1/2
1 1/2	2 1/2	1 11/16	3 1/2	6	7 1/4	....	9 1/2
2	3	2	4				

Extra Heavy Cast Steel Screwed Fittings. — Screwed fittings are made by Walworth Company of cast steel for superheated steam at 350 pounds working pressure and a total temperature of 800 degrees F. or for water working pressures of:

 5,000 pounds for 1 inch and smaller sizes.
 3,500 pounds for 1 1/4 inch to 2 inch.
 2,500 pounds for 2 1/2 inch to 4 1/2 inch.
 2,000 pounds for 5 inch and 6 inch sizes.

Such fittings have a larger radius than ordinary cast iron fittings.

Strength of Fittings. — Some results of tests to determine the average bursting pressures of extra heavy flanged fittings are plotted in Fig. 42. These tests were made by Crane Company who burst several fittings of each size under hydraulic pressure. The average tensile strength of the metal test bars were: Ferrosteel 33,000 pounds per square inch, and cast iron 22,000 pounds per square inch.

The bursting pressure of screwed fittings is from ten to twenty times the working pressure. The internal fluid pressure, however, is not the determining factor, as fittings must withstand the strain of expansion, contraction, weight of piping, settling, and water hammer, and there is also the possibility of non-uniform thickness. For cast iron the bursting pressure is generally in excess of 1000 pounds, and for malleable iron in excess of 2,000 pounds.

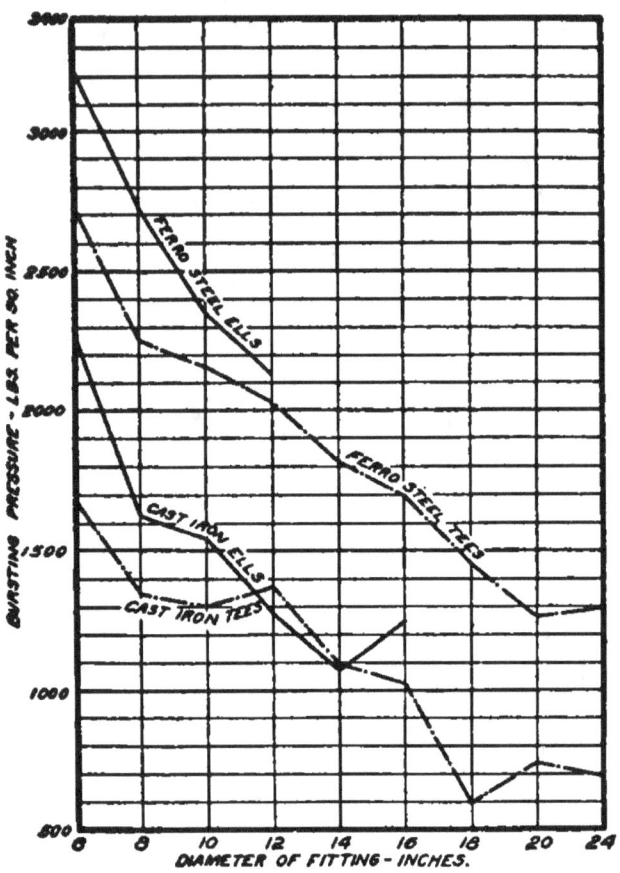

Fig. 42. Bursting Strength of Flanged Fittings.

Flanged Fittings. — Flanged fittings, Fig. 43, are to be preferred for important or high pressure work. Regular fittings are now made with dimensions of the American Standard as devised by a committee of the A. S. M. E., and a Manufacturers' committee. This standard fixes the dimensions for standard weight fittings (125 lbs.) from 1 inch to 100 inches and for extra heavy or high pressure fittings (250 lbs.) from 1 inch to 48 inches. The following tables give the dimensions revised to March 7th and 20th, 1914. The dimensions in Table 35 are common to all fittings for 125 pounds working pressure, and those in Table 36 are common to all fittings for 250 pounds working pressure. Tables 37 and 38 give the thickness of metal, and Tables 39 and 40 the dimensions of pipe flanges.

The following explanatory notes as well as the Tables and data here given are from the A. S. M. E. committee's report.

"(a) Standard and Extra Heavy Reducing Elbows carry same dimensions centre to face as regular Elbows of larger straight size.

"(b) Standard and Extra Heavy Tees, Crosses, and Laterals, reducing on run only, carry same dimensions face to face as larger straight size.

"(c) If Flanged Fittings for lower working pressure than 125 pounds are made, they shall conform in all dimensions except thickness of shell, to this standard and shall have the guaranteed

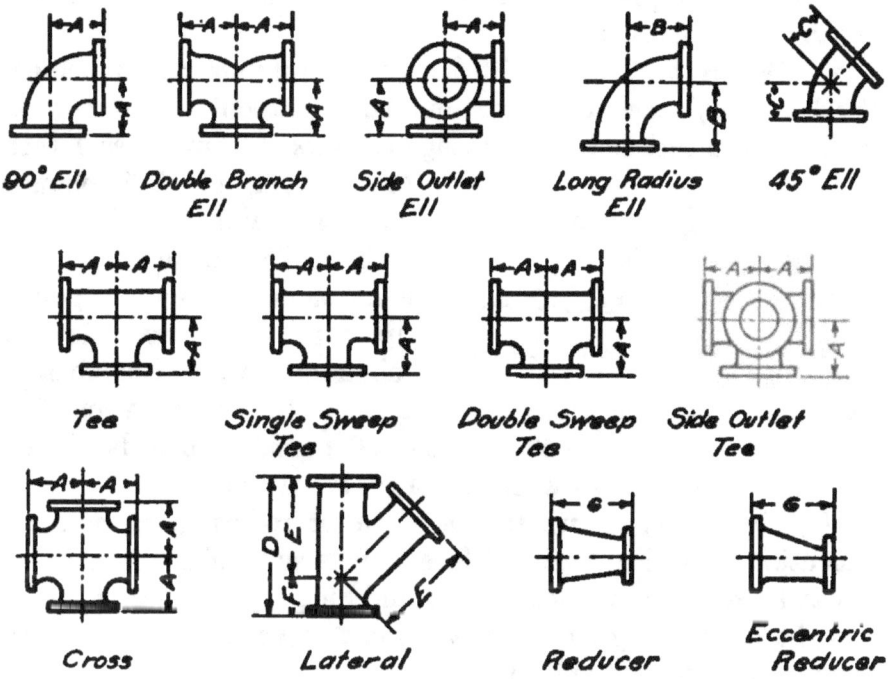

Fig. 43. A. S. M. E. Flanged Fittings.

working pressure cast on each fitting. Flanges for these fittings must be of standard dimensions.

"(d) Where Long Radius Fittings are specified, it has reference only to Elbows which are made in two centre to face dimensions, and to be known as Elbows and Long Radius Elbows, the latter being used only when so specified.

"(e) All standard weight fittings must be guaranteed for 125 pounds working pressure and Extra Heavy Fittings for 250 pounds working pressure, and each fitting must have some mark cast on it indicating the maker and guaranteed working steam pressure.

"(f) All extra heavy fittings and flanges to have a raised surface of $1/16$ inch high inside of bolt holes for gaskets, Fig. 44.

Standard weight fittings and flanges to be plain faced. Bolt holes to be $1/8$ inch larger in diameter than bolts. Bolt holes to straddle centre line.

"(g) Size of all fittings scheduled indicates inside diameter of ports.

"(h) The face to face dimension of reducers, either straight or eccentric, for all pressures, shall be the same face to face as given in table of dimensions.

"(i) Square head bolts with hexagonal nuts are recommended. For bolts, $1^5/8$ inch diameter and larger, studs with a nut on each end are satisfactory. Hexagonal nuts for pipe sizes 1 inch to 46 inch, on 125 pounds standard, and 1 inch to 16 inch on 250 pounds standard can be conveniently pulled up with open wrenches of minimum design of heads.

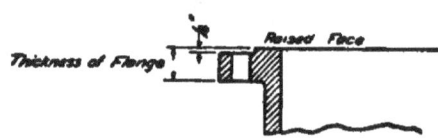

Fig. 44. Raised Face on Flange.

Hexagonal nuts for pipe sizes 48 inch to 100 inch on 125 pounds, and 18 inch to 48 inch on 250 pound standards can be conveniently pulled up with box or socket wrenches.

"(j) Twin Elbows, whether straight or reducing, carry same dimensions centre to face and face to face as regular straight size ells and tees. Side Outlet Elbows and Side Outlet Tees, whether straight or reducing sizes, carry same dimensions centre to face and face to face as regular tees having same reductions.

"(k) Bull Head Tees or Tees increasing on outlet, will have same centre to face and face to face dimensions as a straight fitting of the size of the outlet.

"(l) Tees and Crosses, 16 inches and down, reducing on the outlet, use the same dimensions as straight sizes of the larger port. Size 18 inch and up, reducing on the outlet, are made in two lengths, depending on the size of the outlet as given in the table of dimensions. Laterals, 16 inches and down, reducing on the branch, use the same dimensions as straight sizes of the larger port.

"(m) Sizes 18 inches and up, reducing on the branch, are made in two lengths, depending on the size of the branch, as given in the table of dimensions. The dimensions of reducing flanged fittings are always regulated by the reductions of the outlet or branch. Fittings reducing on the run only, the long body pattern

PIPE FITTINGS

will always be used. Y's are special and are made to suit conditions. Double sweep tees are not made reducing on the run.

"*(n) Steel Flanges, Fittings, and Valves are recommended for Superheated Steam.*"

TABLE 35 (FIG. 43)

AMERICAN STANDARD FLANGED FITTINGS

125 *Pounds Working Pressure*

Size Inches	A-A Inches	A Inches	B Inches	C Inches	D Inches	E Inches	F Inches	G Inches
1	7	$3^1/_2$	5	$1^3/_4$	$7^1/_2$	$5^3/_4$	$1^3/_4$	
$1^1/_4$	$7^1/_2$	$3^3/_4$	$5^1/_2$	2	8	$6^1/_4$	$1^3/_4$	
$1^1/_2$	8	4	6	$2^1/_4$	9	7	2	
2	9	$4^1/_2$	$6^1/_2$	$2^1/_2$	$10^1/_2$	8	$2^1/_2$	
$2^1/_2$	10	5	7	3	12	$9^1/_2$	$2^1/_2$	
3	11	$5^1/_2$	$7^3/_4$	3	13	10	3	6
$3^1/_2$	12	6	$8^1/_2$	$3^1/_2$	$14^1/_2$	$11^1/_2$	3	$6^1/_2$
4	13	$6^1/_2$	9	4	15	12	3	7
$4^1/_2$	14	7	$9^1/_2$	4	$15^1/_2$	$12^1/_2$	3	$7^1/_2$
5	15	$7^1/_2$	$10^1/_4$	$4^1/_2$	17	$13^1/_2$	$3^1/_2$	8
6	16	8	$11^1/_2$	5	18	$14^1/_2$	$3^1/_2$	9
7	17	$8^1/_2$	$12^3/_4$	$5^1/_2$	$20^1/_2$	$16^1/_2$	4	10
8	18	9	14	$5^1/_2$	22	$17^1/_2$	$4^1/_2$	11
9	20	10	$15^1/_4$	6	24	$19^1/_2$	$4^1/_2$	$11^1/_2$
10	22	11	$16^1/_2$	$6^1/_2$	$25^1/_2$	$20^1/_2$	5	12
12	24	12	19	$7^1/_2$	30	$24^1/_2$	$5^1/_2$	14
14	28	14	$21^1/_2$	$7^1/_2$	33	27	6	16
15	29	$14^1/_2$	$22^3/_4$	8	$34^1/_2$	$28^1/_2$	6	17
16	30	15	24	8	$36^1/_2$	30	$6^1/_2$	18
18	33	$16^1/_2$	$26^1/_2$	$8^1/_2$	39	32	7	19
20	36	18	29	$9^1/_2$	43	35	8	20
22	40	20	$31^1/_2$	10	46	$37^1/_2$	$8^1/_2$	22
24	44	22	34	11	$49^1/_2$	$40^1/_2$	9	24
26	46	23	$36^1/_2$	13	53	44	9	26
28	48	24	39	14	56	$46^1/_2$	$9^1/_2$	28
30	50	25	$41^1/_2$	15	59	49	10	30
32	52	26	44	16	...	...	...	32
34	54	27	$46^1/_2$	17	...	...	...	34
36	56	28	49	18	...	...	...	36
38	58	29	$51^1/_2$	19	...	...	...	38
40	60	30	54	20	...	...	...	40

TABLE 35 (Fig. 43) (Continued)

American Standard Flanged Fittings

125 Pounds Working Pressure

Size Inches	A-A Inches	A Inches	B Inches	C Inches	D Inches	E Inches	F Inches	G Inches
42	62	31	56½	21	...	...	...	42
44	64	32	59	22	...	...	...	44
46	66	33	61½	23	...	...	...	46
48	68	34	64	24	...	...	...	48
50	70	35	66½	25	...	...	...	50
52	74	37	69	26	...	...	...	52
54	78	39	71½	27	...	...	...	54
56	82	41	74	28	...	...	...	56
58	84	42	76½	29	...	...	...	58
60	88	44	79	30	...	...	...	60
62	90	45	81½	31	...	...	...	62
64	94	47	84	32	...	...	...	64
66	96	48	86½	33	...	...	...	66
68	100	50	89	34	...	...	...	68
70	102	51	91½	35	...	...	...	70
72	106	53	94	36	...	...	...	72
74	108	54	96½	37	...	...	...	74
76	112	56	99	38	...	...	...	76
78	116	58	101½	39	...	...	...	78
80	118	59	104	40	...	...	...	80
82	120	60	106½	41	...	...	...	82
84	124	62	109	42	...	...	...	84
86	126	63	111½	43	...	...	...	86
88	130	65	114	44	...	...	...	88
90	134	67	116½	45	...	...	...	90
92	136	68	119	46	...	...	...	92
94	138	69	121½	47	...	...	...	94
96	142	71	124	48	...	...	...	96
98	146	73	126½	49	...	...	...	98
100	148	74	129	50	...	...	...	100

PIPE FITTINGS

TABLE 36 (FIG. 43)

EXTRA HEAVY AMERICAN STANDARD FLANGED FITTINGS

250 Pounds Working Pressure

Size Inches	A-A Inches	A Inches	B Inches	C Inches	D Inches	E Inches	F Inches	G Inches
1	8	4	5	2	8¹/₂	6¹/₂	2	
1¹/₄	8¹/₂	4¹/₄	5¹/₂	2¹/₂	9¹/₂	7¹/₄	2¹/₄	
1¹/₂	9	4¹/₂	6	2³/₄	11	8¹/₂	2¹/₂	
2	10	5	6¹/₂	3	11¹/₂	9	2¹/₂	
2¹/₂	11	5¹/₂	7	3¹/₂	13	10¹/₂	2¹/₂	
3	12	6	7³/₄	3¹/₂	14	11	3	6
3¹/₂	13	6¹/₂	8¹/₂	4	15¹/₂	12¹/₂	3	6¹/₂
4	14	7	9	4¹/₂	16¹/₂	13¹/₂	3	7
4¹/₂	15	7¹/₂	9¹/₂	4¹/₂	18	14¹/₂	3¹/₂	7¹/₂
5	16	8	10¹/₄	5	18¹/₂	15	3¹/₂	8
6	17	8¹/₂	11¹/₂	5¹/₂	21¹/₂	17¹/₂	4	9
7	18	9	12³/₄	6	23¹/₂	19	4¹/₂	10
8	20	10	14	6	25¹/₂	20¹/₂	5	11
9	21	10¹/₂	15¹/₄	6¹/₂	27¹/₂	22¹/₂	5	11¹/₂
10	23	11¹/₂	16¹/₂	7	29¹/₂	24	5¹/₂	12
12	26	13	19	8	33¹/₂	27¹/₂	6	14
14	30	15	21¹/₂	8¹/₂	37¹/₂	31	6¹/₂	16
15	31	15¹/₂	22³/₄	9	39¹/₂	33	6¹/₂	17
16	33	16¹/₂	24	9¹/₂	42	34¹/₂	7¹/₂	18
18	36	18	26¹/₂	10	45¹/₂	37¹/₂	8	19
20	39	19¹/₂	29	10¹/₂	49	40¹/₂	8¹/₂	20
22	41	20¹/₂	31¹/₂	11	53	43¹/₂	9¹/₂	22
24	45	22¹/₂	34	12	57¹/₂	47¹/₂	10	24
26	48	24	36¹/₂	13	...	...	...	26
28	52	26	39	14	...	...	...	28
30	55	27¹/₂	41¹/₂	15	...	...	...	30
32	58	29	44	16	...	...	...	32
34	61	30¹/₂	46¹/₂	17	...	...	...	34
36	65	32¹/₂	49	18	...	...	...	36
38	68	34	51¹/₂	19	...	...	...	38
40	71	35¹/₂	54	20	...	...	...	40
42	74	37	56¹/₂	21	...	...	...	42
44	78	39	59	22	...	...	...	44
46	81	40¹/₂	61¹/₂	23	...	...	...	46
48	84	42	64	24	...	...	...	48

TABLE 37

AMERICAN STANDARD CAST IRON PIPE, WALL THICKNESS

125 *Pounds Working Pressure*

Diameter of Pipe Inches	Thickness of Pipe Inches	Minimum Thickness Inches	Stress per Square Inch Pounds	Diameter of Pipe Inches	Thickness of Pipe Inches	Minimum Thickness Inches	Stress per Square Inch Pounds
1	.43	7/16	143	42	1.82	1 13/16	1448
1 1/4	.44	7/16	178	44	1.87	1 7/8	1467
1 1/2	.45	7/16	214	46	1.94	1 15/16	1484
2	.46	7/16	286	48	2.00	2	1500
2 1/2	.48	7/16	357	50	2.07	2 1/16	1515
3	.50	7/16	428	52	2.14	2 1/8	1530
3 1/2	.52	7/16	500	54	2.20	2 3/16	1543
4	.53	1/2	500	56	2.27	2 1/4	1555
4 1/2	.55	1/2	562	58	2.34	2 5/16	1567
5	.56	1/2	625	60	2.41	2 7/16	1538
6	.60	9/16	667	62	2.47	2 1/2	1550
7	.63	5/8	700	64	2.54	2 9/16	1561
8	.66	5/8	800	66	2.61	2 5/8	1572
9	.70	11/16	818	68	2.68	2 11/16	1582
10	.73	3/4	833	70	2.74	2 3/4	1591
12	.80	13/16	923	72	2.81	2 13/16	1600
14	.86	7/8	1000	74	2.88	2 7/8	1609
15	.90	7/8	1072	76	2.94	2 15/16	1617
16	.93	1	1000	78	3.01	3	1625
18	1.00	1 1/16	1059	80	3.08	3 1/16	1633
20	1.07	1 1/8	1111	82	3.15	3 1/8	1640
22	1.13	1 3/16	1158	84	3.21	3 3/16	1647
24	1.20	1 1/4	1200	86	3.28	3 1/4	1653
26	1.27	1 5/16	1238	88	3.35	3 5/16	1660
28	1.33	1 3/8	1273	90	3.41	3 3/8	1667
30	1.40	1 7/16	1304	92	3.48	3 1/2	1643
32	1.47	1 1/2	1333	94	3.55	3 9/16	1649
34	1.54	1 9/16	1360	96	3.62	3 5/8	1655
36	1.60	1 5/8	1385	98	3.68	3 11/16	1661
38	1.67	1 11/16	1407	100	3.75	3 3/4	1667
40	1.73	1 3/4	1428				

PIPE FITTINGS

TABLE 38
Extra Heavy Cast Iron Pipe, Wall Thickness
250 Pounds Working Pressure

Diameter of Pipe Inches	Thickness of Pipe Inches	Minimum Thickness Inches	Stress per Square Inch Pounds	Diameter of Pipe Inches	Thickness of Pipe Inches	Minimum Thickness Inches	Stress per Square Inch Pounds
1	.45	1/2	250				
1 1/4	.47	1/2	312	16	1.27	1 1/4	1600
1 1/2	.49	1/2	375	18	1.37	1 3/8	1636
2	.51	1/2	500	20	1.48	1 1/2	1666
2 1/2	.53	9/16	555	22	1.59	1 9/16	1760
3	.56	9/16	667	24	1.70	1 5/8	1846
3 1/2	.59	9/16	778	26	1.81	1 13/16	1793
4	.61	5/8	800	28	1.91	1 7/8	1866
4 1/2	.64	5/8	900	30	2.02	2	1875
5	.67	11/16	909	32	2.13	2 1/8	1882
6	.72	3/4	1000	34	2.24	2 1/4	1889
7	.78	13/16	1077	36	2.35	2 3/8	1894
8	.83	13/16	1230	38	2.46	2 7/16	1948
9	.89	7/8	1285	40	2.56	2 9/16	1953
10	.94	15/16	1333	42	2.67	2 11/16	1953
12	1.05	1	1500	44	2.78	2 13/16	1955
14	1.16	1 1/8	1555	46	2.89	2 7/8	2000
15	1.21	1 3/16	1579	48	3.00	3	2000

TABLE 39
American Standard Pipe Flanges — 125 Pounds Working Pressure

Size Inches	Diameter of Flanges Inches	Thickness of Flanges Inches	Bolt Circle Inches	Number of Bolts	Size of Bolts Inches	Length of Bolts Inches	Length of Studs with Two Nuts Inches
1	4	7/16	3	4	7/16	1 1/2	
1 1/4	4 1/2	1/2	3 3/8	4	7/16	1 1/2	
1 1/2	5	9/16	3 7/8	4	1/2	1 3/4	
2	6	5/8	4 3/4	4	5/8	2	
2 1/2	7	11/16	5 1/2	4	5/8	2 1/4	
3	7 1/2	3/4	6	4	5/8	2 1/2	
3 1/2	8 1/2	13/16	7	4	5/8	2 1/2	
4	9	15/16	7 1/2	8	5/8	2 3/4	
4 1/2	9 1/4	15/16	7 3/4	8	3/4	3	
5	10	15/16	8 1/2	8	3/4	3	
6	11	1	9 1/2	8	3/4	3	
7	12 1/2	1 1/16	10 3/4	8	3/4	3	
8	13 1/2	1 1/8	11 3/4	8	3/4	3 1/4	
9	15	1 1/8	13 1/4	12	3/4	3 1/4	
10	16	1 3/16	14 1/4	12	7/8	3 1/2	
12	19	1 1/4	17	12	7/8	3 3/4	
14	21	1 3/8	18 3/4	12	1	4 1/4	
15	22 1/4	1 3/8	20	16	1	4 1/4	

TABLE 39 (*Continued*)

AMERICAN STANDARD PIPE FLANGES — 125 *Pounds Working Pressure*

Size Inches	Diameter of Flanges Inches	Thickness of Flanges Inches	Bolt Circle Inches	Number of Bolts	Size of Bolts Inches	Length of Bolts Inches	Length of Studs with Two Nuts Inches
16	23 1/2	1 7/16	21 1/4	16	1	4 1/4	
18	25	1 9/16	22 3/4	16	1 1/8	4 3/4	
20	27 1/2	1 11/16	25	20	1 1/8	5	
22	29 1/2	1 13/16	27 1/4	20	1 1/4	5 1/2	
24	32	1 7/8	29 1/2	20	1 1/4	5 1/2	
26	34 1/4	2	31 3/4	24	1 1/4	5 3/4	
28	36 1/2	2 1/16	34	28	1 1/4	6	
30	38 3/4	2 1/8	36	28	1 3/8	6 1/4	
32	41 3/4	2 1/4	38 1/2	28	1 1/2	6 1/2	
34	43 3/4	2 5/16	40 1/2	32	1 1/2	6 1/2	
36	46	2 3/8	42 3/4	32	1 1/2	7	
38	48 3/4	2 3/8	45 1/4	32	1 5/8	7	9
40	50 3/4	2 1/2	47 1/4	36	1 5/8	7	9
42	53	2 5/8	49 1/2	36	1 5/8	7 1/2	9 1/2
44	55 1/4	2 5/8	51 3/4	40	1 5/8	7 1/2	9 1/2
46	57 1/4	2 11/16	53 3/4	40	1 5/8	7 1/2	9 1/2
48	59 1/2	2 3/4	56	44	1 5/8	8	9 1/2
50	61 3/4	2 3/4	58 1/4	44	1 3/4	8	10
52	64	2 7/8	60 1/2	44	1 3/4	8	10 1/2
54	66 1/4	3	62 3/4	44	1 3/4	8 1/2	10 1/2
56	68 3/4	3	65	48	1 3/4	8 1/2	10 1/2
58	71	3 1/8	67 1/4	48	1 3/4	9	11
60	73	3 1/8	69 1/4	52	1 3/4	9	11
62	75 3/4	3 1/4	71 3/4	52	1 7/8	9	11 1/2
64	78	3 1/4	74	52	1 7/8	9	11 1/2
66	80	3 3/8	76	52	1 7/8	9 1/2	11 1/2
68	82 1/4	3 3/8	78 1/4	56	1 7/8	9 1/2	11 1/2
70	84 1/2	3 1/2	80 1/2	56	1 7/8	10	12
72	86 1/2	3 1/2	82 1/2	60	1 7/8	10	12
74	88 1/2	3 5/8	84 1/2	60	1 7/8	10	12
76	90 3/4	3 5/8	86 1/2	60	7 7/8	10	12
78	93	3 3/4	88 3/4	60	2	10 1/2	12 1/2
80	95 1/4	3 3/4	91	60	2	10 1/2	12 1/2
82	97 1/2	3 7/8	93 1/4	60	2	10 1/2	13
84	99 3/4	3 7/8	95 1/2	64	2	10 1/2	13
86	102	4	97 3/4	64	2	11	13
88	104 1/4	4	100	68	2	11	13
90	106 1/2	4 1/8	102 1/4	68	2 1/8	11 1/2	14
92	108 3/4	4 1/8	104 1/2	68	2 1/8	11 1/2	14
94	111	4 1/4	106 1/4	68	2 1/8	11 1/2	14
96	113 1/4	4 1/4	108 1/2	68	2 1/4	11 1/2	14 1/2
98	115 1/2	4 3/8	110 3/4	68	2 1/4	12	14 1/2
100	117 3/4	4 3/8	113	68	2 1/4	12	14 1/2

TABLE 40

EXTRA HEAVY AMERICAN STANDARD PIPE FLANGES

250 Pounds Working Pressure

Size Inches	Diameter of Flanges Inches	Thickness of Flanges Inches	Bolt Circle Inches	Number of Bolts	Size of Bolts Inches	Length of Bolts Inches	Length of Studs with Two Nuts Inches
1	4 1/2	11/16	3 1/4	4	1/2	2	
1 1/4	5	3/4	3 3/4	4	1/2	2 1/4	
1 1/2	6	13/16	4 1/2	4	5/8	2 1/2	
2	6 1/2	7/8	5	4	5/8	2 1/2	
2 1/2	7 1/2	1	5 7/8	4	3/4	3	
3	8 1/4	1 1/8	6 5/8	8	3/4	3 1/4	
3 1/2	9	1 3/16	7 1/4	8	3/4	3 1/4	
4	10	1 1/4	7 7/8	8	3/4	3 1/2	
4 1/2	10 1/2	1 5/16	8 1/2	8	3/4	3 1/2	
5	11	1 3/8	9 1/4	8	3/4	3 3/4	
6	12 1/2	1 7/16	10 5/8	12	3/4	3 3/4	
7	14	1 1/2	11 7/8	12	7/8	4	
8	15	1 5/8	13	12	7/8	4 1/4	
9	16 1/4	1 3/4	14	12	1	4 3/4	
10	17 1/2	1 7/8	15 1/4	16	1	5	
12	20 1/2	2	17 3/4	16	1 1/8	5 1/2	
14	23	2 1/8	20 1/4	20	1 1/8	5 3/4	
15	24 1/2	2 3/16	21 1/2	20	1 1/4	6	
16	25 1/2	2 1/4	22 1/2	20	1 1/4	6	
18	28	2 3/8	24 3/4	24	1 1/4	6 1/4	
20	30 1/2	2 1/2	27	24	1 3/8	6 3/4	
22	33	2 5/8	29 1/4	24	1 1/2	7	
24	36	2 3/4	32	24	1 5/8	7 1/2	9 1/2
26	38 1/4	2 13/16	34 1/2	28	1 5/8	8	10
28	40 3/4	2 15/16	37	28	1 5/8	8	10
30	43	3	39 1/4	28	1 3/4	8 1/2	10 1/2
32	45 1/4	3 1/8	41 1/2	28	1 7/8	9	11
34	47 1/2	3 1/4	43 1/2	28	1 7/8	9	11 1/2
36	50	3 3/8	46	32	1 7/8	9 1/2	11 1/2
38	52 1/4	3 7/16	48	32	1 7/8	9 1/2	11 1/2
40	54 1/2	3 9/16	50 1/4	36	1 7/8	10	12
42	57	3 11/16	52 3/4	36	1 7/8	10	12
44	59 1/4	3 3/4	55	36	2	10 1/2	12 1/2
46	61 1/2	3 7/8	57 1/4	40	2	10 1/2	13
48	65	4	60 3/4	40	2	11	13

Reducing Fittings. — The sizes for reducing fittings are given in Tables 41, 42, 43, and 44.

On all reducing tees and crosses from 1 inch to 16 inches, inclusive, the centre to face dimension of the various outlets is the same on fittings of the same size run. Thus a 5 × 5 × 1 tee has the same centre to face dimension as a 5 × 5 × 5 tee, and is interchangeable with any combination of 5 inch cross. For sizes 18 inches and up interchangeability exists in two classes, one for short body patterns and one for long body patterns.

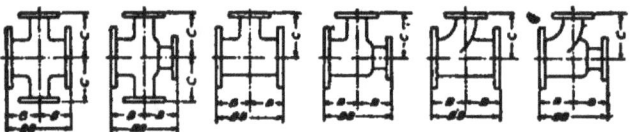

Fig. 45. Short Body Reducing Crosses and Tees.

TABLE 41 (FIG. 45)

AMERICAN STANDARD REDUCING TEES AND CROSSES

Short Body Pattern

125 *Pounds, Working Pressure*

Size Inches	Size of Outlet and Smaller Inches	B-B Inches	B Inches	C Inches	Size Inches	Size of Outlet and Smaller Inches	B-B Inches	B Inches	C Inches
18	12	26	13	15½	60	40	66	33	41
20	14	28	14	17	62	40	66	33	42
22	15	28	14	18	64	42	68	34	44
24	16	30	15	19	66	44	70	35	45
26	18	32	16	20	68	44	70	35	46
28	18	32	16	21	70	46	64	37	47
30	20	36	18	23	72	48	80	40	48
32	20	36	18	24	74	48	80	40	49
34	22	38	19	25	76	50	84	42	50
36	24	40	20	26	78	52	86	43	52
38	24	40	20	28	80	52	86	43	53
40	26	44	22	29	82	54	88	44	54
42	28	46	23	30	84	56	94	47	56
44	28	46	23	31	86	56	94	47	57
46	30	48	24	33	88	58	96	48	58
48	32	52	26	34	90	60	100	50	61
50	32	52	26	35	92	60	100	50	62
52	34	54	27	36	94	62	104	52	63
54	36	58	29	37	96	64	106	53	64
56	36	58	29	39	98	64	106	53	65
58	38	62	31	40	100	66	110	55	67

PIPE FITTINGS

TABLE 42 (FIG. 45)

EXTRA HEAVY AMERICAN STANDARD REDUCING TEES AND CROSSES
Short Body Pattern

250 Pounds. Working Pressure

Size Inches	Size of Outlet and Smaller Inches	B-B Inches	B Inches	C Inches	Size Inches	Size of Outlet and Smaller Inches	B-B Inches	B Inches	C Inches
18	12	28	14	17	34	22	44	22	28
20	14	31	15½	18½	36	24	47	23½	29½
22	15	33	16½	20	38	24	47	23½	30½
24	16	34	17	21½	40	26	50	25	31½
26	18	38	19	23	42	28	53	26½	33½
28	18	38	19	24	44	28	53	26½	34½
30	20	41	20½	25½	46	30	55	27½	35½
32	20	41	20½	26½	48	32	58	29	37½

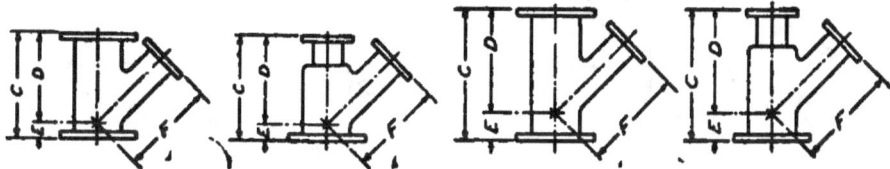

Fig. 46. Short Body Reducing Laterals.   Fig. 47. Long Body Reducing Laterals.

TABLE 43

AMERICAN STANDARD REDUCING LATERALS
Short Body Pattern (FIG. 46)

125 Pounds Working Pressure

Size Inches	Size of Branch and Smaller Inches	C Inches	D Inches	E Inches	F Inches
18	9	26	25	1	27½
20	10	28	27	1	29½
22	10	29	28½	½	31½
24	12	32	31½	½	34½
26	12	35	35	0	38
28	14	37	37	0	40
30	15	39	39	0	42

Long Body Pattern (FIG. 47)

Size Inches	Size of Branch and Larger Inches	C Inches	D Inches	E Inches	F Inches
18	10	39	32	7	32
20	12	43	35	8½	35
22	12	46	37½	8	37½
24	14	49½	40½	9	40½
26	14	53	44	9	44
28	15	56	46½	9½	46½
30	16	59	49	10	49

TABLE 44

EXTRA HEAVY AMERICAN STANDARD REDUCING LATERALS

Short Body Pattern (FIG. 46)

250 Pounds Working Pressure

Size Inches	Size of Branch and Smaller Inches	C Inches	D Inches	E Inches	F Inches
18	9	34	31	3	32½
20	10	37	34	3	
22	10	40	37	3	
24	12	44	41	3	43

Long Body Pattern (FIG. 47)

Size Inches	Size of Branch and Larger Inches	C Inches	D Inches	E Inches	F Inches
18	10	45½			
20	12	49			
22	12	53			
24	14	57½			

Cast Steel Fittings. — Walworth Company list cast steel fittings for steam pressures up to 350 pounds working pressure, and total temperature of 800 degrees, or working water pressures of

1,000 pounds for 2 inch to 4 inch sizes.
800 pounds for 4½ inch to 8 inch sizes.
500 pounds for 9 inch to 24 inch sizes.

PIPE FITTINGS

These fittings have the same dimensions as extra heavy cast iron fittings but are made from steel, having a tensile strength of 60,000 pounds.

Ammonia Fittings. — For ammonia piping malleable iron screwed fittings are made with a recess for soldering to insure tightness. Flange fittings are made tongued and grooved and provided with gaskets. The flanges may be round, square, or

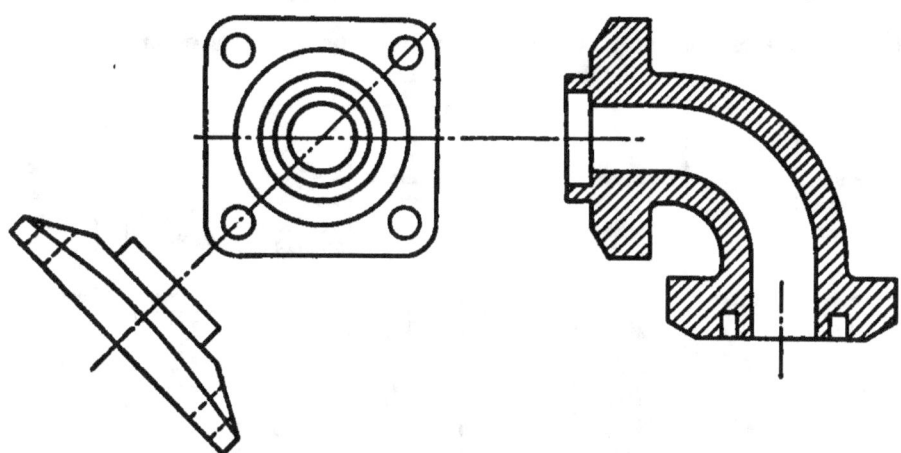

Fig. 48. Flanged Ammonia Fittings.

oval. Fig. 48 shows some flanged ammonia fittings, and Table 45 gives the sizes of lead or rubber gaskets for tongued and grooved ammonia joints, as made by the Walworth Company.

TABLE 45

AMMONIA GASKETS FOR TONGUED AND GROOVED JOINTS

Size Inches	Outside Diameter Inches	Inside Diameter Inches	Size Inches	Outside Diameter Inches	Inside Diameter Inches
$1/4$	$31/32$	$19/32$	2	$3^3/_{32}$	$2^7/_{16}$
$3/8$	$1^3/_{32}$	$23/32$	$2^1/_2$	$3^{21}/_{32}$	$2^{15}/_{16}$
$1/2$	$1^7/_{32}$	$7/8$	3	$4^9/_{32}$	$3^9/_{16}$
$3/4$	$1^{15}/_{32}$	$1^1/_8$	$3^1/_2$	$4^{27}/_{32}$	$4^1/_{16}$
1	$1^{25}/_{32}$	$1^3/_8$	4	$5^{15}/_{32}$	$4^9/_{16}$
$1^1/_4$	$2^7/_{32}$	$1^{11}/_{16}$	5	$6^{27}/_{32}$	$5^5/_8$
$1^1/_2$	$2^{15}/_{32}$	$1^{15}/_{16}$	6	$7^{27}/_{32}$	$6^{11}/_{16}$

72 A HANDBOOK ON PIPING

British Standard Pipe Flanges and Fittings. — The dimensions for standard pipe flanges used in England are given in Tables 46 and 47.

TABLE 46

BRITISH STANDARD PIPE FLANGES

For Working Steam Pressures up to 55 Pounds per Square Inch, and for Water Pressure up to 200 Pounds per Square Inch

This table does not apply to boiler feed pipes, or other water pipes subject to exceptional shocks.

Internal Diameter of Pipe	Diameter of Flange	Diameter of Bolt Circle	Number of Bolts	Diameter of Bolts	Thickness of Flanges		
					Cast Iron and Steel or Iron Welded on	Cast Steel and Bronze	Stamped or Forged Wrought Iron or Steel
Inches	Inches	Inches		Inches	Inches	Inches	Inches
$1/2$	$3^3/4$	$2^5/8$	4	$1/2$	$1/2$	$3/8$	$3/16$
$3/4$	4	$2^7/8$	4	$1/2$	$1/2$	$3/8$	$3/16$
1	$4^1/2$	$3^1/4$	4	$1/2$	$1/2$	$3/8$	$3/16$
$1^1/4$	$4^3/4$	$3^7/16$	4	$1/2$	$5/8$	$1/2$	$1/4$
$1^1/2$	$5^1/4$	$3^7/8$	4	$1/2$	$5/8$	$1/2$	$1/4$
2	6	$4^1/2$	4	$5/8$	$3/4$	$9/16$	$5/16$
$2^1/2$	$6^1/2$	5	4	$5/8$	$3/4$	$9/16$	$5/16$
3	$7^1/4$	$5^3/4$	4	$5/8$	$3/4$	$9/16$	$3/8$
$3^1/2$	8	$6^1/2$	4	$5/8$	$3/4$	$9/16$	$3/8$
4	$8^1/2$	7	4	$5/8$	$7/8$	$11/16$	$3/8$
$4^1/2$	9	$7^1/2$	8	$5/8$	$7/8$	$11/16$	$7/16$
5	10	$8^1/4$	8	$5/8$	$7/8$	$11/16$	$1/2$
6	11	$9^1/4$	8	$5/8$	$7/8$	$11/16$	$1/2$
7	12	$10^1/4$	8	$5/8$	1	$3/4$	$1/2$
8	$13^1/4$	$11^1/2$	8	$5/8$	1	$3/4$	$1/2$
9	$14^1/2$	$12^3/4$	8	$5/8$	1	$3/4$	$5/8$
10	16	14	8	$3/4$	1	$3/4$	$5/8$
12	18	16	12	$3/4$	$1^1/8$	$7/8$	$5/8$
14	$20^3/4$	$18^1/2$	12	$7/8$	$1^1/4$	1	$3/4$
15	$21^3/4$	$19^1/2$	12	$7/8$	$1^1/4$	1	$3/4$
16	$22^3/4$	$20^1/2$	12	$7/8$	$1^1/4$	1	$3/4$
18	$25^1/4$	23	12	$7/8$	$1^3/8$	$1^1/8$	$7/8$
20	$27^3/4$	$25^1/4$	16	$7/8$	$1^1/2$	$1^1/4$	1
24	$32^1/2$	$29^3/4$	16	1	$1^5/8$	$1^3/8$	$1^1/8$

Bolt-holes. — For $1/2$-inch and $5/8$-inch bolts the diameters of the holes to be $1/16$-inch larger than the diameters of the bolts, and for larger sizes of bolts, $1/8$ inch. Bolt-holes to be drilled off centre lines.

PIPE FITTINGS

TABLE 47

British Standard Pipe Flanges

For Working Pressures up to 125 Pounds, 225 Pounds, and 325 Pounds per Square Inch

Internal Diam. of Pipe	Diam. of Flange	Diam. of Bolt Circle	Number of Bolts	Diameter of Bolts		Thickness of Flanges					
						Cast Iron, and Steel or Iron Welded on			Steel (Cast or Riveted on) and Bronze		
	125 Lbs. 225 Lbs. 325 Lbs.	125 Lbs. 225 Lbs. 325 Lbs.	125 Lbs. 225 Lbs. 325 Lbs.	125 Lbs. 225 Lbs.	325 Lbs.	125 Lbs.	225 Lbs.	325 Lbs.	125 Lbs.	225 Lbs.	325 Lbs.
Inches	Inches	Inches		Inches	Inches	In.	In.	In.	In.	In.	In.
1/2	3 3/4	2 5/8	4	1/2	1/2	1/2	1/2	5/8	3/8	7/16	5/8
3/4	4	2 7/8	4	1/2	1/2	1/2	1/2	5/8	3/8	7/16	5/8
1	4 5/8	3 7/16	4	5/8	5/8	1/2	5/8	3/4	3/8	1/2	11/16
1 1/4	5 1/4	3 7/8	4	5/8	5/8	5/8	5/8	3/4	1/2	1/2	11/16
1 1/2	5 1/2	4 1/8	4	5/8	5/8	5/8	3/4	7/8	1/2	9/16	3/4
2	6 1/2	5	4	5/8	3/4	3/4	7/8	1	5/8	11/16	7/8
2 1/2	7 1/4	5 5/8	8	5/8	3/4	3/4	7/8	1	5/8	11/16	7/8
3	8	6 1/2	8	5/8	3/4	3/4	1	1 1/4	5/8	3/4	1
3 1/2	8 1/2	7	8	5/8	3/4	7/8	1	1 1/4	3/4	3/4	1
4	9	7 1/2	8	5/8	3/4	7/8	1 1/8	1 3/8	3/4	7/8	1 1/8
4 1/2	10	8 1/4	8	3/4	7/8	7/8	1 1/8	1 3/8	3/4	7/8	1 1/8
5	11	9 1/4	8	3/4	7/8	1	1 1/4	1 1/2	7/8	1	1 1/4
6	12	10 1/4	12	5/8	7/8	1	1 1/4	1 1/2	7/8	1	1 1/4
7	13 1/4	11 1/2	12	3/4	7/8	1	1 3/8	1 5/8	7/8	1 1/8	1 3/8
8	14 1/2	12 3/4	12	3/4	7/8	1 1/8	1 3/8	1 3/4	1	1 1/8	1 3/8
9	16	14	12	7/8	1	1 1/8	1 1/2	1 3/4	1	1 1/4	1 1/2
10	17	15	12	7/8	1	1 1/8	1 1/2	1 7/8	1	1 1/4	1 5/8
12	19 1/4	17 1/4	16	7/8	1	1 1/4	1 5/8	2	1 1/8	1 3/8	1 3/4
14	21 3/4	19 1/2	16	1	1 1/8	1 3/8	1 3/4	2 1/4	1 1/4	1 1/2	1 7/8
15	22 3/4	20 1/2	16	1	1 1/8	1 3/8	1 7/8	2 1/4	1 1/4	1 5/8	1 7/8
16	24	21 3/4	20	1	1 1/8	1 3/8	1 7/8	2 1/4	1 1/4	1 5/8	2
18	26 1/2	24	20	1 1/8	1 1/4	1 1/2	2	2 3/8	1 3/8	1 3/4	2 1/8
20	29	26 1/2	24	1 1/8	1 1/4	1 5/8	2 1/8	2 1/2	1 1/2	1 7/8	2 1/4
22	31	28 1/2	24	1 1/8	1 1/4	1 5/8	2 1/4	2 5/8	1 1/2	2	2 3/8
24	33 1/2	30 3/4	24	1 1/8	1 3/8	1 3/4	2 3/8	2 3/4	1 5/8	2 1/8	2 1/2

Bolt-holes. — For 1/2-inch and 5/8-inch bolts the diameters of the holes to be 1/16-inch larger than the diameters of the bolts, and for larger sizes of bolts, 1/8-inch. Bolt-holes to be drilled off center lines.

The Engineering Standards Committee gives four classes according to pressure, as follows: low pressure for steam up to 55 pounds, and water up to 200 pounds per square inch; intermediate pressure for steam over 55 pounds and not exceeding 125 pounds; high pressure for steam pressure over 125 pounds, and not exceeding 225 pounds; extra high pressure for steam pressure over 225 pounds and not exceeding 325 pounds.

74　A HANDBOOK ON PIPING

General dimensions for British Standard Flanged Fittings are given in Tables 48 and 49 for short tees and bends of cast material and for long bends of wrought iron and steel.

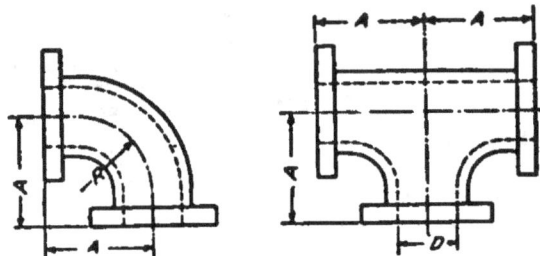

Fig. 49. British Standard Short Tees and Bends.

TABLE 48 (FIG. 49)

BRITISH STANDARD SHORT BENDS AND TEES

325 Pounds Working Pressure

Size D Inches	A Inches	R Inches	Size D Inches	A Inches	R Inches
1/2	3 1/2	2 1/2	7	10	7 1/4
3/4	3 3/4	2 3/4	8	11	8 1/4
1	4	2 3/4	9	12	9
1 1/4	4 1/4	3	10	13	10
1 1/2	4 1/2	3	12	15	11 3/4
2	5	3 1/2	14	17	13 1/2
2 1/2	5 1/2	3 3/4	15	18	14 1/4
3	6	4	16	19	15 1/4
3 1/2	6 1/2	4 1/2	18	21	17
4	7	4 3/4	20	23	18 3/4
5	8	5 1/2	21	24	19 1/2
6	9	6 1/2	24	27	22 1/4

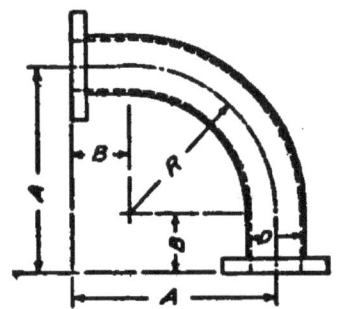

Fig. 50. British Standard Long Bends.

TABLE 49 (FIG. 50)
BRITISH STANDARD LONG BENDS OF WROUGHT IRON AND STEEL

Size D Inches	A Inches	B Inches	R Inches	Size D Inches	A Inches	B Inches	R Inches
1/2	4 1/2	2 1/2	2	6	25	7	18
3/4	5	2 1/2	2 1/2	7	31 1/2	7	24 1/2
1	6	3	3	8	36	8	28
1 1/4	6 3/4	3	3 3/4	9	39 1/2	8	31 1/2
1 1/2	7 1/2	3	4 1/2	10	49	9	40
2	9 1/2	3 1/2	6	12	58	10	48
2 1/2	11 1/2	4	7 1/2	14	74	11	63
3	13	4	9	15	79 1/2	12	67 1/2
3 1/2	15 1/2	5	10 1/2	16	93	13	80
4	17	5	12	18	104	14	90
5	21	6	15	20	126	16	110

CHAPTER V

PIPE JOINTS

There are a great many kinds of joints used for connecting pieces of pipe. Some forms are described in this chapter but there are many others which space does not permit showing. The ideal arrangement would be to have the pipe in one continuous piece, but this is not practicable, although the number of joints

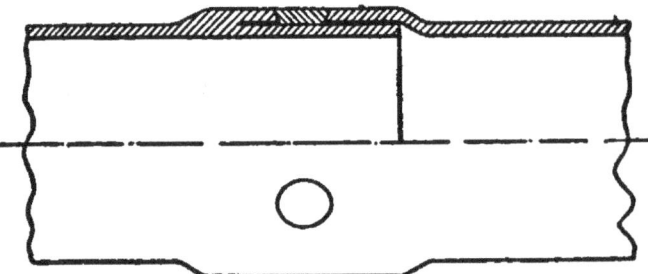

Fig. 51. Atwood Line Weld.

can be greatly reduced by using welded joints. The question of joints should receive very careful attention and the type selected which will best meet the conditions involved.

Welded Joints. — Any means of reducing the number of joints to be made in pipe lines is distinctly worth while as it makes

Fig. 52. Interlock Welded Necks.

fewer chances for leakage, lessens repairs, and is generally commendable. The oxy-acetylene blow torch is used by the Pittsburg Valve, Foundry, and Construction Company for doing

PIPE JOINTS

welded work, as illustrated in the patented joints shown in Figs. 51 and 52. The "Atwood line weld," Fig. 51, allows the fabrication of pipes into lengths as long as can be handled for shipment, with a consequent reduction by about fifty per cent of the number

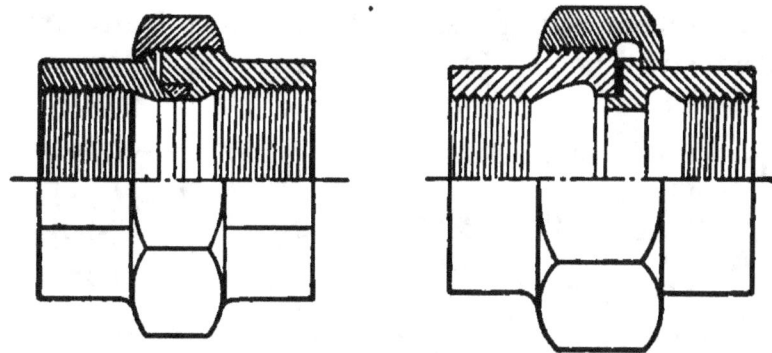

Figs. 53 and 54. Screwed Unions.

of flange joints in the line. For connecting branch lines of wrought pipe in mains of the same material, "interlock welded necks" are made use of to eliminate cast fittings. This appears to good advantage in welded headers where the weight is reduced in addition to doing away with a large number of joints. The method of making this connection is shown in Fig. 52.

Screw Unions. — For joining two lengths of small screwed pipe, couplings are in general use, as described in Chapter IV,

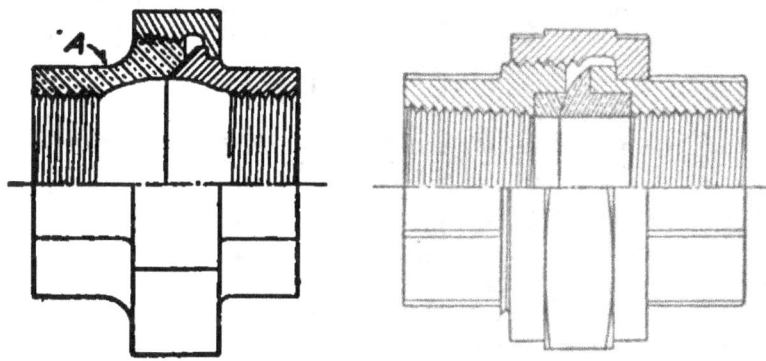

Figs. 55 and 56. Screwed Unions.

Table 20. When the joint must be unmade frequently, or for making the last joint in a line, unions may be used. Fig. 53 shows a union made of malleable iron with a brass seat forced into place so that contact is between iron and brass. Both ends are ground together, making a tight joint. Fig. 54 shows a union made of malleable iron, using a metallic gasket to make a tight joint. The Kewanee Union shown in Fig. 55 is made by the National Tube Company. Part A is made of brass, giving a brass to iron thread connection, and a brass to iron ball joint seat. Fig. 56 shows the Dart Union, having inserted brass seats. Unions are also made entirely of brass. Table 50 gives the dimensions of Crane Company Unions.

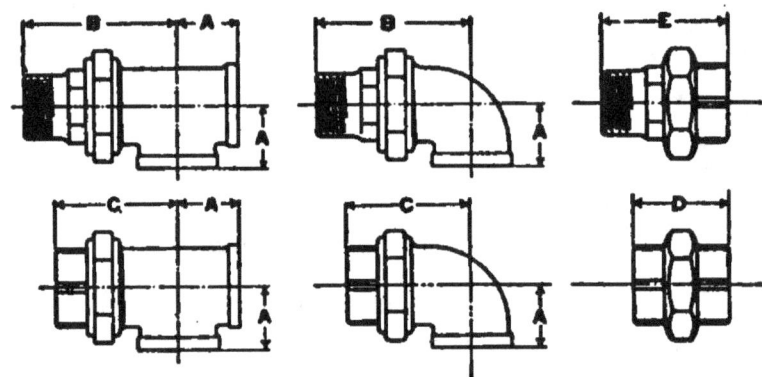

Figs. 57. Screwed Unions.

TABLE 50 (FIG. 57)

CRANE MALLEABLE IRON UNIONS, UNION ELLS, UNION TEES

Size Inches	Elbows and Tees A Inches	Union, Male B Inches	Union, Female C Inches	Standard Union D Inches	Crane and Navy Unions D Inches	Standard and Railroad Unions with Male and Female Ends E Inches
1/8	...	...	...	1 1/2		
1/4	13/16	2 7/16	1 13/16	1 5/8	2 3/16	2 1/4
3/8	15/16	2 3/4	2 1/16	1 3/4	2 1/4	2 7/16
1/2	1 1/8	3 1/16	2 5/16	1 7/8	2 5/16	2 11/16
3/4	1 5/16	3 1/2	2 5/8	2 1/8	2 1/2	3 1/16
1	1 7/16	3 15/16	3	2 3/8	2 5/8	3 5/16
1 1/4	1 3/4	4 7/16	3 7/16	2 5/8	2 29/32	3 11/16
1 1/2	1 15/16	4 3/4	3 13/16	2 15/16	3 1/8	4
2	2 1/4	5 3/8	4 3/8	3 1/4	3 9/16	4 5/16
2 1/2	2 11/16	6	4 7/8	3 9/16	4 1/16	4 5/8
3	...	...	...	3 15/16	4 7/8	
3 1/2	...	...	...	4 3/8		
4	...	...	...	4 1/2		

Flange Unions. — For many purposes, especially for the larger sizes, flange unions, Figs. 58 and 59, are to be preferred. These are made in a large variety of forms. The object of using them is to facilitate the erection and disassembling of the piping. The Kewanee Flange Union is shown in Fig. 59.

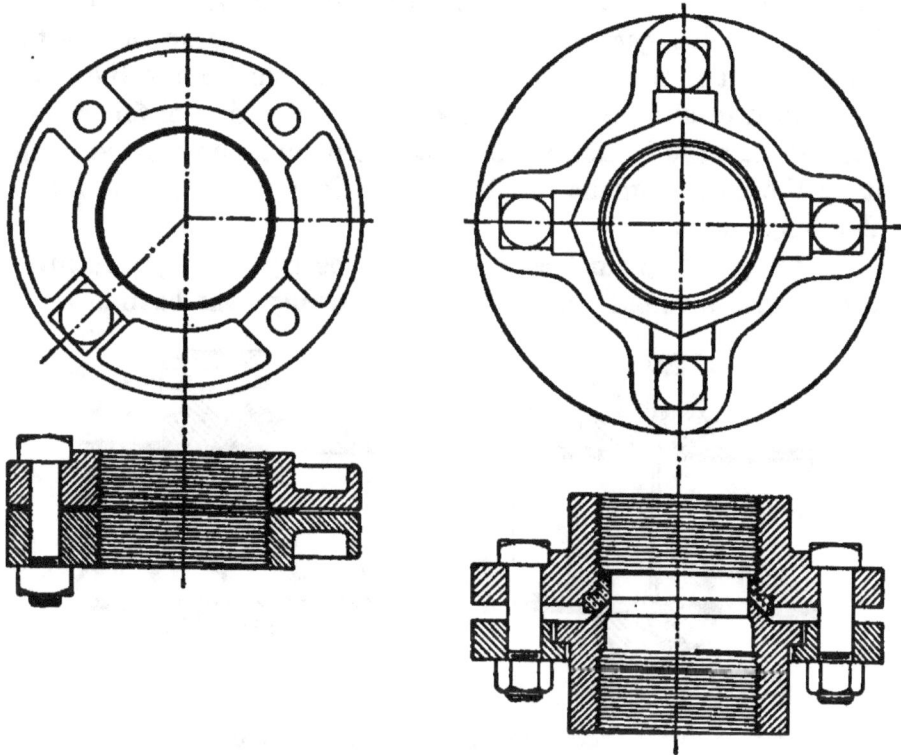

Figs. 58 and 59. Flanged Unions.

Bolt Circles and Drilling. — The diameters of bolt circles, sizes of bolts and bolt holes, number of bolts, etc., are given in Tables 39 and 40, Chapter IV, for the American Standard which is generally used in the United States. When cast steel flanges are used the bolt holes are spot faced. This is done by facing off around the bolt holes on the back side of the flange, where the nut or head of the bolt bears. This gives a truer and firmer bearing than can be had with a rough casting.

Flange Facing. — There are a large number of methods of facing flanges and providing for the holding of gaskets to make tight joints. These may be listed as

80 A HANDBOOK ON PIPING

Straight plain face Raised face for gasket
Corrugated, plain face Raised face for ground joint
Scored, plain face Tongue and groove
Grooved, plain face Male and female

Fig. 60 shows the plain straight-faced flange commonly employed for pressures up to 125 pounds on steam and water. Either a full face or ring gasket is used. The full face gasket is a little easier to put in place and to centralize with the bore of the pipe. Very good results can be obtained by a ring gasket of fair thickness, so that the gasket will have sufficient pressure exerted upon it by the bolts to make a tight joint, before the outside edges of the flange meet.

A corrugated, plain face flange is made by cutting concentric curves with a round nosed tool. The corrugations have a tend-

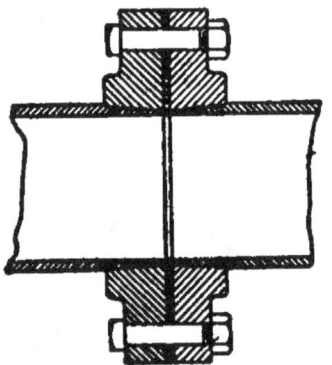
Fig. 60. Straight Faced Flange.

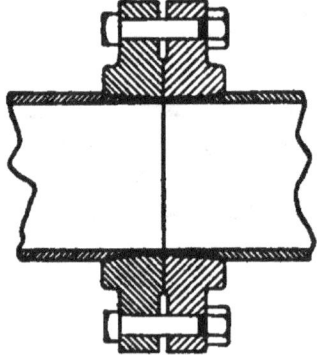
Fig. 61. Raised Face Flange.

ency to prevent the gaskets from blowing out. Their use is desirable when the fluid conveyed requires extra thick gaskets. A scored, plain face flange is one which has concentric rings scored upon the face by a diamond-pointed tool. When lead gaskets must be used, as on oil and acid lines, this form of flange is desirable. The lead gasket squeezes into the scores and helps to maintain a tight joint without bringing undue strain on the bolts. Same forms of grooved flanges are used in which contact is made by a copper or lead wire pressed into a groove cut into both flanges. This joint is effective, but the flanges must be strong to withstand the stresses set up when the bolts are tightened.

A very satisfactory joint for high pressure steam lines is made by raising the face of the flange between the inside of the bolt

PIPE JOINTS

holes and the bore $1/16$ inch above the rest of the flange, Fig. 61. The entire force exerted by the bolts is concentrated at the joint without danger of the edges of the flanges coming together, making an efficient joint. Such flange faces are advised by the A. S. M. E. Committee for all flanges and fittings for use with pressures above 125 pounds.

It is essential that no organic matter should be in contact with superheated steam as it will carbonize. For such use the raised faces of Fig. 61 may be ground, giving a metal to metal joint. Special gaskets may be had for superheated steam.

The tongued and grooved flange shown in Fig. 62 provides a recess to hold the packing in place so that it cannot blow out.

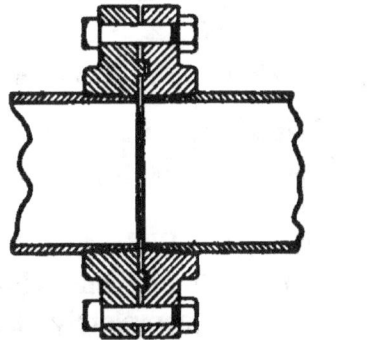   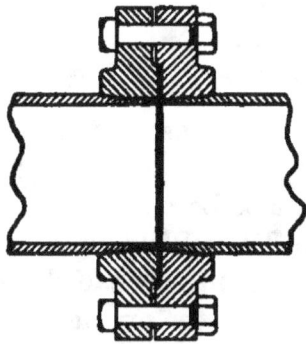

Fig. 62. Tongued and Grooved Flanges.   Fig. 63. Male and Flanges.

The male and female flanges shown in Fig. 63 are used considerably on high pressure hydraulic lines and to some extent on high pressure steam lines. The gasket is held securely in place, but both Figs. 62 and 63 are difficult to take down as they must be separated a distance equal to the projection before the pipe can be moved.

Flange Joints for Steel Pipe.—For making joints with wrought pipe various forms of flange joints are made, of which the following may be mentioned:

Screwed	Rolled joint
Screwed and calked	Riveted and shrunk
Screwed and welded	Swivel
Welded	Shrunk joint

The screwed joint shown in Fig. 60 is a common method of attaching flanges. The flange is screwed on until the pipe projects through, then the flange and pipe are faced off together. It is advisable to have the gasket bear on the end of the pipe to insure tightness. The threading weakens the pipe so that for high pressures some of the following types are advisable. Flanges

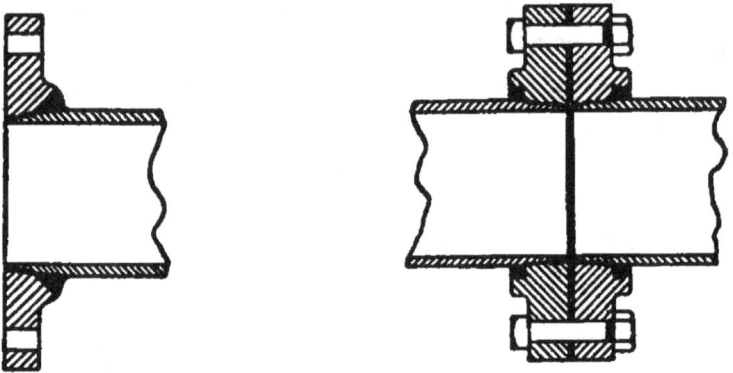

Fig. 64. Walco-Weld Flange.   Fig. 65. Flange with Calking Recess.

are made of cast iron, semi steel, malleable iron, cast steel, and forged steel suitable for the method of joining to the pipe and the pressure to be met.

The Walco-Weld flange, Fig. 64, made by the Walworth Company, is made by half-threading on the flange and then welding the back by the oxy-acetylene method, thereby completely eliminating the possibility of an imperfect or incomplete weld, as sometimes occurs with the furnace-welded flange. Flanges with a calking recess, Fig. 65, are made by the Crane Company by cutting a recess in the hubs on the backs of the flanges. This recess is $1/2$ inch in depth, $1/4$ inch wide at top, and $5/16$ inch wide at bottom. It can be applied to extra heavy flanges in sizes from 2 to 24 inch. Flanges so fitted are $1/2$ inch higher than the regular flanges. When the flanges are used on cold water, the recesses are filled with lead, and when used on steam the recesses are filled with soft copper, which is

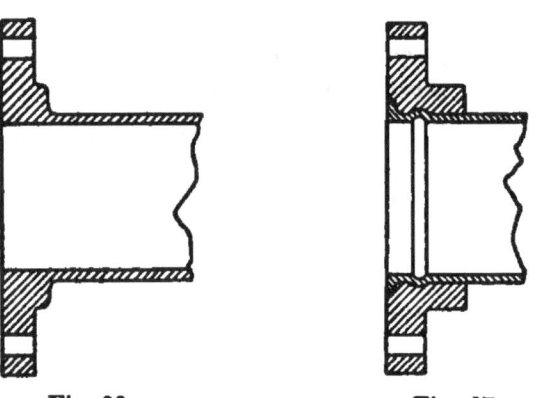

Fig. 66. Welded Flange.   Fig. 67. Rolled Joint.

calked in firmly to keep the flanges from leaking where they are made on pipe.

Welded joints are made by welding a wrought steel flange to the pipe, making them into one piece, as shown in Fig. 66. Fig. 67 shows a form of rolled joint. A groove is turned into the flange and the pipe rolled into it. The shrunk joint is shown in Fig. 68. The flange is first bored to a shrink fit, and then heated and placed over the end of the pipe which is peened into the

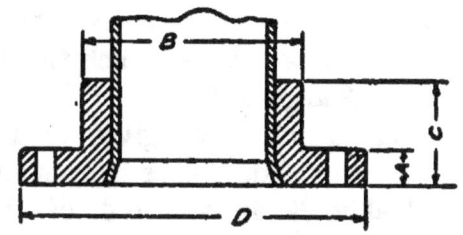

Fig. 68. Shrink Joint.

recess in the flange. Afterwards a facing cut is taken across the end of the pipe and flange. The gasket should bear on the end of the pipe as the joint between pipe and flange may not be absolutely tight. Shrunk joints are also made with either single or double riveting.

The Walmanco joint, Fig. 69, was developed in the Walworth shops in 1897. Some of the advantages of this form as stated by the makers are: first — the pipe is not weakened by cutting into the wall; second — the gasket bears on the face of the lap, and

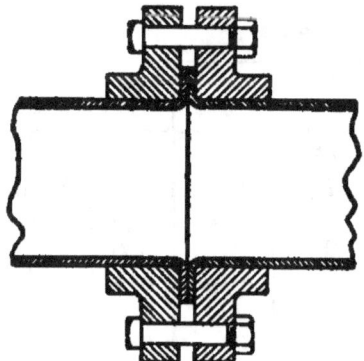
Fig. 69. Walmanco Joint.

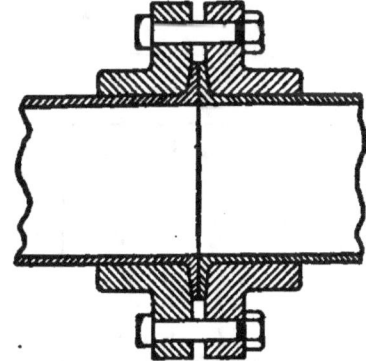
Fig. 70. Cranelap Joint.

absolutely prevents leakage through the bore of the flange; third — the advantage of the flange swiveling on the pipe is obvious to the fitter; fourth — the flange has maximum strength, and is not subject to torsional strains in attaching.

The Cranelap joint made by Crane Company is shown in Fig. 70. The face of the flange is bevelled to the width of the lap, to compensate for the difference in the thickness of the pipe between the inside and outside portions of the lap, caused by drawing over the pipe, and the lap is made with a square corner so that the inside of the pipe runs straight to the face of the joint, as illustrated in Fig. 70. The flanges in these joints are loose and swivel. This is a great convenience when it is necessary to change the position of bolt holes, which this makes possible.

Pipe Flange Tables. — The principal dimensions for the various flange joints are given in Tables 51 to 56 inclusive. For American standard pipe flanges and British standard pipe flanges see Tables 39, 40, 46 and 47 of Chapter IV.

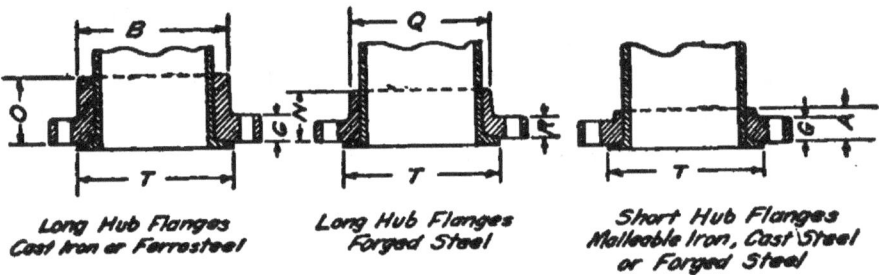

Fig. 71. Cranelap Flanges.

TABLE 51 (FIG. 71)

EXTRA HEAVY CRANELAP PIPE JOINTS

250 Pounds Working Pressure

Size Inches	B Inches	Q Inches	G Inches	R Inches	T Inches	O Inches	N Inches	A Inches
4	6³/₁₆	5³/₄	1¹/₄	1¹/₈	6⁵/₈	3³/₄	3¹/₈	1³/₄
4¹/₂	6¹¹/₁₆	6¹/₄	1⁵/₁₆	1¹/₄	7¹/₄	3¹⁵/₁₆	3¹/₄	1¹³/₁₆
5	7³/₈	7	1³/₈	1¹/₄	7³/₄	4¹/₈	3¹/₄	1⁷/₈
6	8¹/₂	7¹⁵/₁₆	1⁷/₁₆	1¹/₄	9	4¹/₄	3¹/₄	2
7	9³/₄	9¹/₈	1¹/₂	1⁵/₁₆	10	4⁷/₁₆	3³/₈	2¹/₁₆
8	10⁷/₈	10⁵/₁₆	1⁵/₈	1³/₈	11	4⁵/₈	3¹/₂	3³/₁₆
9	11⁷/₈	11³/₈	1³/₄	1⁷/₁₆	12¹/₄	4¹³/₁₆	3⁵/₈	2¹/₄
10	13¹/₈	12⁵/₈	1⁷/₈	1¹/₂	13¹/₂	4¹⁵/₁₆	3³/₄	2³/₈
12	15³/₈	14³/₄	2	1⁵/₈	15³/₄	5⁵/₁₆	4	2⁹/₁₆
14	16³/₄	16³/₁₆	2¹/₈	1³/₄	17	5¹/₂	4³/₈	2¹¹/₁₆
15	17⁷/₈	17¹/₄	2³/₁₆	1¹³/₁₆	18	5⁵/₈	4¹/₂	2¹³/₁₆
16	19¹/₄	18¹/₂	2¹/₄	1⁷/₈	19	6	4³/₄	2⁷/₈
18	21¹/₂	20³/₄	2³/₈	2	21¹/₂	6¹/₄	5	3¹/₁₆
20	23³/₄	22¹/₂	2¹/₂	2¹/₄	23¹/₂	6¹/₂	5¹/₂	3¹/₄
22	26	24³/₄	2⁵/₈	2¹/₄	25¹/₂	6⁷/₈	5¹/₂	3⁷/₁₆
24	28¹/₄	27	2³/₄	2⁷/₁₆	27¹/₂	7¹/₄	6¹/₄	3⁵/₈

PIPE JOINTS

Cast Iron
Semi Steel

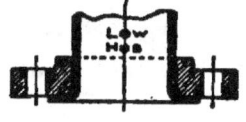
Malleable Iron
Forged Steel

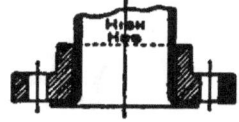
Cast Iron, Semi Steel,
Forged Steel

Fig. 72. Walmanco Flanges.

TABLE 52 (Fig. 72)

Standard Weight Walmanco Flanges

Pipe Size	New Style		Low Hub		High Hub	
	Diameter of Flange	Thickness through Hub	Thickness at Edge	Thickness through Hub	Thickness at Edge	Thickness through Hub
Inches	Inches	Inches	Inches	Inches	Inches	Inches
4	9	$1^7/_8$	$^{15}/_{16}$	$1^3/_{16}$	$^{15}/_{16}$	$2^5/_8$
$4^1/_2$	$9^1/_4$	$1^7/_8$	$^{15}/_{16}$	$1^1/_4$		
5	10	$1^7/_8$	$^{15}/_{16}$	$1^5/_{16}$	$^{15}/_{16}$	$2^5/_8$
6	11	2	1	$1^7/_{16}$	1	$2^7/_8$
7	$12^1/_2$	$2^1/_{16}$	$1^1/_{16}$	$1^1/_2$	$1^1/_{16}$	$2^7/_8$
8	$13^1/_2$	$1^1/_8$	$1^1/_8$	$1^5/_8$	$1^1/_8$	3
9	15	$2^1/_8$	$1^1/_8$	$1^3/_4$	$1^1/_8$	$3^1/_4$
10	16	$2^3/_{16}$	$1^3/_{16}$	$1^7/_8$	$1^3/_{16}$	$3^1/_2$
12	19	$2^1/_4$	$1^1/_4$	$2^1/_{16}$	$1^1/_4$	$4^1/_4$
14	21	$2^3/_8$	$1^3/_8$	$2^3/_{16}$	$1^3/_8$	$4^1/_4$
15	$22^1/_4$	$2^3/_8$	$1^3/_8$	$2^5/_{16}$	$1^3/_8$	$4^1/_4$
16	$23^1/_2$	$2^7/_{16}$	$1^7/_{16}$	$2^7/_{16}$	$1^7/_{16}$	$4^1/_4$
18	25	$2^9/_{16}$	$1^9/_{16}$	$2^5/_8$	$1^9/_{16}$	$4^1/_2$
20	$27^1/_2$	$2^{11}/_{16}$	$1^{11}/_{16}$	$2^3/_4$	$1^{11}/_{16}$	$4^1/_2$
22	$29^1/_2$	$2^{13}/_{16}$	$1^{13}/_{16}$	$2^7/_8$		
24	32	$2^7/_8$	$1^7/_8$	3	$1^7/_8$	$5^1/_2$
26	$34^1/_2$	3				
28	$36^1/_2$	$3^1/_{16}$				
30	$38^3/_4$	$3^1/_8$				

TABLE 53 (FIG. 72)

EXTRA HEAVY WALMANCO FLANGES

Pipe Size Inches	Outside Diameter Inches	New Style		High Hub			Forged Steel High Hub
		Diameter of Face	Thickness of Flange	Diameter of Hub	Thickness through Hub	Thickness of Flange	Thickness of Flange
4	10	$6^3/_4$	2	$5^7/_8$	$2^{13}/_{16}$	$1^1/_4$	$1^1/_8$
$4^1/_2$	$10^1/_2$	$7^1/_2$	2	$6^3/_{16}$	$2^7/_8$	$1^5/_{16}$	$1^1/_4$
5	11	8	$2^1/_8$	$6^{13}/_{16}$	3	$1^3/_8$	$1^1/_4$
6	$12^1/_2$	$9^1/_4$	$2^1/_4$	$8^1/_{16}$	$3^1/_{16}$	$1^7/_{16}$	$1^1/_4$
7	14	$10^1/_4$	$2^3/_8$	9	$3^7/_{16}$	$1^1/_2$	$1^5/_{16}$
8	15	$11^1/_4$	$2^1/_2$	10	$3^7/_{16}$	$1^5/_8$	$1^3/_8$
9	$16^1/_4$	$12^1/_2$	$2^5/_8$	$11^1/_4$	$3^{13}/_{16}$	$1^3/_4$	$1^7/_{16}$
10	$17^1/_2$	$13^3/_4$	$2^3/_4$	$12^5/_{16}$	$3^{15}/_{16}$	$1^7/_8$	$1^1/_2$
12	$20^1/_2$	16	3	$14^9/_{16}$	$4^5/_{16}$	2	$1^5/_8$
14	23	$17^1/_4$	$3^1/_8$	$15^{11}/_{16}$	$4^{11}/_{16}$	$2^1/_8$	$1^3/_4$
15	$24^1/_2$	$18^1/_4$	$3^1/_4$	17	$4^5/_8$	$2^3/_{16}$	$1^{13}/_{16}$
16	$25^1/_2$	$19^1/_4$	$3^1/_2$	18	5	$2^1/_4$	$1^7/_8$
18	28	$21^3/_4$	$3^5/_8$	20	$5^3/_{16}$	$2^3/_8$	2
20	$30^1/_2$	$23^3/_4$	$3^3/_4$	$22^1/_2$	$5^1/_2$	$2^1/_2$	$2^1/_4$
22	33	$25^3/_4$	$3^7/_8$	$24^1/_2$	$5^5/_8$	$2^5/_8$	
24	36	$27^3/_4$	4	$26^1/_2$	$6^{11}/_{16}$	$2^3/_4$	

TABLE 54 (FIG. 68)

SHRUNK AND PEENED FLANGES — EXTRA HEAVY

Size Inches	Diameter of Flange Inches	Cast Flanges			Forged Flanges		
		A	B	C	A	B	C
4	10	$1^1/_4$	$6^3/_{16}$	$3^3/_4$	$1^1/_8$	$5^3/_4$	$3^1/_8$
$4^1/_2$	$10^1/_2$	$1^5/_{16}$	$6^{11}/_{16}$	$3^{15}/_{16}$	$1^1/_4$	$6^1/_4$	$3^1/_4$
5	11	$1^3/_8$	$7^3/_8$	$4^1/_8$	$1^1/_4$	7	$3^1/_4$
6	$12^1/_2$	$1^7/_{16}$	$8^1/_2$	$4^1/_4$	$1^1/_4$	$7^{15}/_{16}$	$3^1/_4$
7	14	$1^1/_2$	$9^3/_4$	$4^7/_{16}$	$1^5/_{16}$	$9^1/_2$	$3^3/_8$
8	15	$1^5/_8$	$10^7/_8$	$4^5/_8$	$1^3/_8$	$10^3/_{16}$	$3^1/_2$
9	$16^1/_4$	$1^3/_4$	$11^7/_8$	$4^{13}/_{16}$	$1^7/_{16}$	$11^3/_8$	$3^5/_8$
10	$17^1/_2$	$1^7/_8$	$13^1/_8$	$4^{15}/_{16}$	$1^1/_2$	$12^5/_8$	$3^3/_4$
12	$20^1/_2$	2	$15^3/_8$	$5^5/_{16}$	$1^5/_8$	$14^3/_4$	4
14	23	$2^1/_8$	$16^3/_4$	$5^1/_2$	$1^3/_4$	$16^3/_{16}$	$4^1/_8$
15	$24^1/_2$	$2^3/_{16}$	$17^7/_8$	$5^5/_8$	$1^{13}/_{16}$	$17^1/_4$	$4^1/_2$
16	$25^1/_2$	$2^1/_4$	$19^1/_4$	6	$1^7/_8$	$18^1/_2$	$4^3/_4$
18	28	$2^3/_8$	$21^1/_2$	$6^1/_4$	2	$20^3/_4$	5
20	$30^1/_2$	$2^1/_2$	$23^3/_8$	$6^1/_2$	$2^1/_4$	$22^1/_2$	$5^1/_2$
22	33	$2^5/_8$	26	$6^7/_8$	$2^1/_4$	$24^3/_4$	$5^1/_2$
24	36	$2^3/_4$	$28^1/_4$	$7^1/_4$	$2^7/_{16}$	27	$6^1/_4$

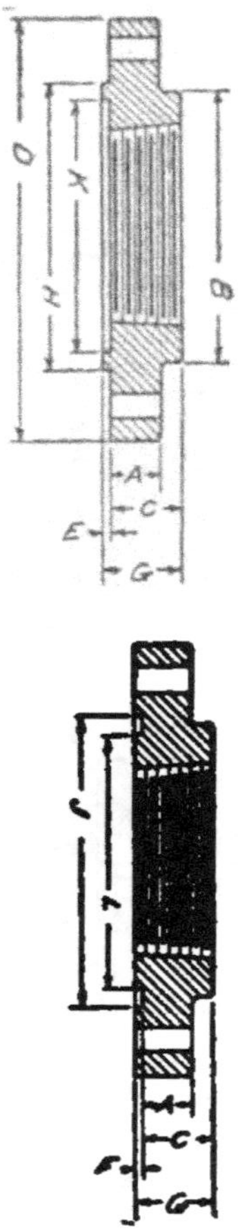

Fig. 73. Tongued and Grooved Flanges.

TABLE 55 (Fig. 73)

Extra Heavy Tongued and Grooved Flanges

Size Inches	D	E	F	H	J	K	L	Cast Flanges A	Cast Flanges B	Cast Flanges C	Forged Flanges A	Forged Flanges B	Forged Flanges G
1	4 1/2	3/16	1/8	2 1/2	2 9/16	1 3/4	1 11/16	...	...	...	9/16	2	1
1 1/4	5	3/16	1/8	3	3 1/16	2 1/8	2 1/16	...	...	...	5/8	2 1/2	1 1/8
1 1/2	6	3/16	1/8	3 5/8	3 11/16	2 3/4	2 11/16	...	...	...	3/4	2 7/8	1 1/4
2	6 1/2	3/16	1/8	4 1/8	4 3/16	3 1/8	3 1/16	7/8	3 5/16	1 3/8	7/8	3 3/8	1 5/8
2 1/2	7 1/2	3/16	1/8	4 5/8	4 11/16	3 5/8	3 9/16	1	4	1 7/16	1	4 1/16	1 7/16
3	8 1/4	3/16	1/8	5 1/4	6 5/16	4 1/4	4 3/16	1 1/8	4 5/8	1 9/16	1	4 11/16	1 9/16
3 1/2	9	3/16	1/8	5 3/4	5 13/16	4 3/4	4 11/16	1 3/16	5 1/4	1 5/8	1 1/8	5 5/16	1 5/8
4	10	3/16	1/8	6 1/4	6 5/16	5 1/4	5 3/16	1 1/4	5 3/4	1 3/4	1 1/8	5 13/16	1 3/4
4 1/2	10 1/2	3/16	1/8	6 3/4	6 13/16	5 3/4	5 11/16	1 5/16	6 3/16	1 13/16	1 1/4	6 1/4	1 13/16
5	11	3/16	1/8	7 1/4	7 5/16	6 1/4	6 3/16	1 3/8	6 3/4	1 7/8	1 1/4	6 13/16	1 7/8
6	12 1/2	3/16	1/8	8 1/2	8 9/16	7 1/2	7 7/16	1 7/16	7 15/16	2	1 1/4	7 7/8	2
7	14	1/4	3/16	9 5/8	9 11/16	8 5/8	8 9/16	1 1/2	9	2 1/16	1 5/16	9 1/8	2 1/16
8	15	1/4	3/16	10 7/8	10 15/16	9 5/8	9 9/16	1 5/8	10 1/8	2 3/16	1 3/8	10 1/8	2 3/16
9	16 1/4	1/4	3/16	11 7/8	11 15/16	10 5/8	10 9/16	1 3/4	11 1/16	2 1/4	1 7/16	11 3/16	2 1/4
10	17 1/2	1/4	3/16	13 1/8	13 3/16	11 5/8	11 9/16	1 7/8	12 3/8	2 3/8	1 1/2	12 9/16	2 3/8
12	20 1/2	1/4	3/16	15 1/8	15 3/16	13 5/8	13 9/16	2	14 5/8	2 9/16	1 5/8	14 5/8	2 9/16
14	23	1/4	3/16	17 3/8	17 7/16	15 7/8	15 13/16	2 1/2	15 7/8	2 11/16	1 3/4	15 13/16	2 11/16
15	24 1/2	1/4	3/16	18 5/8	18 11/16	17 1/8	17 1/16	2 3/16	16 15/16	2 13/16	1 13/16	17 3/16	2 13/16
16	25 1/2	1/4	3/16	20 1/8	20 5/16	18 3/8	18 5/16	2 1/4	18	2 7/8	1 7/8	20 1/4	3 1/16
18	28	1/4	3/16	22 3/8	22 7/16	20 3/8	20 5/16	2 3/4	20 1/8	3 1/16	2	20 3/8	3 1/16
20	30 1/2	1/4	3/16	24 5/16	24 3/8	22 5/16	22 1/4	2 1/2	22 5/16	3 1/4	2 1/8	22 1/2	3 1/4
22	33	1/4	3/16	26 1/2	26 9/16	24 1/2	24 7/16	2 5/8	24 1/2	3 7/16	2 1/4	24 3/4	3 3/8
24	36	1/4	3/16	28 1/2	28 9/16	26 1/2	26 7/16	2 3/4	26 3/4	3 5/8	2 3/8	26 15/16	3 5/8

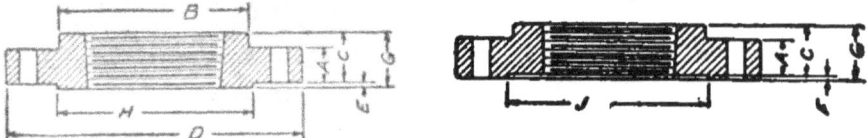

Fig. 74. Male and Female Flanges.

TABLE 56 (FIG. 74)

EXTRA HEAVY MALE AND FEMALE FLANGES

Size Inches	D	E	F	H	J	Cast Flanges			Forged Flanges		
						A	B	C	A	B	G
1	4¹/₂	³/₁₆	¹/₁₆	2⁵/₁₆	2⁵/₈	...	...	...	⁹/₁₆	2	1
1¹/₄	5	³/₁₆	¹/₈	2³/₄	2¹³/₁₆	...	...	...	⁵/₈	2¹/₈	1¹/₈
1¹/₂	6	³/₁₆	¹/₈	3¹/₈	3³/₁₆	...	...	...	³/₄	2⁷/₈	1¹/₄
2	6¹/₂	³/₁₆	¹/₈	3⁵/₈	3¹¹/₁₆	⁷/₈	3⁵/₁	1³/₈	⁷/₈	3³/₈	1³/₈
2¹/₂	7¹/₂	³/₁₆	¹/₈	4¹/₈	4³/₁₆	1	4	1⁷/₁₆	1	4¹/₁₆	1⁷/₁₆
3	8¹/₄	³/₁₆	¹/₈	5	5¹/₁₆	1¹/₈	4⁵/₈	1⁹/₁₆	1	4¹¹/₁₆	1⁹/₁₆
3¹/₂	9	³/₁₆	¹/₈	5¹/₂	5⁹/₁₆	1³/₁₆	5¹/₄	1⁵/₈	1¹/₈	5⁵/₁₆	1⁵/₈
4	10	³/₁₆	¹/₈	6	6¹/₁₆	1¹/₄	5³/₄	1³/₄	1¹/₈	5¹¹/₁₆	1³/₄
4¹/₂	10¹/₂	³/₁₆	¹/₈	6¹/₂	6⁹/₁₆	1⁵/₁₆	6³/₁₆	1¹³/₁₆	1¹/₄	6¹/₄	1¹³/₁₆
5	11	³/₁₆	¹/₈	7¹/₄	7⁵/₁₆	1³/₈	6³/₄	1⁷/₈	1¹/₄	6¹¹/₁₆	1⁷/₈
6	12¹/₂	³/₁₆	¹/₈	8³/₈	8⁷/₁₆	1⁷/₁₆	7¹⁵/₁₆	2	1¹/₄	7⁷/₈	2
7	14	¹/₄	⁵/₁₆	9³/₈	9⁷/₁₆	1¹/₂	9	2¹/₁₆	1⁵/₁₆	9¹/₈	2¹/₁₆
8	15	¹/₄	⁵/₁₆	10⁵/₈	10¹¹/₁₆	1⁵/₈	10¹/₈	2³/₁₆	1⁵/₈	10¹/₈	2³/₁₆
9	16¹/₄	¹/₄	⁵/₁₆	11⁵/₈	11¹¹/₁₆	1³/₄	11¹/₁₆	2¹/₄	1⁷/₁₆	11³/₁₆	2¹/₄
10	17¹/₂	¹/₄	⁵/₁₆	12³/₄	12¹³/₁₆	1⁷/₈	12³/₈	2³/₈	1¹/₂	12⁹/₁₆	2³/₈
12	20¹/₂	¹/₄	⁵/₁₆	15¹/₄	15⁵/₁₆	2	14⁵/₈	2⁹/₁₆	1⁵/₈	14⁵/₈	2⁹/₁₆
14	23	¹/₄	⁵/₁₆	16¹/₂	16⁹/₁₆	2¹/₈	15⁷/₁₆	2¹¹/₁₆	1³/₄	15¹³/₁₆	2¹¹/₁₆
15	24¹/₂	¹/₄	⁵/₁₆	17¹/₂	17⁹/₁₆	2³/₁₆	16¹⁵/₁₆	2¹³/₁₆	1¹³/₁₆	17³/₁₆	2¹³/₁₆
16	25¹/₂	¹/₄	⁵/₁₆	18¹/₂	18⁹/₁₆	2¹/₄	18	2⁷/₈	1⁷/₈	20¹/₄	3¹/₁₆
18	28	¹/₄	⁵/₁₆	21	21¹/₁₆	2³/₈	20¹/₈	3¹/₁₆	2	20³/₈	3¹/₁₆
20	30¹/₂	¹/₄	⁵/₁₆	23	23¹/₁₆	2¹/₂	22⁵/₁₆	3¹/₄	2¹/₈	22¹/₂	3¹/₄
22	33	¹/₄	⁵/₁₆	25¹/₂	25⁹/₁₆	2⁵/₈	24¹/₂	3⁷/₁₆	2¹/₄	24³/₄	3³/₈
24	36	¹/₄	⁵/₁₆	27¹/₂	27⁹/₁₆	2³/₄	26³/₄	3⁵/₈	2³/₈	26¹⁵/₁₆	3⁵/₈

Special Connections. — Several forms of special connections for lap-welded steel pipe, as made by the American Spiral Pipe Works, are shown in Figs. 75 to 80. The flanges are all made of forged steel. Fig. 75 shows a riveted steel flange connection which is made in different standards for high and low pressure work. Fig. 76 shows a welded steel flange with follower rings, a form of connection especially suited for high pressure work. A field riveted joint suitable for long lines where facilities are ample for riveting up at destination, is shown in Fig. 77. It possesses many advantages over the ordinary field joint as the taper end may be inserted into the flared end without difficulty, thus enabling holes to be brought quickly into alignment.

PIPE JOINTS

A form of bell and spigot lead joint for low pressure water lines is shown in Fig. 78. It requires a less amount of lead than

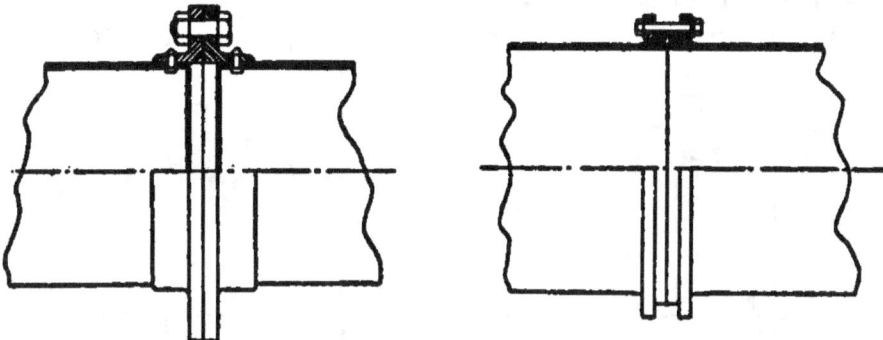

Fig. 75. Riveted Flanges.   Fig. 76. Flanges with Follower Rings.

ordinary cast pipe. The bolted socket joint shown in Fig. 79 is especially suited for long line work or for connections on submerged

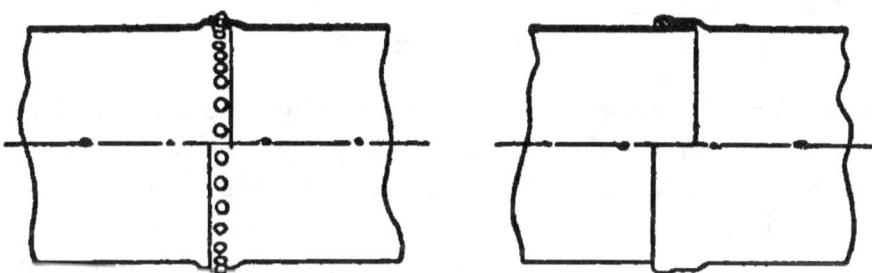

Fig. 77. Field Riveted Joint.   Fig. 78. Bell and Spigot Joint.

pipe lines as it allows for a slight deflection at each joint. The standard bolted joint connection shown in Fig. 80 forms an ex-

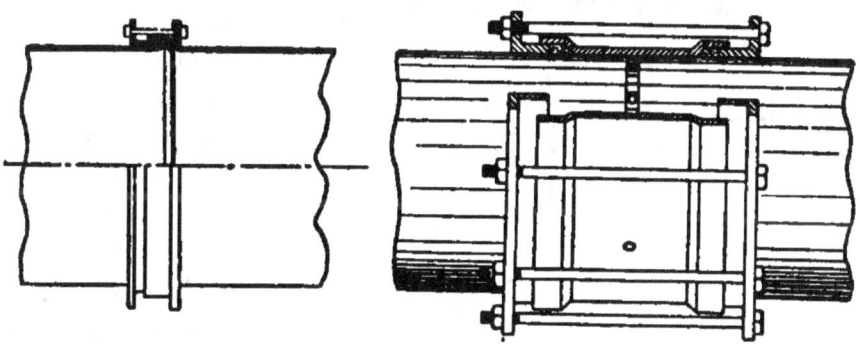

Fig. 79. Bolted Socket Joint.   Fig. 80. Bolted Joint.

pansion joint and permits a deflection or slight angle to be made at each joint.

Converse Joints. — The Converse lock joint pipe, Fig. 81, and the Matheson joint pipe, Fig. 82, are made by the National Tube Company in sizes ranging from 2 inches to 30 inches outside diameter, and about 18 feet long. The joints are made with

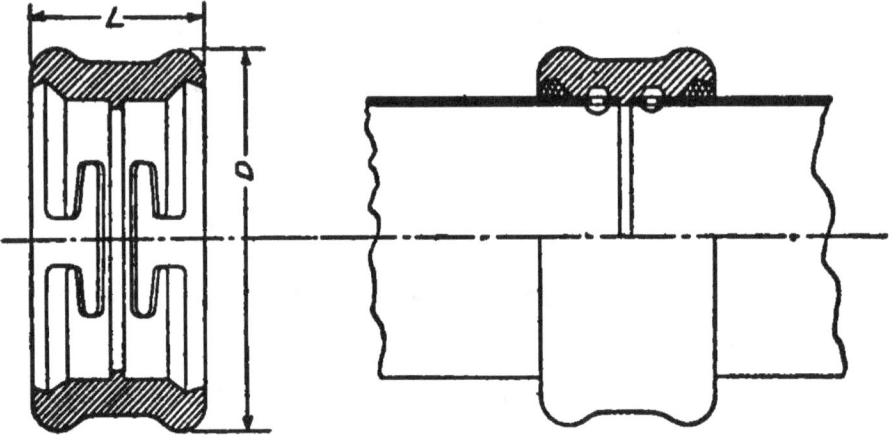

Fig. 81. Converse Joint.

lead. The Converse Lock Joint is made by means of a cast iron hub whose inner surface has an inwardly projecting ring at midlength; on each side of this ring are two wedge-shaped pockets, diametrically opposite; near each mouth of the hub is a recess for lead. Close to each end of the pipe are two strong rivets, placed at such distance from the end that when the pipe is inserted into the hub and slightly rotated, the rivets engage the slopes of the wedge-shaped pockets and force the end of the pipe against the central ring of the hub. Lead is then poured into the recess provided for it,

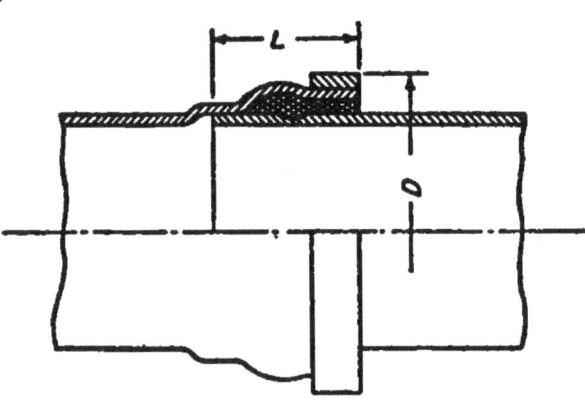

Fig. 82. Matheson Joint.

and securely calked. Table 57 gives standard sizes, thicknesses, etc., for Converse joint pipe.

Matheson Joints. — Matheson joint pipe is a pipe with a joint of a bell and spigot type, very similar in appearance to a cast iron

TABLE 57 (Fig. 81)
Converse Lock Joint Pipe

External Diameter	Thickness	Weight per Foot Plain Ends	Hub — Cast Iron			Weight of Lead for Field End	Weight per Foot Complete, Including Hub Leaded on Mill End	Mill Test
			Diameter D	Length L	Weight			
2.00	.095	1.932	3 3/4	3 1/2	4.25	1.00	2.207	700
3.00	.109	3.365	5 1/8	3 3/4	8.50	2.25	3.931	700
4.00	.128	5.293	6 1/4	4	10.50	3.00	5.991	600
5.00	.134	6.963	7 1/4	4 1/4	15.00	3.75	7.932	600
6.00	.140	8.762	8 1/4	4 1/2	19.00	4.50	9.969	600
7.00	.149	10.902	9 1/2	4 1/2	24.00	5.50	12.419	600
8.00	.158	13.233	10 1/2	4 3/4	28.25	6.50	15.008	600
9.00	.167	15.754	11 3/4	4 3/4	34.50	8.50	17.958	500
10.00	.175	18.363	12 3/4	5	39.00	9.00	20.801	500
11.00	.185	21.368	13 3/4	5	41.50	10.00	23.963	500
12.00	.194	24.461	15	5 1/2	55.00	11.00	27.795	500
13.00	.202	27.610	16 1/8	5 1/2	59.00	12.00	31.179	500
14.00	.210	30.928	17 1/8	5 3/4	67.00	14.50	35.013	500
15.00	.222	35.038	18 3/8	5 3/4	78.00	15.50	39.731	500
16.00	.234	39.401	19 3/4	6 1/4	102.00	25.00	45.847	500
17.00	.240	42.959	20 7/8	6 1/4	110.00	26.00	49.850	450
18.00	.245	46.458	22 1/8	6 3/4	140.00	30.00	55.123	450
19.00	.259	51.840	23 3/16	6 3/4	150.00	32.00	61.081	450
20.00	.272	57.309	24 7/16	7 1/4	180.00	37.00	68.337	450
22.00	.301	69.765	26 5/8	7 3/4	215.00	45.00	82.868	450
24.00	.330	83.423	29	8 1/4	275.00	50.00	99.789	450
26.00	.362	99.122	31 5/8	8 3/4	360.00	64.00	120.555	450
28.00	.396	116.746	33 15/16	9 1/4	425.00	77.00	142.000	450
30.00	.432	136.421	36 3/16	10	525.00	82.00	166.828	450

pipe joint. The joint is made by belling out or expanding one end of the pipe in such a manner as to permit the bell end to slip over the plain or spigot end of the next length of pipe, leaving enough space between the two for the lead which is to make the joint. After the end of the pipe has been shaped a wrought band is shrunk on the outside of the bell to reinforce it at this point and to keep it in shape to withstand the calking of the lead. The spigot end of the pipe has a recess turned in it which prevents the lead from blowing out or the pipe from pulling out. This pipe is extensively used for water service in the west. Table 58 gives standard sizes, thicknesses, etc., for Matheson pipe.

TABLE 58 (Fig. 82)

Matheson Joint Pipe

External Diameter	Thickness	Outside Diameter of Reinforcing Ring D	Length of Joint L	Weight per Foot		Weight of Lead per Joint	Mill Test
				Plain Ends	Complete		
2.00	.095	2.966	2.16	1.932	1.952	1.00	700
3.00	.109	4.034	2.26	3.365	3.392	1.75	700
4.00	.128	5.236	2.32	5.293	5.339	2.75	600
5.00	.134	6.268	2.38	6.963	7.019	3.50	600
6.00	.140	7.446	2.50	8.762	8.872	4.75	600
7.00	.149	8.484	2.58	10.902	11.028	5.50	600
8.00	.158	9.646	2.73	13.233	13.405	6.75	600
9.00	.167	10.684	2.73	15.754	15.945	8.25	500
10.00	.175	11.846	2.82	18.363	18.610	9.50	500
11.00	.185	12.886	2.91	21.368	21.638	11.00	500
12.00	.194	14.048	3.00	24.461	24.880	13.25	500
13.00	.202	15.084	3.07	27.610	28.060	15.25	500
14.00	.210	16.370	3.15	30.928	31.536	17.25	500
15.00	.222	17.394	3.24	35.038	35.686	19.25	500
16.00	.234	18.438	3.32	39.401	40.089	22.00	500
17.00	.240	19.470	3.41	42.959	43.687	23.75	450
18.00	.245	20.730	3.50	46.458	47.384	25.75	450
19.00	.259	21.778	3.57	51.840	52.815	29.00	450
20.00	.272	22.804	3.64	57.309	58.332	31.00	450
22.00	.301	24.882	4.06	69.756	71.098	40.25	450
24.00	.330	26.980	4.26	83.423	84.882	48.00	450
26.00	.362	29.064	4.40	99.122	100.697	55.25	450
28.00	.396	31.672	4.58	116.746	119.021	65.00	450
30.00	.432	33.764	4.75	136.421	138.851	75.00	450

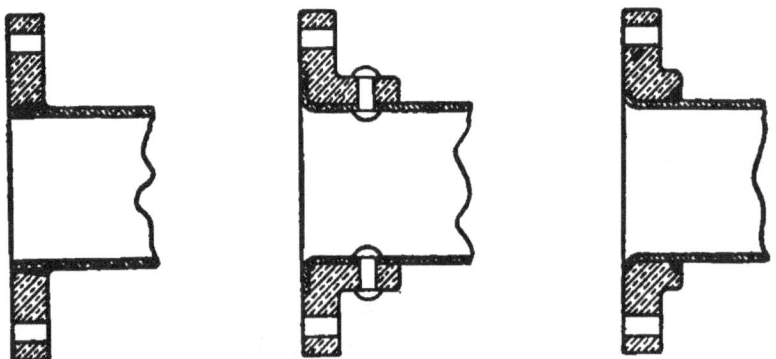

Figs. 83, 84, and 85. Flanges for Copper Pipe.

PIPE JOINTS

Flanges for Copper Pipe. — For copper pipe the flanges are made of composition and are attached by brazing or brazing and riveting. Figs. 83, 84, and 85 show three methods of attaching flanges to copper pipe, the first form is a plain flange brazed on, the second is brazed and riveted, and the third is peened and brazed.

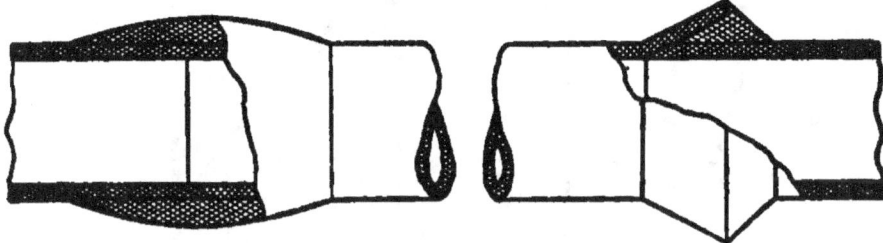

Fig. 86. Wiped Joint.   Fig. 87. Blown Joint.

Lead Pipe Joints. — Lead pipe may be joined by means of flanges bolted together or by wiped or blown joints as shown in Figs. 86 and 87. The flanges may be of lead integral with the pipe or separate cast iron flanges may be used as in Figs. 88 and 89 respectively. The amount of lead required for making lead joints is given in the following tabulation. The thickness of the joint ranges from $1/4$ inch on small sizes to $1/2$ inch on larger sizes.

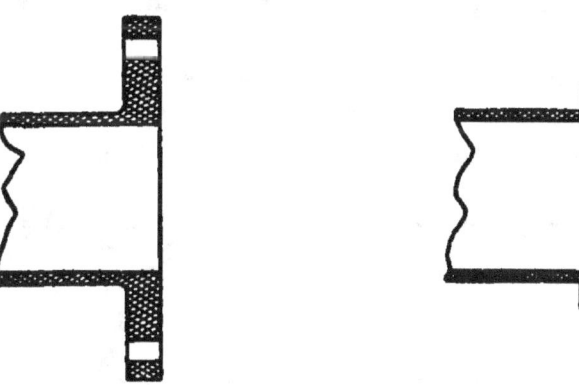

Fig. 88. Lead Flanges.   Fig. 89. Iron Flanges for Lead Pipe.

Diameter Inches	Weight Pounds	Diameter Inches	Weight Pounds
2	2½	12	15
3	3½	14	18
4	4½	16	22
6	6½	18	26
8	9	20	33
10	13		

Joints for Riveted Pipe. — Straight riveted pipe may be joined by riveting while in the course of erection, by flanges riveted on to the end of the pipe, or in some cases by a slip joint. Spiral riveted pipe may be joined by flanges riveted to the ends of the

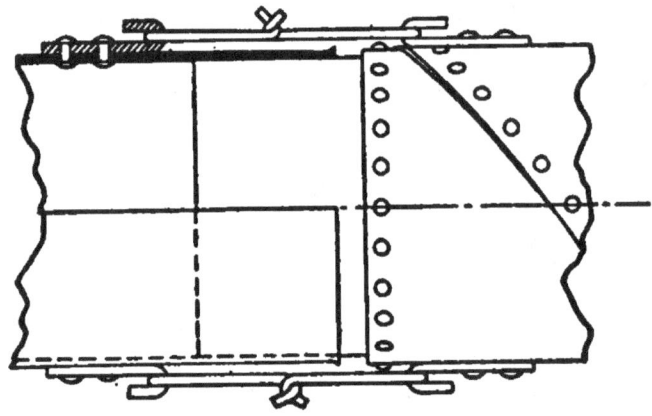

Fig. 90. Slip Joint.

pipe, by a slip joint, Fig. 90, by means of a crimped end and sleeve, Fig. 91, or by bolting, Figs. 80 and 92. The makers have their own standard for dimensions of flanges and drilling so that the American Standard is not supplied unless called for. Table 59 gives the spiral pipe manufacturers' standard dimensions for flanges.

The Root bolted joint, Fig. 92, is recommended for both asphalted and galvanized pipe when used to convey water. The joints shown in Figs. 90 and 93 are from literature of the American Spiral Pipe

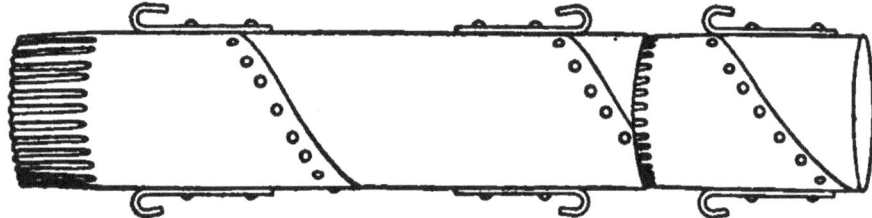

Fig. 91. Crimped End and Sleeve.

Works. The lugs shown in Figs. 90 and 91 are for the purpose of drawing up the pipe. Calking is necessary to obtain a tight joint. Differences in temperature cause a large amount of expansion and contraction on long lines of flanged pipe. Either bolted joints, Figs. 80 and 92 or an expansion joint, Fig. 93, may be

PIPE JOINTS

used at intervals of about 400 feet to take care of these changes in length. The expansion joint consists of a cast body and brass sleeve, with a gland and packing as shown in the figure.

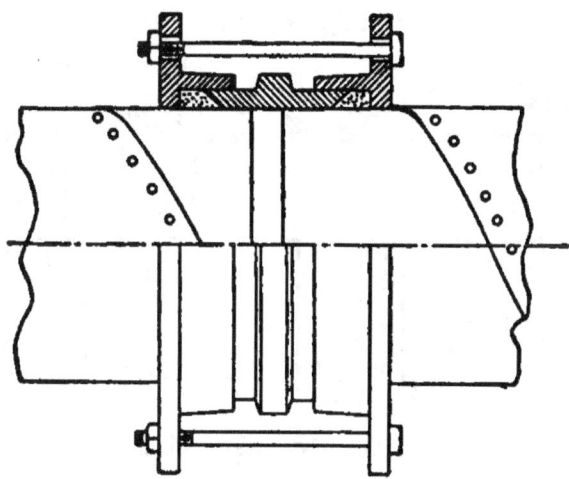

Fig. 92. Bolted Joint.

TABLE 59

FLANGES FOR RIVETED PIPE

Riveted Pipe Manufacturers' Standard

Inside Diameter Inches	Outside Diameter Inches	Diameter of Bolt Circle Inches	Sizes of Bolts Inches	Number of Bolts	Diameter of Bolt Holes Inches
3	6	$4^3/_4$	$7/_{16}$	4	$1/_2$
4	7	$5^{15}/_{16}$	$7/_{16}$	8	$1/_2$
5	8	$6^{15}/_{16}$	$7/_{16}$	8	$1/_2$
6	9	$7^7/_8$	$1/_2$	8	$5/_8$
7	10	9	$1/_2$	8	$5/_8$
8	11	10	$1/_2$	8	$5/_8$
9	13	$11^1/_4$	$1/_2$	8	$5/_8$
10	14	$12^1/_4$	$1/_2$	8	$5/_8$
11	15	$13^3/_8$	$1/_2$	12	$5/_8$
12	16	$14^1/_4$	$1/_2$	12	$5/_8$
13	17	$15^1/_4$	$1/_2$	12	$5/_8$
14	18	$16^1/_4$	$1/_2$	12	$5/_8$
15	19	$17^7/_{16}$	$1/_2$	12	$5/_8$
16	$21^1/_4$	$19^1/_4$	$1/_2$	12	$5/_8$
18	$23^1/_4$	$21^1/_4$	$5/_8$	16	$3/_4$
20	$25^1/_4$	$23^1/_8$	$5/_8$	16	$3/_4$
22	$28^1/_4$	26	$5/_8$	16	$3/_4$
24	30	$27^3/_4$	$5/_8$	16	$3/_4$

Joints for Cast Iron Pipe. — Two forms of joints for cast iron pipe are mentioned and illustrated in Chapter I. Dimensions for cast iron bell and spigot joints are given in Tables 1 and 2, Chapter II. For flanges the dimensions for the American Standard are given in Tables 39 and 40, Chapter IV.

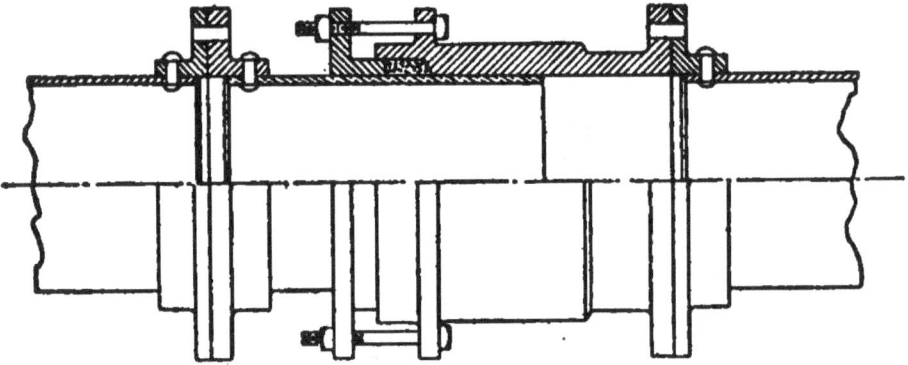

Fig. 93. Expansion Joint.

The form of joint shown in Fig. 94 is used on "universal" cast iron pipe made by the Central Foundry Company. The contact surfaces are machined on a taper at slightly different angles and drawn together by bolts, giving an iron to iron joint. The different tapers permit a deflection of three degrees so that the joint allows for expansion and uneven ground settlement.

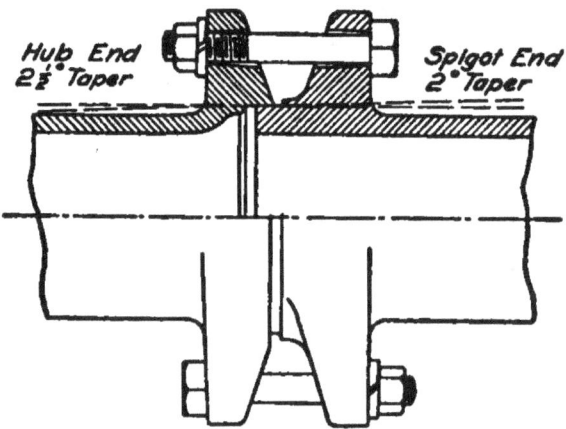

Fig. 94. Universal Cast Iron Joint.

Straight lengths may be laid on a curve of 150 feet radius. Two bolts per joint are sufficient for pressures up to 175 pounds. Table 60 gives the thicknesses and weights of "universal" pipe. Lengths lay a full six feet.

TABLE 60 (FIG. 94)

UNIVERSAL CAST IRON PIPE

Nominal Inside Diameter	Class No. 100, 100 Lbs. Pressure			Class No. 130, 130 Lbs. Pressure			Class No. 175, 175 Lbs. Pressure			Class No. 250, 250 Lbs. Pressure			Bolt Sizes
	Approx. Thickness Inches	Estimated Weight Pounds per Foot	Estimated Weight Pounds per 6-Foot Length	Approx. Thickness Inches	Estimated Weight Pounds per Foot	Estimated Weight Pounds per 6-Foot Length	Approx. Thickness Inches	Estimated Weight Pounds per Foot	Estimated Weight Pounds per 6-Foot Length	Approx. Thickness Inches	Estimated Weight Pounds per Foot	Estimated Weight Pounds per 6-Foot Length	
2	....	....	....	....	....	....	.35	$8\frac{1}{2}$	51	.39	$9\frac{1}{2}$	57	$\frac{1}{2}$ x $3\frac{1}{2}$
3	....	....	....	....	....	....	.37	13	78	.42	$14\frac{1}{2}$	87	$\frac{1}{2}$ x 4
4	.37	18	108	.40	$18\frac{1}{4}$	$112\frac{1}{2}$	.43	$20\frac{1}{4}$	$121\frac{1}{2}$	.45	$21\frac{1}{4}$	$127\frac{1}{2}$	$\frac{5}{8}$ x 5
5	.40	24	144	.425	25	150	.45	26	156	.49	29	174	$\frac{5}{8}$ x $5\frac{1}{2}$
6	.43	30	180	.45	31	186	.47	32	192	.51	$35\frac{1}{2}$	213	$\frac{3}{4}$ x 6
8	.47	$44\frac{1}{4}$	$265\frac{1}{2}$	.49	46	276	.525	$49\frac{1}{4}$	$295\frac{1}{2}$	.58	$53\frac{1}{4}$	$319\frac{1}{2}$	$\frac{7}{8}$ x $\frac{1}{2}$
10	.50	$60\frac{1}{2}$	363	.53	$63\frac{1}{2}$	381	.58	$67\frac{3}{4}$	$406\frac{1}{2}$	.64	74	444	1 x $7\frac{1}{2}$
12	.53	$75\frac{1}{2}$	453	.57	$80\frac{1}{2}$	483	.62	87	522	.70	$97\frac{1}{2}$	585	1 x 8
14	.565	$94\frac{1}{2}$	567	.60	$99\frac{1}{2}$	597	.66	$107\frac{1}{2}$	645	.76	124	741	$1\frac{1}{8}$ x 9
16	.60	$115\frac{1}{2}$	693	.65	123	738	.72	134	804	.83	156	936	$1\frac{1}{4}$ x $9\frac{1}{2}$
20	.67	166	996	.73	178	1068	.82	196	1176	.94	223	1338	$1\frac{5}{8}$ x $11\frac{3}{4}$

CHAPTER VI

STANDARD VALVES

Valves. — Valves of many forms are used to control the conveying of fluids in pipes. It will be impossible to describe all of the valves made for the different purposes for which they are required. The general classes and types, however, will be illustrated and described. The figures have been chosen to illustrate these types, and it does not follow that the particular design or make shown is the best of its class, as it would be difficult to make such a selection from the many reliable valves now manufactured.

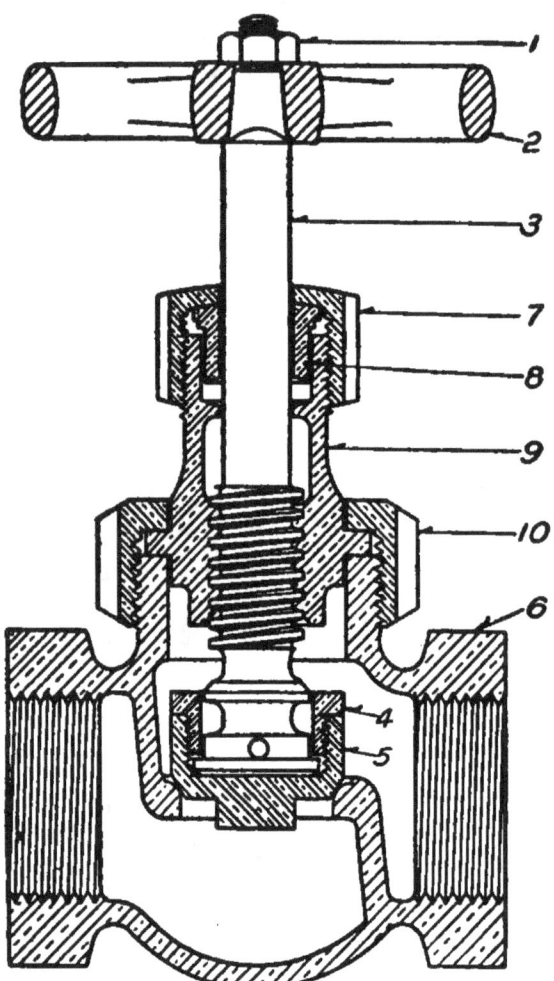

Fig. 95. Sectional View of Globe Valve.

Valves are made with either screwed or flanged ends. It is not desirable to use screwed ends for sizes larger than six inches for steam pressure. For high pressures and superheated steam flanged end fittings should be used for all sizes. It is good practice to call for flanged fittings and valves in all cases for sizes larger than 2½ inches.

STANDARD VALVES

Materials. — Valves are made of various materials suited to the purpose in view. Brass or bronze valves are ordinarily made in sizes up to and including three inches. These valves are used for steam up to $1\frac{1}{2}$ inches and the larger sizes on boiler feed lines. Valves with cast iron bodies are suitable for water or saturated steam. For steam under high pressure and superheat other materials are necessary, such as Ferrosteel and cast steel.

Globe and Gate Valves. — There are two general classes of valves, globe valves and gate valves. The globe valve has a spherical body and a circular opening at right angles to the axis of the pipe. A section of a globe valve, together with the names of the principal parts, is shown in Fig. 95.

NAMES OF PARTS OF GLOBE VALVE

1. Stem nut
2. Hand wheel
3. Valve stem
4. Valve nut
5. Valve (swivel)
6. Valve body
7. Gland Nut
8. Gland
9. Bonnet
10. Bonnet ring

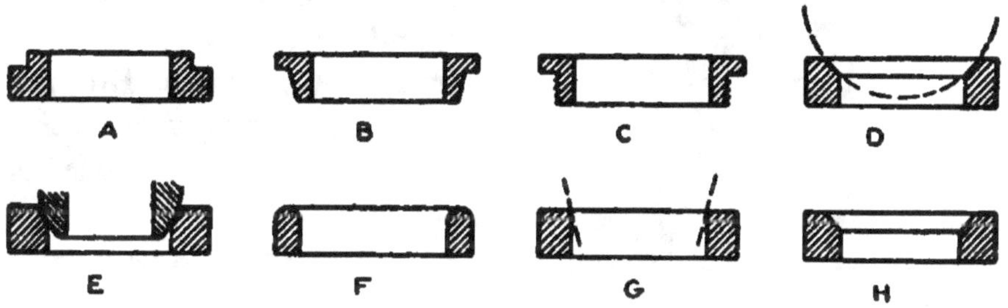

Fig. 96. Forms of Valve Seats.

A valve may be used in place of an elbow and a globe valve, in which case it is called an angle valve, Fig. 107. A cross valve is shown in Fig. 107. There are several objections to the use of globe valves, among which are the resistance which they offer to the fluid, and the water pocket which is present when they are used for steam lines. They are desirable, however, when throttling is necessary.

Valve Seats. — A variety of valve seats are shown in Figs. 96, 97, and 98. In Fig. 96 A, B, and C are plain flat seats; D is a concave or spherical seat; E and F are rounded seats; G is a square seat, and H is a bevel seat. Any of these forms may be made as a part of the valve body or separate, and either screwed or forced into place. The forms of valve discs differ, as shown

100 A HANDBOOK ON PIPING

in the various figures. The valve seat shown in Fig. 97 is made up of two conical surfaces and a groove. The disc is made in

Fig. 97. Spring Valve Seat.

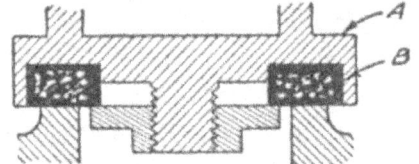

Fig. 98. Removable Disc Valve Seat.

similar form. The grooves permit a certain amount of spring and insure tightness when the valve is closed. This form of seat is made by the Crosby Steam Gage and Valve Company. Fig. 98 shows the use of a removable disc instead of a solid disc. The disc holder A is of brass or other suitable material and the disc B of softer material. When leakage takes place the disc can be removed and replaced by a new one. Discs are made by Jenkins Brothers of various compounds suiting them to different kinds of service.

Gate Valves.—A gate valve is shown in section in Fig. 99, and as will be observed has its openings parallel to the cross section of the pipe, so there is little or no resistance to the flow, making it preferable for most purposes. The valve disc which closes the passage way may be a solid tapered wedge, as in Fig. 99, may be in two parts, as in Fig. 100, or may have parallel faces, as in Fig. 101.

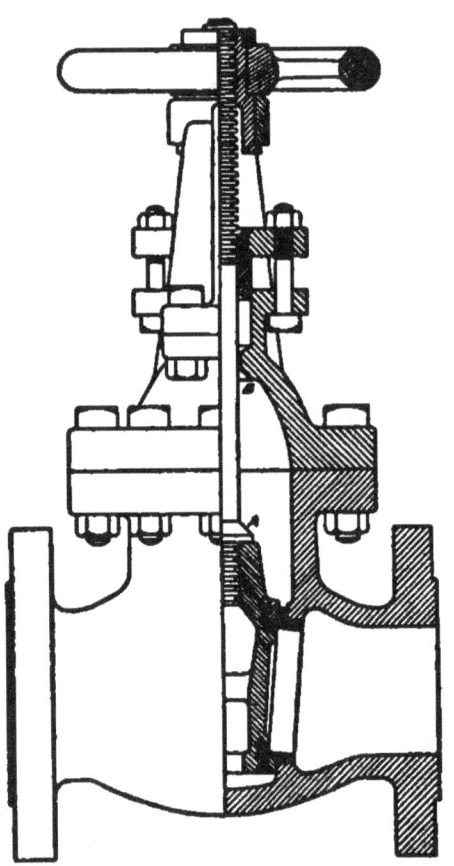

Fig. 99. Gate Valve — Solid Tapered Wedge — Rising Stem.

The gate valve shown in Fig. 99 is made by Walworth Company. It is of the solid wedge gate type, in which the disc consists of a single piece faced with hard metal. This disc slides on ribs in the valve body. This valve may be packed while under pressure by screwing the stem out until the beveled collar A on the stem engages with the beveled recess B of the bonnet, forming a tight joint. The valve shown in Fig. 100 is made by the

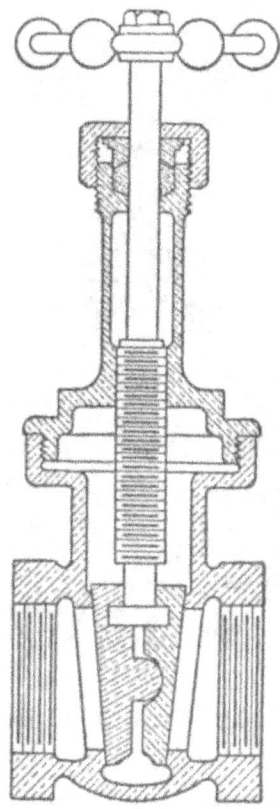

Fig. 100. Gate Valve — Two Part Wedge.

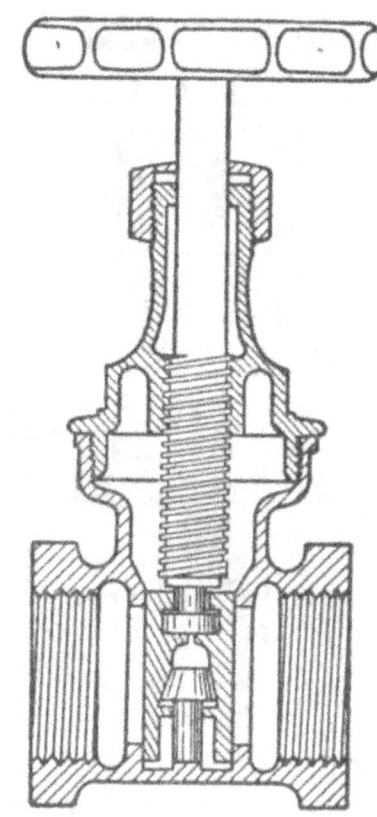

Fig. 101. Gate Valve — Parallel Seat.

Lunkenheimer Company. The principle upon which the discs are seated makes them self-adjusting and they will accommodate themselves to scale or sediment which may lodge on one of the seats, so that at least one disc will close tightly. This is accomplished by the ball and socket bearing between the discs, which permits sufficient play in any direction. The stuffing boxes can be packed when the valve is wide open and under pressure, as a shoulder on the stem directly above the threads forms a seat beneath the stuffing box.

The parallel seat double disc valve shown in Fig. 101 is con-

102 A HANDBOOK ON PIPING

structed so that the discs do not bear on the seats in opening or closing. The discs are hung on the stem and are seated by a wedge which bears on the centres of the discs. The lug on the bottom of the valve body brings the wedge into action just before the discs reach their lowest position. At the instant of starting

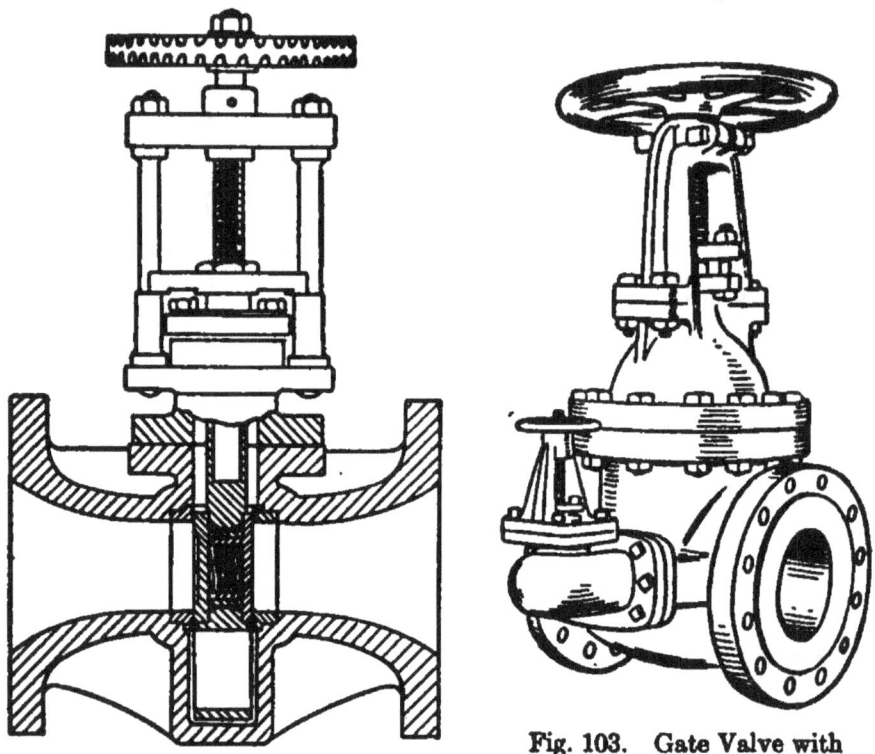

Fig. 102. Hopkinson-Ferranti Valve.

Fig. 103. Gate Valve with By-pass.

to open the valve the wedge is released. This valve is made by the National Tube Company.

A form of gate valve based upon the principle of the Venturi meter is the Hopkinson-Ferranti valve shown in Fig. 102. This form of valve is widely used in England and has found application in American practice. The velocities are increased in the central portion of the valve which is only one half the size of the pipe in which the valve is used. The contour of the delivery side is such as to reduce the velocity and restore the pressure. The reduced size of the valve faces makes them less liable to distortion. The small valve seats make a by-pass valve unnecessary as the steam can be throttled when opening. A throat piece is drawn up when the valve is opened, and this forms a continuous

STANDARD VALVES

Venturi tube. These valves are adaptable for steam lines where velocities of less than 6000 feet per minute are used.

By-pass Valves. — The effort required to open a large valve with the steam acting upon one side is considerable, and some means of equalizing the pressure on the two sides of the disc as well as to permit "warming up" is desirable. This is accomplished by means of a small auxiliary valve in the passage joining the two ends of the valve, called a by-pass. Fig. 103 shows an extra heavy Walworth valve with a by-pass.

Valve Stem Arrangements. — There are two general arrangements of the valve stem known as inside screw and outside screw. The inside screw may be either rising stem or non-rising stem. Fig. 104 shows a valve with an inside screw, non-rising stem; Fig. 101 an inside screw, rising stem; and Fig. 99 an outside screw, rising stem. When the screw is outside it is protected from stem corrosion, and can be kept oiled. The rising stem is desirable as its position clearly indicates whether the valve is open or closed. In some parts of the country laws require the use of the rising stem on boiler stop valves, and certain classes of work. The valve stem on small sizes is generally made of bronze and larger sizes of steel, nickel plated. The valve shown in Fig. 104 is made by Crane Company for steam working pressures up to 250 pounds. It has an inside screw, non-rising stem. The seats are made of hard brass and screwed to shoulders in the body of the valve. They are renewable. The gate is faced with hard brass. This valve may be packed while under pressure by opening the valve wide

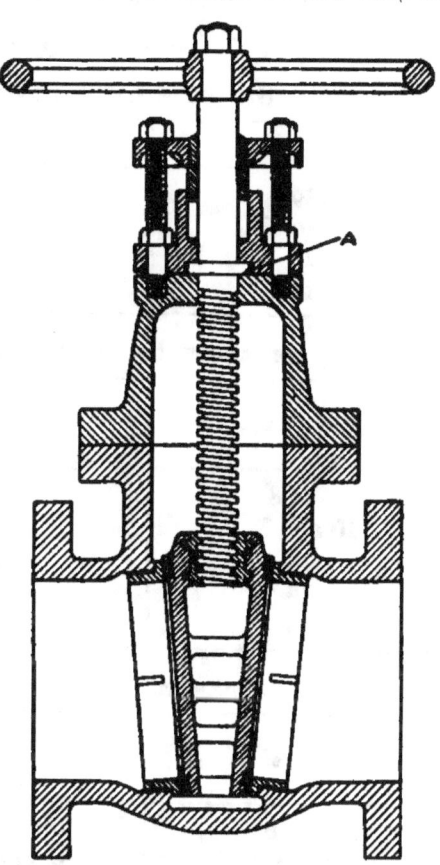

Fig. 104. Gate Valve — Inside Screw — Non-rising Stem.

and running the wedge tightly up to the top of the bonnet, which draws the collar of the stem down tightly to the flange of the bonnet, forming a steam or water tight joint at *A*.

Strength of Gate Valves. — Some actual bursting pressures for gate valves as tested by Crane Company are given in Table 61.

TABLE 61

STRENGTH OF STANDARD IRON GATE VALVES

Sizes, Inches	Pressure in Pounds per Square Inch
4 to 8	1000 to 1500
10 to 16	900
18	450 without breaking
20 to 30	300 " "

STRENGTH OF MEDIUM PRESSURE GATE VALVES

Sizes, Inches	Pressure in Pounds per Square Inch	
	Cast Iron	Ferro Steel
4 to 8	1200 to 1900	1900 to 2600
10 and 12	850	1400 to 1500
14	...	O.K. at 1000
16	...	O.K. at 750
18	...	O.K. at 700

STRENGTH OF EXTRA HEAVY GATE VALVES

Sizes, Inches	Pressure in Pounds per Square Inch	
	Cast Iron	Ferro Steel
4 to 8	1600 to 1900	2450 to 2600
10 and 12	1350 to 1550	1750 to 1900
14 to 16	1100	1200 to 1350
18	....	O.K. at 850
20 to 24	....	O.K. at 600

Standard Pressures and Dimensions. — Valves are generally constructed for three pressures, standard, medium, and extra heavy. Standard pressure generally means 125 pounds, medium 175 pounds, and extra heavy 250 pounds, when referring to steam. When used for water these values may be greatly increased, depending upon the conditions of service. Valves are made of cast steel suitable for steam pressures up to 350 pounds per square inch. The following tables give some of the dimensions for various kinds of valves as made by different companies.

STANDARD VALVES

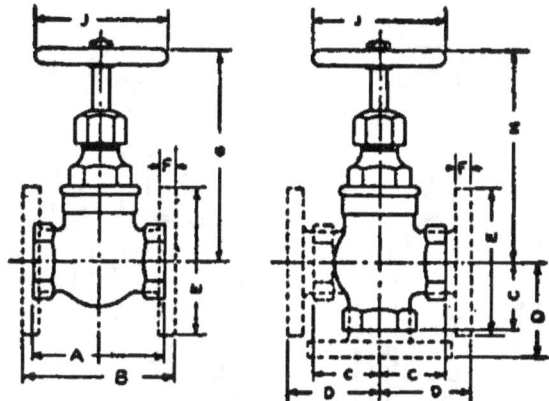

Fig. 105. Jenkins Valves, Globe, Angle, and Cross.

TABLE 62 (Fig. 105)
JENKINS STANDARD GLOBE, ANGLE, AND CROSS VALVES. BRASS—SCREWED AND FLANGED

150 Pounds Working Pressure

Size Inches	A Inches	B Inches	C Inches	D Inches	E Inches	F Inches	G Inches	H Inches	J Inches
1/8	1 9/16	..	13/16	..	..	..	2 5/8	2 3/4	1 1/2
1/4	2 1/8	2 7/16	1 1/16	1 7/8	2 3/4	9/32	3 5/8	3 7/8	2 1/16
3/8	2 3/8	3	1 3/16	2 1/16	2 7/8	11/32	4 1/8	4	2 1/16
1/2	2 3/4	3 1/16	1 3/8	2 1/16	3	3/8	4 7/8	5	2 7/16
3/4	3 3/16	3 5/8	1 1/2	2 3/8	3 1/2	13/32	5 1/4	5 5/8	2 13/16
1	3 13/16	4	1 3/4	2 5/8	4	7/16	5 3/4	6	3
1 1/4	4 1/4	4 5/8	2 1/16	2 15/16	4 1/2	15/32	7	7 1/4	3 1/2
1 1/2	4 7/8	4 7/8	2 1/4	3 3/16	5	1/2	7 1/4	7 7/8	4 1/8
2	5 3/4	6	2 7/8	3 3/4	6	9/16	9 1/8	9 1/4	4 5/8
2 1/2	6 5/8	6 3/4	3 1/8	4 1/4	7	5/8	9 3/8	9 3/4	5
3	8 1/2	7 1/2	4 1/4	4 9/16	7 1/2	11/16	10 1/2	11 1/4	6

TABLE 63 (Fig. 105)
JENKINS EXTRA HEAVY GLOBE, ANGLE, AND CROSS VALVES. BRASS—SCREWED AND FLANGED

250 Pounds Working Pressure

Size Inches	A Inches	B Inches	C Inches	D Inches	E Inches	F Inches	G Inches	H Inches	J Inches
1/2	2 15/16	3 3/4	1 1/2	2 1/3	3 1/4	13/32	4 3/4	4 7/8	2 7/8
3/4	3 1/2	4 1/4	1 3/4	2 1/2	3 3/4	15/32	6	6 1/8	3 1/8
1	4 1/8	4 3/4	2 1/16	2 7/8	4 1/2	1/2	6 7/8	7 1/8	3 1/2
1 1/4	4 5/8	5 1/2	2 5/16	3 1/4	5	17/32	7 1/2	8	4 1/8
1 1/2	5 3/8	6 1/4	2 11/16	3 11/16	6	9/16	8 3/8	8 7/8	4 3/8
2	6 1/4	7 1/4	3 1/8	4 1/8	6 1/2	5/8	9 3/4	10 3/8	5
2 1/2	7 1/2	8 1/4	3 3/4	4 3/4	7 1/2	11/16	11 3/8	12 1/4	6 1/2
3	8 3/4	9 1/2	4 3/8	5 1/8	8 1/4	3/4	12 5/8	13 5/8	7 1/2

106 A HANDBOOK ON PIPING

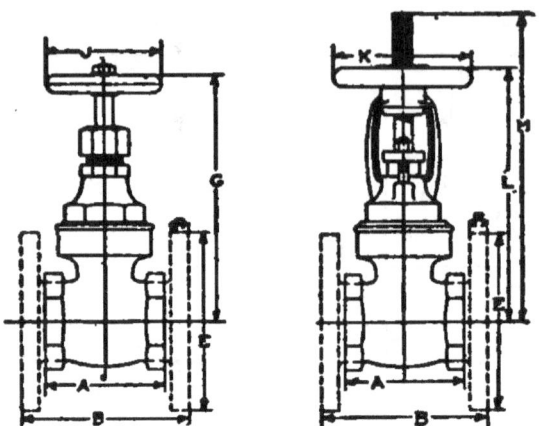

Fig. 106. Jenkins Gate Valves.

TABLE 64 (FIG. 106)
JENKINS STANDARD GATE VALVES. BRASS — SCREWED AND FLANGED
125 Pounds Working Pressure

Size Inches	A Inches	B Inches	E Inches	F Inches	G Inches	J Inches	K Inches	L Inches	M Inches
1/4	1 3/4	2 1/2	2 3/4	9/32	3 1/2	1 3/4			
3/8	1 3/4	2 5/8	2 7/8	11/32	3 1/2	1 3/4			
1/2	1 15/16	2 15/16	3	3/8	3 15/16	2			
3/4	2 5/16	3 3/8	3 1/2	13/32	4 11/16	2 7/16	2 3/4	4 3/4	5 13/16
1	2 11/16	3 15/16	4	7/16	5 9/16	2 13/16	3 5/16	5 1/2	6 7/8
1 1/4	3	4 1/2	4 1/2	15/32	6 1/4	3	3 5/8	6 3/8	8 1/16
1 1/2	3 3/8	4 7/8	5	1/2	6 3/4	3 1/2	3 7/8	7 1/4	9 1/16
2	4	5 3/4	6	9/16	7 3/4	4 1/8	4 3/8	8 1/4	10 5/8
2 1/2	4 11/16	6 1/2	7	5/8	9	4 3/8	4 13/16	9 13/16	12 11/16
3	5 3/8	7 3/8	7 1/2	11/16	10 3/8	5	5 3/4	11 1/2	15

TABLE 65 (FIG. 106)
JENKINS MEDIUM PRESSURE GATE VALVES. BRASS — SCREWED AND FLANGED
175 Pounds Working Pressure

Size Inches	A Inches	B Inches	E Inches	F Inches	G Inches	J Inches	K Inches	L Inches	M Inches
1/4	2 5/16	2 13/16	2 3/4	9/32	3 7/8	2 1/16			
3/8	2 5/16	2 15/16	3	11/32	3 7/8	2 1/16			
1/2	2 1/2	3 1/4	3	3/8	4 3/8	2 7/16			
3/4	2 7/8	3 11/16	3 1/2	13/32	5 1/4	2 13/16	3 5/16	5 1/4	6 3/8
1	3 1/2	4 1/4	4	7/16	6	3	3 5/8	6	7 3/8
1 1/4	3 15/16	4 3/4	4 1/2	15/32	6 5/8	3 1/2	3 7/8	7	8 1/2
1 1/2	4 9/32	5 1/16	5	1/2	7	4 1/8	4 3/8	7 5/8	9 1/2
2	5 1/16	6	6	9/16	8 1/4	4 3/8	4 13/16	8 5/8	11 1/8
2 1/2	5 3/4	6 13/16	7	5/8	10 1/4	5	5 3/4	10 1/2	13 1/2
3	7	8 1/8	7 1/2	11/16	11 1/2	6	6 11/16	12 1/4	15 3/4

STANDARD VALVES

TABLE 66 (FIG. 106)

JENKINS EXTRA HEAVY GATE VALVES. BRASS — SCREWED AND FLANGED

250 Pounds Working Pressure

Size Inches	A Inches	B Inches	E Inches	F Inches	G Inches	J Inches	K Inches	L Inches	M Inches
1/2	2¹⁵/₁₆	3¹⁵/₁₆	3¹/₄	¹³/₃₂	4³/₄	2⁷/₈			
3/4	3⁵/₁₆	4⁷/₁₆	3³/₄	¹⁵/₃₂	5³/₄	3¹/₈	3⁵/₈	5⁵/₈	6³/₄
1	3³/₄	5	4¹/₂	¹/₂	6¹/₂	3¹/₂	3⁷/₈	6⁷/₁₆	7¹³/₁₆
1¹/₄	4¹/₄	5¹/₂	5	¹⁷/₃₂	7¹/₄	4¹/₈	4³/₈	7¹/₄	9
1¹/₂	4³/₄	6	6	⁹/₁₆	8	4³/₈	4¹³/₁₆	8	10
2	5⁵/₈	7¹/₈	6¹/₂	⁵/₈	9¹/₄	5	5³/₄	9⁵/₈	12¹/₈
2¹/₂	6³/₄	8¹/₈	7¹/₂	¹¹/₁₆	11	6¹/₂	6¹¹/₁₆	10⁵/₈	13³/₄
3	7⁷/₈	9	8¹/₄	³/₄	12¹/₈	7¹/₂	7⁵/₈	12⁵/₁₆	15⁷/₈

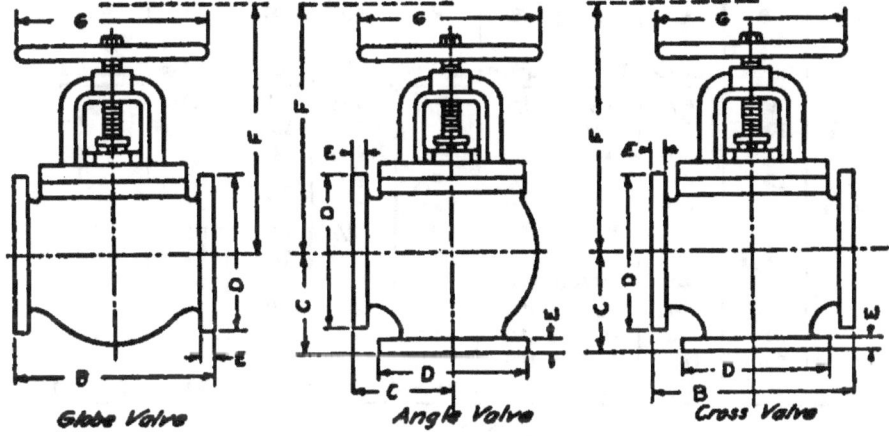

Fig. 107. Crane Globe, Angle, and Cross Valves.

TABLE 67 (FIG. 107)

CRANE STANDARD WEIGHT GLOBE, ANGLE, AND CROSS VALVES

125 Pounds Working Pressure, Iron Body

Size A Ins.	B Ins.	C Ins.	D Ins.	E Ins.	F Ins.	G Ins.	Size A Ins.	B Ins.	C Ins.	D Ins.	E Ins.	F Ins.	G Ins.
2	8	4	6	⁵/₈	10³/₄	6¹/₂	7	16	8	12¹/₂	1¹/₁₆	20¹/₂	14
2¹/₂	8¹/₂	4¹/₄	7	¹¹/₁₆	11¹/₄	6¹/₂	8	17	8¹/₂	13¹/₂	1¹/₈	23³/₄	16
3	9¹/₂	4³/₄	7¹/₂	³/₄	12³/₄	7¹/₂	10	20	10	16	1⁹/₁₆	28	18
3¹/₂	10¹/₂	5¹/₄	8¹/₂	¹³/₁₆	13	7¹/₂	12	24	12	19	1¹/₄	34	20
4	11¹/₂	5³/₄	9	¹³/₁₆	15¹/₄	9	14	28	14	21	1³/₈	38¹/₂	24
4¹/₂	12	6	9¹/₄	¹³/₁₆	15¹/₄	9	15	30	15	22¹/₄	1¹/₂	38¹/₂	24
5	13	6¹/₂	10	¹⁵/₁₆	17¹/₄	10	16	32	16	23¹/₂	1⁷/₁₆	41¹/₂	27
6	14	7	11	1	19	12							

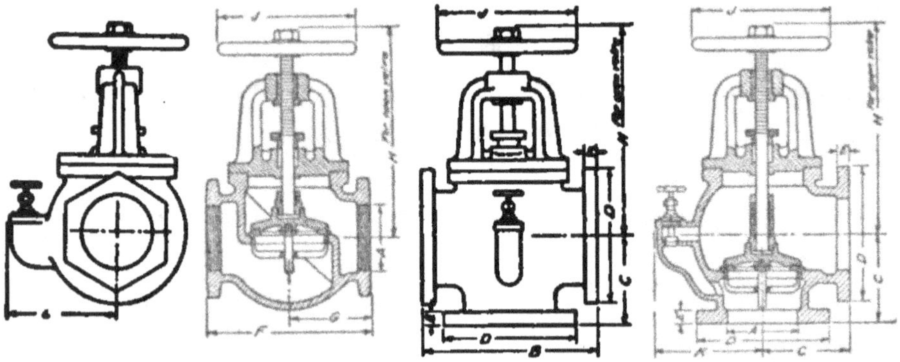

Fig. 108. Crane Globe, Cross, and Angle Valves.

TABLE 68 (Fig. 108)

Crane Medium Pressure Globe, Angle and Cross Valves

175 Pounds Working Pressure — Iron Body

Size A Inches	B Inches	C Inches	D Inches	E Inches	F Inches	G Inches	H Inches	J Inches	Size of By Pass Inches	L Inches
2	9	4 1/2	6 1/2	7/8	7 3/4	3 7/8	11 3/4	6 1/2		
2 1/2	10	5	7 1/2	1	8	4	12 5/8	7 1/2		
3	11	5 1/2	8 1/4	1 1/8	8 1/4	4 1/8	4 1/8	9		
3 1/2	12	6	9	1 3/16	9 1/2	4 3/4	15 3/8	10		
4	13	6 1/2	10	1 1/4	10 1/2	5 1/4	16 3/8	10		
4 1/2	13 1/2	6 3/4	10 1/2	1 5/16	11 1/4	5 5/8	17 3/8	12		
5	14 1/2	7 1/4	11	1 3/8	12 1/4	6 1/8	18 1/4	12		
6	16	8	12 1/2	1 7/16	14	7	20 1/4	14		
7	17 1/2	8 3/4	14	1 1/2	17	8 1/2	21 1/4	14		
8	20	10	15	1 5/8	18 1/2	9 1/4	24 1/2	16	1 1/2	12
10	22 1/2	11 1/4	17 1/2	1 7/8	22 1/2	11 1/4	28 1/2	20	1 1/2	13 3/4
12	25 1/2	12 3/4	20	2	25 1/2	12 3/4	31	20	2	15 3/4

TABLE 69 (Fig. 108)

Crane Extra Heavy Globe, Angle, and Cross Valves

250 Pounds Working Pressure, Iron Body

Size A Inches	B Inches	C Inches	D Inches	E Inches	F Inches	G Inches	H Inches	J Inches	Size of By Pass Inches	Angle Cross K Inches	Globe L Inches
2	10 1/2	5 1/4	6 1/2	7/8	9 1/2	4 3/4	13 3/4	7 1/2	...	...	...
2 1/2	11 1/2	5 3/4	7 1/2	1	10 3/4	5 3/8	14 1/2	9	...	...	...
3	12 1/2	6 1/4	8 1/4	1 1/8	11 3/4	5 7/8	17 1/2	10	...	...	...
3 1/2	13 1/4	6 5/8	9	1 3/16	12 1/4	6 1/8	17 1/2	10	...	...	...
4	14	7	10	1 1/4	13	6 1/2	19 1/2	14	...	...	...
4 1/2	15	7 1/2	10 1/2	1 5/16	14	7	19 1/2	14	...	...	...
5	15 3/4	7 7/8	11	1 3/8	15	7 1/2	21 1/2	16	...	...	...
6	17 1/2	8 3/4	12 1/2	1 7/16	16 1/2	8 1/4	25	18	...	...	...
7	19 1/4	9 5/8	14	1 1/2	18 1/4	9 1/8	26 1/4	20	...	...	...
8	21	10 1/2	15	1 5/8	20	10	29 1/2	24	1 1/2	13 3/8	13 3/8
10	24 1/2	12 1/4	17 1/2	1 7/8	23 1/4	11 5/8	33 1/2	27	1 1/2	14 3/4	14 5/4
12	28	14	20 1/2	2	...	...	39	30	2	17 1/4	17 1/4
14	33	16 1/2	23	2 1/8	...	...	42	36	2	19 3/8	18 3/4
12	33	16 1/2	24 1/2	2 3/16	...	...	42	36	2	19 3/8	18 3/4

STANDARD VALVES

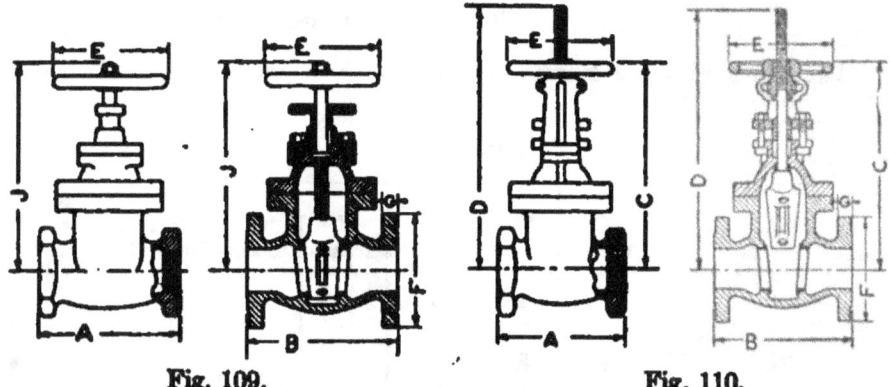

Fig. 109. Fig. 110.
Walworth Gate Valves.

TABLE 70 (FIGS. 109 and 110)
WALWORTH STANDARD GATE VALVES — IRON BODY
125 Pounds Working Pressure

Size Inches	A Inches	B Inches	C Inches	D Inches	E Inches	F Inches	G Inches	J Inches
2	5¹/₂	7	10⁵/₈	12⁷/₈	6	6	⁵/₈	10⁵/₈
2¹/₂	5⁷/₈	7¹/₂	12¹/₈	15	6	7	¹¹/₁₆	11⁷/₈
3	6¹/₈	8	14	17³/₈	8	7¹/₂	³/₄	13³/₈
3¹/₂	6³/₄	8¹/₂	15⁵/₈	19¹/₂	8	8¹/₂	¹³/₁₆	14³/₈
4	7¹/₄	9	17¹/₂	21⁷/₈	9	9	¹⁵/₁₆	16
4¹/₂	7³/₄	9¹/₂	19	23⁷/₈	9	9¹/₄	¹⁵/₁₆	16⁷/₈
5	8	10	20³/₄	26¹/₈	10	10	¹⁵/₁₆	18⁵/₈
6	9	10¹/₂	23⁷/₈	30³/₈	12	11	1	20¹/₂
7	...	11	27¹/₂	34⁷/₈	14	12¹/₂	1¹/₁₆	22⁵/₈
8	...	11¹/₂	29⁷/₈	38¹/₂	14	13¹/₂	1¹/₈	24¹/₂
9	...	12	...	...	14	15	1¹/₈	
10	...	13	35³/₈	46¹/₈	16	16	1³/₁₆	29³/₈
12	...	14	41¹/₂	54¹/₄	16	19	1¹/₄	34
14	...	15	51	65³/₄	18	21	1³/₈	38¹/₈
15	...	15	...	...	20	22¹/₄	1³/₈	
16	...	16¹/₄	56⁵/₈	73¹/₈	20	23¹/₂	1⁷/₁₆	42¹/₈
18	...	17¹/₂	64	83	20	25	1⁹/₁₆	47
20	...	18¹/₂	69	90	24	27¹/₂	1¹¹/₁₆	50³/₈
22	...	19	...	...	27	29¹/₂	1¹³/₁₆	
24	...	21	81	106	30	32	1⁷/₈	58

TABLE 71 (FIGS. 109, 110, AND 111)

WALWORTH MEDIUM PRESSURE GATE VALVES, WITH BY-PASS, IRON BODY

175 Pounds Working Pressure

Size Inches	A Inches	B Inches	C Inches	D Inches	E Inches	F Inches	G Inches	H Inches	Size of By-Pass Inches	J Ins.
2	5$^1/_2$	7$^1/_2$	11$^1/_2$	14	6$^1/_2$	6$^1/_2$	$^7/_8$	...	...	11
2$^1/_2$	6	8	12$^1/_2$	15$^1/_2$	6$^1/_2$	7$^1/_2$	1	...	...	12
3	7$^1/_4$	9$^1/_2$	15	18$^1/_2$	7$^1/_2$	8$^1/_4$	1$^1/_8$	...	...	14
3$^1/_2$	7$^1/_2$	10	16$^3/_2$	20$^1/_2$	7$^1/_2$	9	1$^3/_{16}$	...	...	15
4	7$^3/_4$	10$^1/_2$	19	23$^3/_4$	9	10	1$^1/_4$	...	...	16
4$^1/_2$	8$^1/_4$	11	20	25	9	10$^1/_2$	1$^5/_{16}$	...	...	17
5	8$^1/_2$	11$^1/_2$	22	28	10	11	1$^3/_8$	...	...	19
6	8$^3/_4$	12	25$^1/_4$	32	12	12$^1/_2$	1$^7/_{16}$	14	1$^1/_4$	21
7	...	12$^1/_2$	28	36	12	14	1$^1/_2$	15	1$^1/_4$	23
8	...	13$^1/_2$	32	41	14	15	1$^5/_8$	16	1$^1/_2$	26
9	...	14	34	44	14	16$^1/_4$	1$^3/_4$	16$^1/_2$	1$^1/_2$	28
10	...	15	39	50	16	17$^1/_2$	1$^7/_8$	17$^1/_2$	1$^1/_2$	30
12	...	16	43$^1/_2$	57	18	20$^1/_2$	2	18$^1/_2$	2	34
14	...	18	49$^1/_2$	65	20	23	2$^1/_8$	20	2	
15	...	18$^3/_4$	52$^1/_2$	69	20	24$^1/_2$	2$^3/_{16}$	21	2	
16	...	19$^1/_2$	57$^1/_2$	75	22	25$^1/_2$	2$^1/_4$	23	3	

Table 72 gives the dimensions of Walworth extra heavy iron gate valves for both screwed and flanged ends. The dimensions for sizes from 6 inches to 12 inches are the same with or without a by-pass valve. The dimensions given in Table 72 hold for non-rising stem valves except the distances from centre of valve to top of wheel and diameter of handwheel above the 6 inch size. The values for these two dimensions are given in Table 73.

The dimension D is to the top of the valve when it is wide open. The arrangement of the valve stem and the kind of ends, whether screwed or flanged, is shown in the figures.

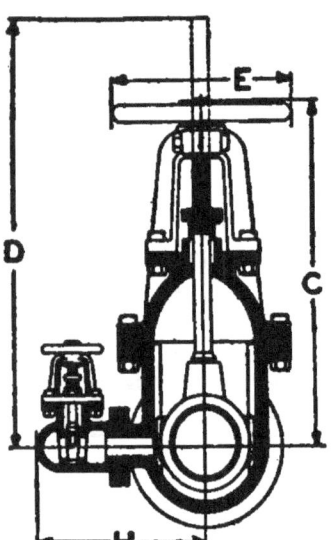

Fig. 111. Walworth Gate Valve.

STANDARD VALVES

TABLE 72 (FIGS. 110 AND 111)

WALWORTH EXTRA HEAVY GATE VALVES WITH BY-PASS RISING STEM, OUTSIDE SCREW AND YOKE, IRON BODY — SCREWED AND FLANGED

250 Pounds Working Pressure

Size Inches	A Inches	B Inches	C Inches	D Inches	E Inches	F Inches	G Inches	H Inches	Size of By-Pass Inches
$2^1/_2$	$8^1/_2$	$9^1/_2$	$13^1/_2$	$16^1/_4$	8	$7^1/_2$	1	...	...
3	$9^1/_2$	$11^1/_8$	$15^1/_8$	$18^3/_8$	10	$8^1/_4$	$1^1/_8$	...	...
$3^1/_2$	$11^3/_8$	$11^7/_8$	$17^1/_8$	21	10	9	$1^3/_{16}$	...	...
4	$12^3/_8$	12	$18^7/_8$	$23^3/_8$	11	10	$1^1/_4$	...	...
$4^1/_2$	14	$13^1/_4$	$23^3/_8$	$29^1/_4$	11	$10^1/_2$	$1^5/_{16}$	...	...
5	$15^3/_8$	15	$23^3/_8$	$29^1/_4$	12	11	$1^3/_8$	...	...
6	$16^1/_4$	$15^7/_8$	$25^3/_8$	32	13	$12^1/_2$	$1^7/_{16}$	14	$1^1/_2$
7	...	$16^1/_4$	$29^3/_4$	38	15	14	$1^1/_2$	15	$1^1/_2$
8	...	$16^1/_2$	$32^1/_2$	41	15	15	$1^5/_8$	16	$1^1/_2$
9	...	17	$36^1/_2$	46	16	$16^1/_4$	$1^3/_4$	$16^1/_2$	$1^1/_2$
10	...	18	$39^3/_8$	50	16	$17^1/_2$	$1^7/_8$	$17^1/_2$	$1^1/_2$
12	...	$19^3/_4$	$45^1/_4$	$58^1/_2$	18	$20^1/_2$	2	20	2
14	...	$21^1/_2$	$50^1/_2$	66	22	23	$2^1/_8$	21	2
15	...	$22^1/_2$	$52^1/_2$	69	22	$24^1/_2$	$2^3/_{16}$	$21^1/_2$	2
16	...	24	58	$75^1/_2$	24	$25^1/_2$	$2^1/_4$	27	3
18	...	26	...	$82^1/_4$	27	28	$2^3/_8$	...	3
20	...	28	...	$91^1/_2$	30	$30^1/_2$	$2^1/_2$	...	4
24	...	31	...	109	36	36	$2^3/_4$	...	4

TABLE 73 (FIG. 109)

WALWORTH EXTRA HEAVY GATE VALVES WITH BY-PASS, NON-RISING STEM, IRON BODY — SCREWED AND FLANGED

250 Pounds Working Pressure

Size Inches	J Inches	E Inches	Size Inches	J Inches	E Inches
$2^1/_2$	$12^3/_4$	8	6	$22^3/_4$	13
3	$14^1/_2$	10	7	25	14
$3^1/_2$	$15^1/_4$	10	8	28	14
4	$16^1/_2$	11	9	$29^3/_4$	15
$4^1/_2$	21	11	10	33	15
5	21	12	12	$37^1/_2$	18

Check Valves. — There are a large variety of special forms of valves, some of which will be mentioned. When necessary to permit flow in one direction and to prevent it in the opposite direction a check or non-return valve is used. These are made in many forms; Fig. 112 shows a swing check valve, Fig. 113 shows a ball check valve, Fig. 114 a lift check valve, and Fig.

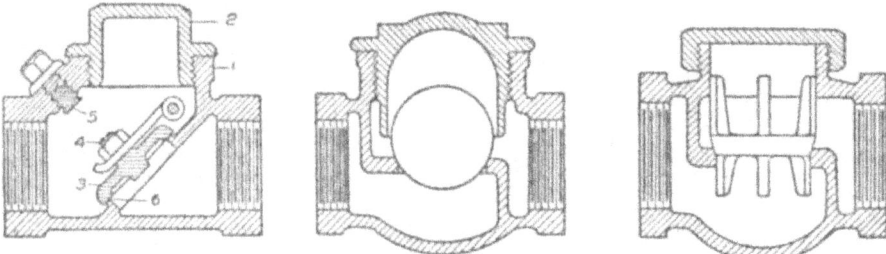

Figs. 112, 113, and 114. Swing Check Valve, Ball Check Valve, and Lift Check Valve.

115 a large flanged check valve having a relief gate, as made by Walworth Company. It is desirable that there should be provision for regrinding. The swing check valve shown in Fig. 112 is made with or without the stop plug *5*. The purpose of the stop plug is to allow for re-grinding in the following manner. Unscrew the cap *2* and the stop plug *5*, place a small amount of abrasive moistened with soap or oil on the valve seat *6*. By inserting a screw driver through the stop plug opening and engaging the slot in the clapper stud *4*, the disc *3* can be rotated and re-ground upon its seat.

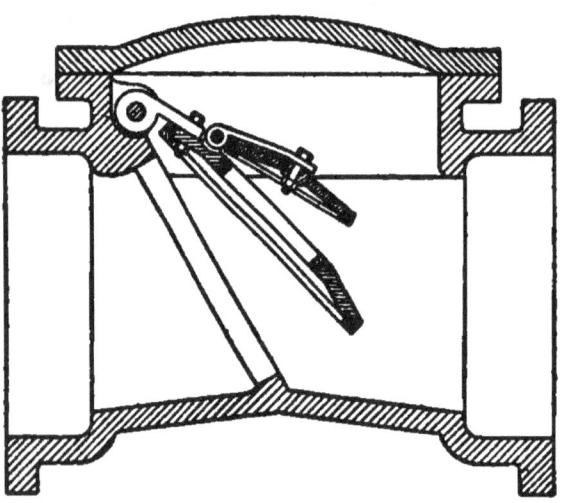

Fig. 115. Large Swing Check Valve with Gate.

The iron body swing check valve shown in Fig. 115 is for water pressure up to 150 pounds. The relief gate shown is used on sizes larger than 16 inch. These valves are made with screwed ends, flanged ends, and hub ends, and in sizes from 2½ to 24 inches.

Operation of Valves. — While the purpose of this book is not to deal with operation of valves and piping, there are a few points which are worth setting down. A steam valve should never be opened quickly as the rush of steam is likely to bring about a dangerous condition, especially if there is any water present. A leaky valve cannot be made tight except by re-grinding. Screwing down the valve excessively will only result in damage to the valve. In attaching screwed end valves the wrench should always be applied to the end nearest the pipe, as valve bodies are not designed to transmit the forces required in "making up" lines. When the wrench is applied to the opposite end of the valve it produces distortion. A valve should always be closed tightly when being put into place. Cement or graphite should not be put into the valve threads, but on to the pipe so that it will not get into the valve and hold such grit and dirt as may come through the pipe. A new pipe line should always be thoroughly blown out after construction, and it is well if possible to examine the valves after this blowing out and before closing them.

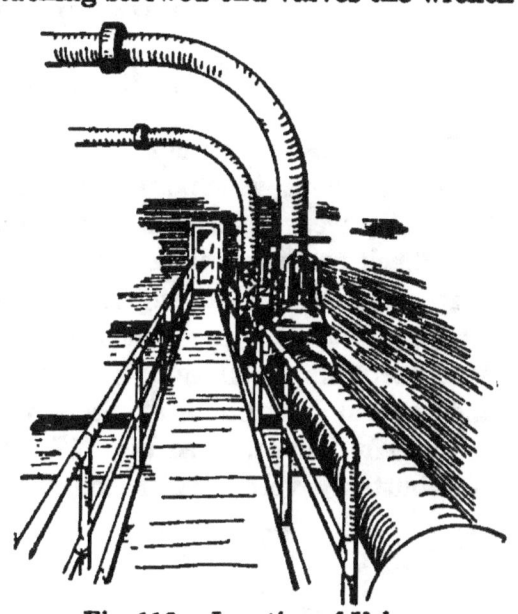

Fig. 116. Location of Valves.

Location. — The location of valves should receive careful attention, as many accidents have occurred through the placing of valves in inconvenient places. Sometimes the valve stem can be placed in a horizontal position and operated from the floor by means of a chain or similar device. The operator should not be required to open and close valves when they are in such a position that he places his life in danger should there be an accident of any kind. Where a gallery or platform is used near valves it should be placed to one side of the line, as shown in Fig. 116 rather than directly over the steam line with the valve stems extending through the platform. In the latter case the workman is directly over the line and in case of breakage is in great danger of being scalded by the escaping steam.

CHAPTER VII

SPECIAL VALVES

The purpose of this chapter is to describe some rather special forms of valves which are used for various purposes, such as blow-off valves, boiler stop valves, reducing valves, pump governors, back pressure valves, and relief valves. The large number of special forms and arrangements make it impossible to do more than suggest the types that are available and some of the uses. Manufacturers' catalogs should be consulted for more complete and detailed descriptions of special valves that are regularly made.

Butterfly Valves. — In Fig. 117 is shown a cross-sectional view of a butterfly valve, which consists of a disc which may be revolved either in line with or across the opening, very much like the damper in an ordinary stove pipe. These valves can be used only for regulating purposes where absolute tightness is not essential.

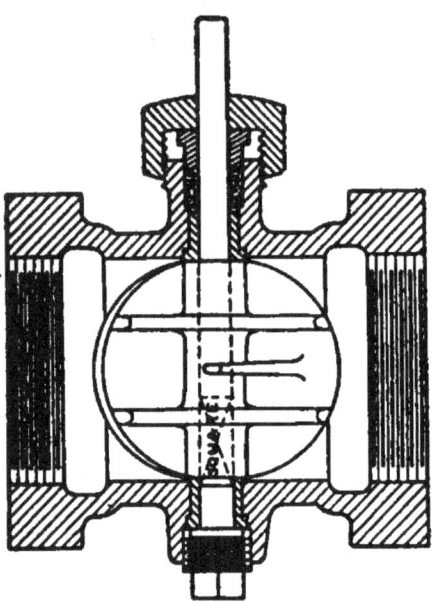

Fig. 117. Butterfly Valve.

Blow-off Valves. — Special valves are made for use in the blow-off pipes of boilers. Such valves require as clear a passage way as possible, and that it shall be without interfering parts. Several designs are shown in Figs. 118, 119, and 120, where the construction of each is clearly shown. The objection to ordinary valves is that they afford an opportunity for scale or sediment to obtain lodgment and prevent closing. The severe conditions of service require that blow-off valves be of heavy construction. Blow-off valves are made either straight, angle,

SPECIAL VALVES 115

or Y form. Fig. 118 is a Y blow-off valve, made by Walworth Company. Often two valves are used together in the blow-off pipe to make sure of a tight blow-off. Fig. 119 shows a Crane blow-off cock with a compensating spring *2* located between the

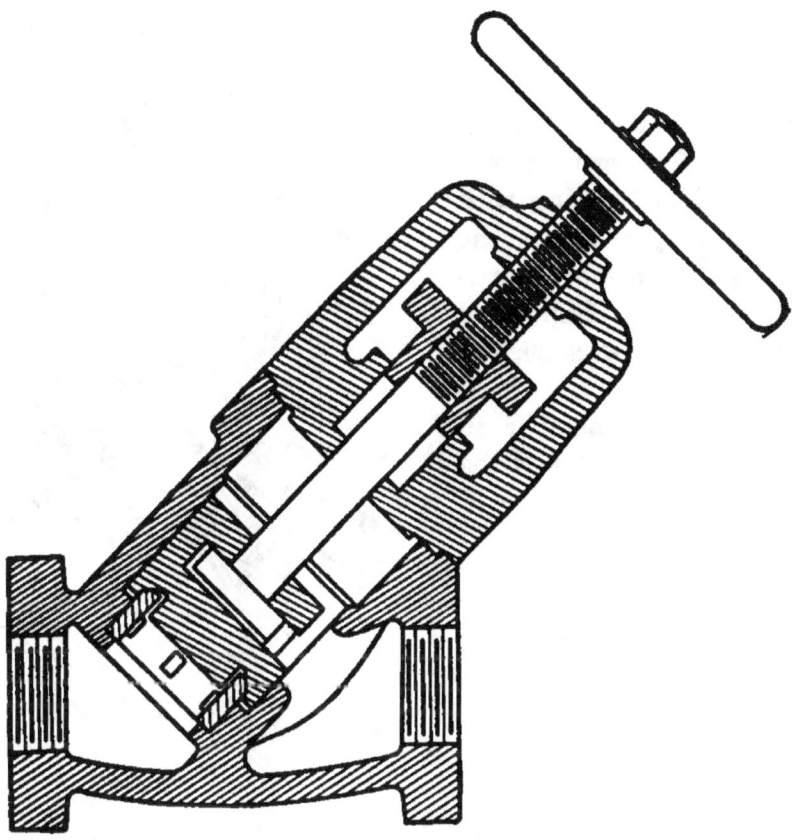

Fig. 118. Y — Blow-off Valve.

plug *1* and the cap *3* which automatically takes up wear and holds the plug securely in place at all times, preventing the accumulation of scale, sediment, etc., which would tend to impair the ground surfaces of the plug and body. The Simplex seatless blow-off valve as made by the Yarnell-Waring Company is illustrated in Fig. 120. This valve has no seat but closes by moving the plunger *3* down past the port. In closing the valve the shoulder *1* on the plunger *3* engages the loose follower gland *2* and so compresses the packing *4* above and below the port, thus making the valve tight. There are many other worthy forms which space will not permit describing.

116 A HANDBOOK ON PIPING

Plug Valves. — The plug valve shown in Fig. 121 is made by the Homestead Valve Manufacturing Company for steam, compressed air, and hydraulic service. This valve is so constructed that when it is closed it is at the same time forced firmly to its seat. This result is secured by means of the traveling cam A through which the stem passes. The cam is prevented from turning with the stem by means of the lugs B which move vertically in slots. Supposing the valve to be open, the cam will be in the lower part of the chamber in which it is placed, and the

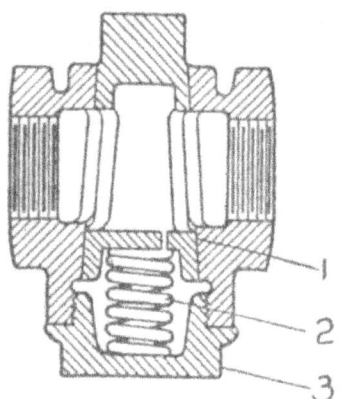

Fig. 119. Crane Cock.

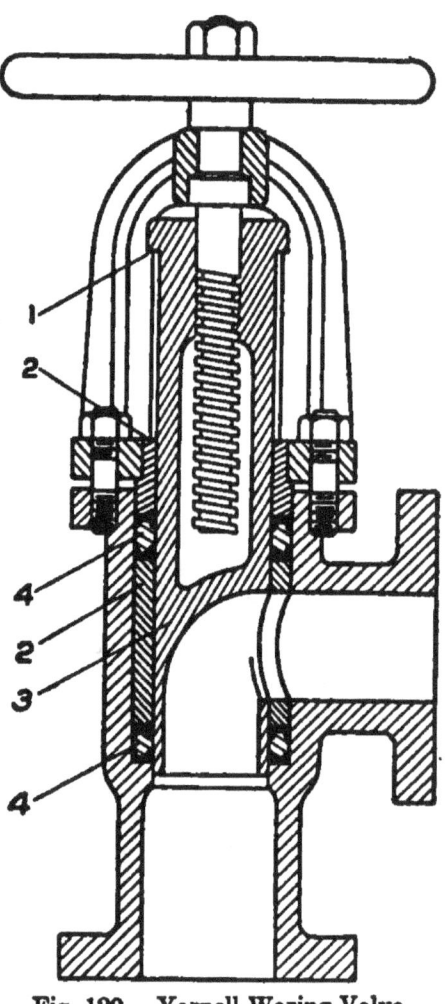

Fig. 120. Yarnell-Waring Valve.

plug will be free to be easily moved. A quarter of a turn in the direction for closing it causes the cam to rise and take a bearing on the upper surface of the chamber, and the only effect of further effort to turn the stem in that direction is to force the plug more firmly to the seat. A slight motion in the other direction immediately releases the cam and the plug turns easily, being arrested at the proper open position by contact of the fingers of the cam at the other end of its travel. The balancing ports E and D allow the pressure to predominate at the top of the plug,

SPECIAL VALVES

holding it gently in its seat while the valve is open. This valve is made in sizes up to six inches, and for pressures up to 5000 pounds.

Boiler Stop Valves. — A boiler stop valve is a valve in the connection of the boiler to the steam main, and may be of the globe or angle type, hand operated. The larger sizes should be fitted with a by-pass. When a plant consists of two or more boilers some form of automatic non-return valve in addition to the stop valve should be provided. The purpose of the automatic valve is to prevent back flow from the main steam pipe when the pressure in one boiler is lower, due to the bursting of a tube or other causes. Such valves are made by many of the valve companies, and advantages are claimed for each design.

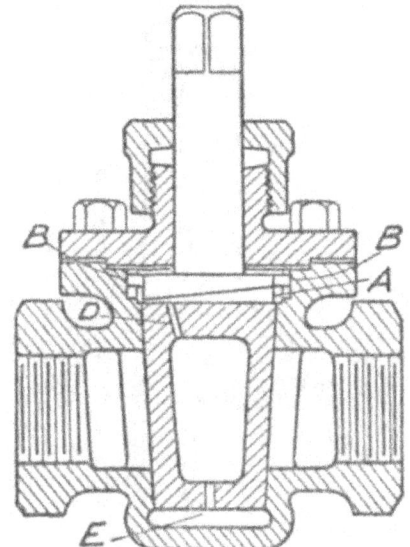

Fig. 121. Homestead Cock.

Foster Automatic Valve. — The automatic non-return stop valve shown in Fig. 122 is made by the Foster Engineering Company. When installed between the boiler and header it will equalize the pressure between the units of a battery of boilers, remaining closed so long as the pressure is lower than that of the header. The valve will open and remain in that position when the boiler pressure is equal to the pressure in the header. It automatically prevents the back flow of steam into a disabled boiler and acts as a safety stop valve to prevent steam being turned into a cold boiler while men are working inside — the pressure in the header making it impossible to open the valve. The valve may be closed in the same manner as an ordinary stop valve by screwing down the stem.

The operation of the valve is described as follows: Inlet A is connected to the boiler nozzle, and outlet side B to the header. When the pressure at A is one pound or more greater than the pressure at B the valve C lifts and is held open by the flow of steam passing through the valve. If the pressure at A should fall below that at B, due to the blowing out or weaning of a tube, a cock blowing off, or from other cause, the back flow of steam from B acting on the upper side of clapper C plus its weight, forces the valve automatically to its seat. The clapper is then held to its seat until there is an equalization of pressure on both sides of it.

Emergency Stop Valves. — As a further protection against accidents, and to safeguard the lives of operators, emergency valves have been devised, combining the duties of the automatic non-return stop valve, automatic safety stop valve, automatic emergency stop valve, and hand stop valve.

The Foster automatic non-return emergency stop valve is shown in Fig. 123 and described as follows: in the event of a rupture in the main line or a break in fittings causing a sudden escape of steam, it will close automatically and prevent further flow of steam from the boiler or boilers. Small emergency pipes may be run to different parts of the plant, and when desired, steam may be shut off by opening a small globe valve, which should be placed at convenient points, in the emergency lines, permitting the isolating of any boiler in a battery at will from a distant point if necessary. The valve may also be closed in the same manner as an ordinary stop valve. The operation of this valve as a *non-return valve* is the same as for Fig. 122. *As an automatic and emergency stop valve.* The pilot or governing valve Fig. 124 may be placed near the main valve, or located at any point desired, Fig. 125. A ⅜-inch pipe connection is made from

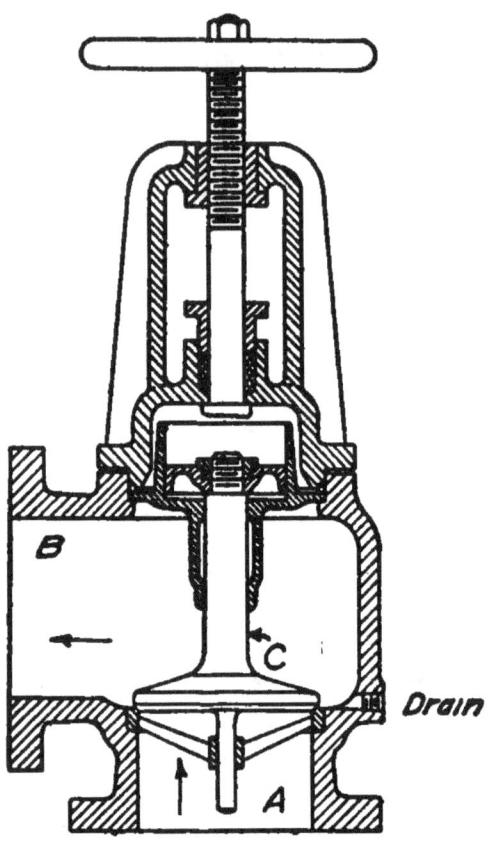

Fig. 122. Foster Automatic Valve.

SPECIAL VALVES 119

the boiler to the pilot valve at *C*, and from the pilot, at *E* to the chamber *D* of the main valve at *F*. The diaphragm chamber

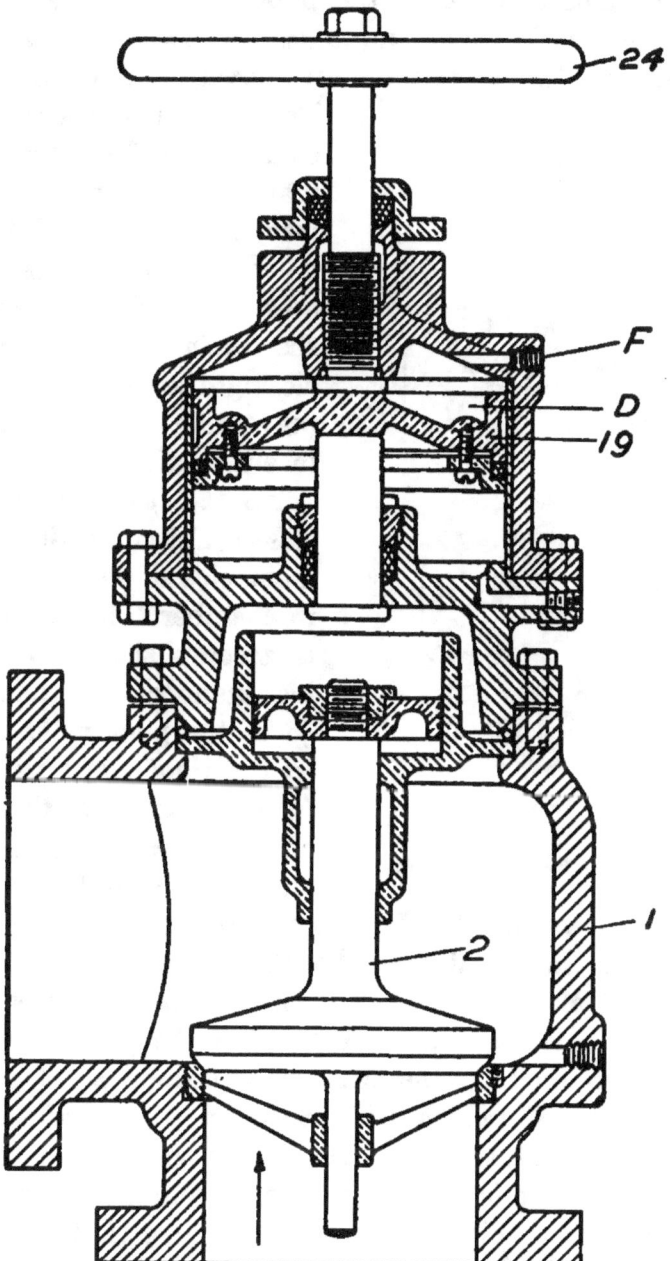

Fig. 123. Automatic Non-return Emergency Stop Valve.

J of this pilot valve is also connected to the header or at any point on the main steam line beyond the outlet of the main valve.

Whenever, from rupture or other causes, the pressure in the main lines falls abruptly, a corresponding effect is experienced upon the upper diaphragm *48* of the pilot valve, thus allowing the boiler pressure acting upon and under the lower diaphragm *48'* to open valve *36* (which is normally closed) and close valve *37*. The full boiler pressure then is enabled to flow through the main port of the pilot valve into chamber *D* of the main valve, against piston *19*, the area of which is greater than the main valve *2*, instantly closing the latter to its seat, preventing the flow of steam in either direction. The main valve *2*, then having been closed automatically, will remain closed until the pressure in chamber *D* is relieved. This is accomplished in the following manner: the hand wheel *46* of the pilot valve is turned to the right until valve *36* is forced to its seat, thus cutting off live steam chamber *D* of main valve, and at the same time forcing valve *37* off its seat, exhausting the steam in chamber *D* of main valve to the atmosphere, through the pipe connection at *M*. After sufficient steam pressure has been raised to hold down the upper diaphragm *48* of the pilot valve, which may be determined by the exhaust connection at *M* not blowing, the alarm *K* (which will otherwise give notice) is then closed by turning the hand wheel *46* to the left, in which (its normal position) it is again ready for automatic action.

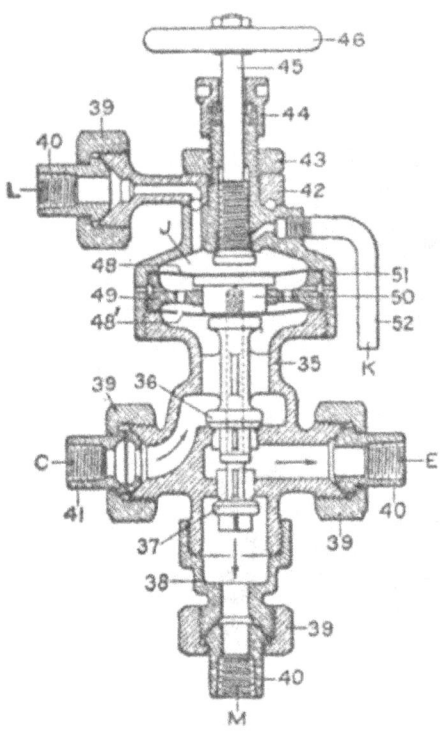

Fig. 124. Foster Pilot Valve.

The Pilot Valve, Fig. 124, is constructed so that variations or fluctuating conditions of the boiler pressure between maximum and minimum loads will not influence the pilot, which requires no adjustment to meet these conditions. The valve is automatic and will respond only to any drop in line pressure for which it is designed and intended. A number of $3/8$-inch branch pipes may

SPECIAL VALVES

be run to and located at any desired point from the line leading to the diaphragm chamber J of the pilot valve — on each of these laterals a small globe valve is mounted. The mere cracking of one of these globe valves obtains the same result as a break in the main lines, in that the steam is in this way bled from the

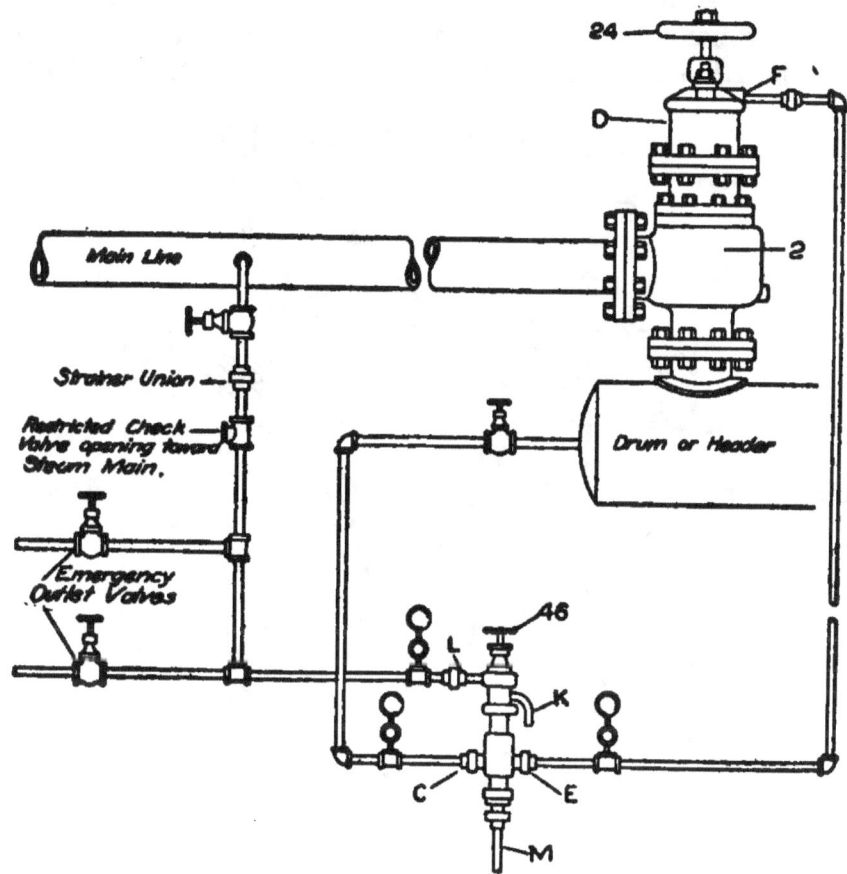

Fig. 125. Arrangement of Piping for Pilot Valve.

diaphragm chamber J, functioning both the pilot and the main valve. By the use of these emergency valves a boiler may be cut out from a battery at will, from a distant point without the necessity of access to the boiler.

Crane-Erwood Automatic Valve. — The automatic double acting non-return and emergency cut-out valve shown in Fig. 126 is made by Crane Company. Some of the claims for this valve are as follows: The valve will close automatically if any part of the header or distributing lines fail; the valve will open when the boiler to which it is connected reaches the full pressure in the main; the valve will prevent back-flow of steam from the main in the event of a tube blowing out or other accident to the boiler; the valve may be used as an emergency valve by attaching a cord to the lever so that it can be closed by hand at a distance, or may be operated electrically. The levers on the outside of the valve are in line with the discs, and indicate their position and operation. The separating link connecting the outside lever may be adjusted to suit the load carried. Shortening the link

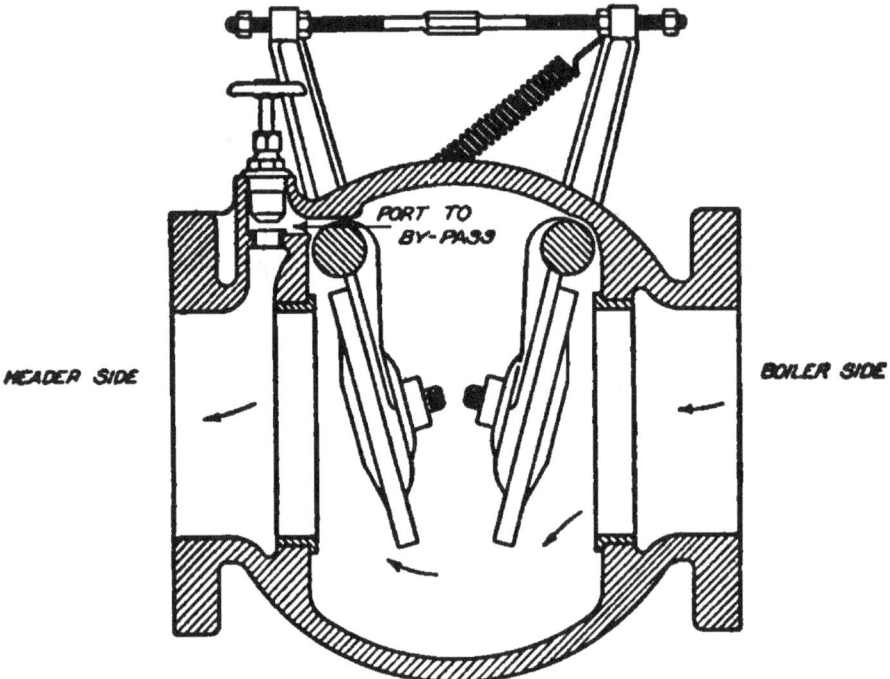

Fig. 126. Crane-Erwood Valve.

decreases the volume of steam passing through the valve; lengthening the link increases the volume. Such adjustments do not interfere with the operation of the valve. The valve may be adjusted to close at any desired velocity. The purpose of the by-pass is to provide for the valve to open automatically when the pressure in the header equals the pressure in the boiler after the valve has been closed due to a break or reduction in pressure beyond the outlet of the valve.

Reducing Valves. — Reducing valves are valves made to reduce and maintain automatically a constant pressure of steam or air with variable initial pressures. Such valves are employed

SPECIAL VALVES

for reducing boiler pressure for use with all kinds of steam heating systems, central station heating, paper machines, engines, kettles and cooking apparatus, and other conditions necessitating a reduced pressure.

A reducing valve used to supply a steam engine should be placed some distance from the engine in order to provide as large a reservoir as possible for the engine to draw from. A receiver may be placed between the valve and steam cylinder to serve the same purpose. It should have a capacity greater than the volume of the steam cylinder. When a reducing valve is to be placed in a pipe line, the piping should be thoroughly blown out. With new pipe sufficient time should be allowed for the oil or grease to be completely burned out.

The reducing valve shown in Fig. 127 is made by the Mason Regulator Company. This valve is controlled by the variation of the reduced pressure acting through the port A, on the diaphragm 1. This diaphragm is resisted by a spring 2, which is adjusted to the reduced pressure. The auxiliary valve 3 is held in contact with the diaphragm by the auxiliary valve spring 4, and moves up and down freely with the diaphragm. As soon as the valve 3 is open, steam passes through into the port B, and under piston 5. By raising piston 5, the main valve 6 opens against the initial pressure because the area of valve 6 is only one-half of that of piston 5; steam is thus admitted to the system. When

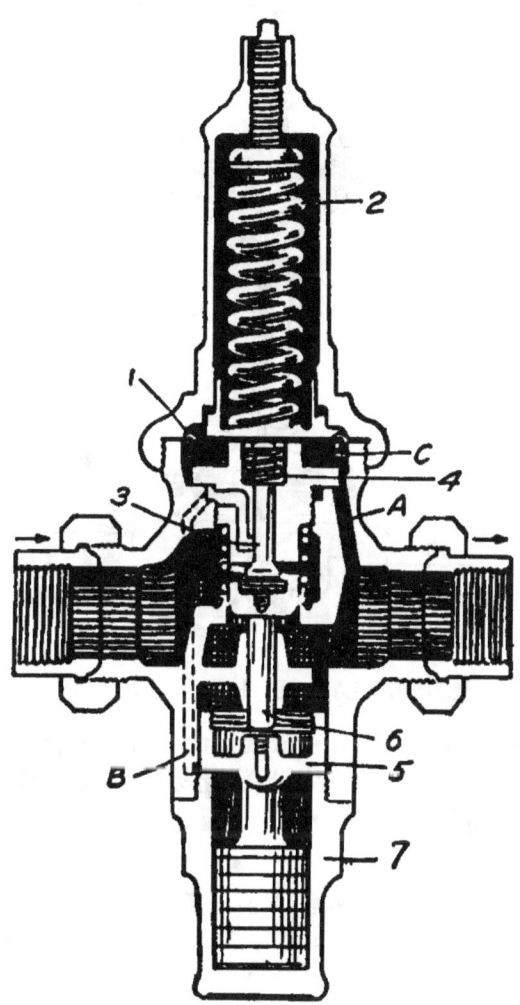

Fig. 127. Mason Reducing Valve.

the pressure in the system has reached the required point, which is determined by the spring *2*, the diaphragm is forced upward by the low pressure which passes up through port *A* to chamber *C* under the diaphragm, allowing valve *3* to close, shutting off the steam from piston *5*. The main valve *6* is now forced to

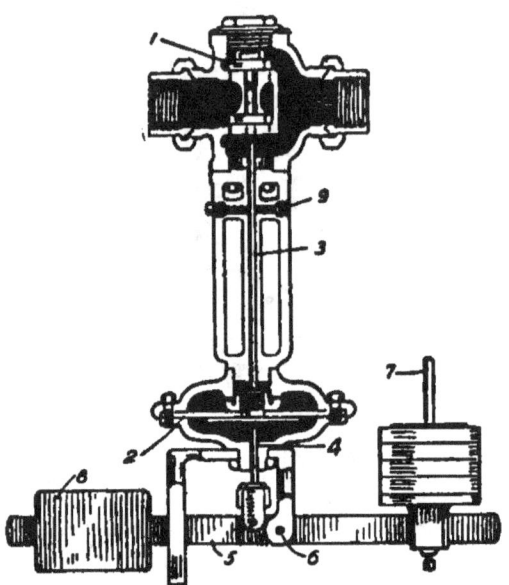

Fig. 128. Lever Style Reducing Valve.

its seat by the initial pressure shutting off steam from the system and pushing the piston *5* down to the bottom of its stroke. The steam beneath piston *5* exhausts freely around the piston, being fitted loosely for this purpose, and passes off into the system. In practice the main valve does not open or close entirely with each slight variation of pressure, but assumes a position which furnishes just the steam required to maintain the required pressure. Piston *5* is fitted with dashpot *7* which prevents chattering or pounding.

Where low pressures of from zero to 25 pounds per square inch are employed, as on low pressure heating systems, central station heating, and similar conditions where the initial pressure may be high, the form of valve shown in Fig. 128 is often used. The valve illustrated is the Mason lever style, and consists of a balanced valve *1*, which is under the control of the diaphragm *2*, by means of the stem *3* and an extension stem *4* which is connected to lever *5*. This lever is pivoted at *6*. The reduced pressure is determined by the amount of weights *7*, and for very low pressures the weight *8* is used to counterbalance the weight of the lever. In action, the reduced pressure from the low pressure system passes through a small pipe to connection *9*, and then down around the stem *3* into the diaphragm chamber where it exerts its pressure on the diaphragm. This pressure, balanced by weights *7*, causes the valve *1* to assume the proper position to supply the

SPECIAL VALVES

necessary volume of steam to maintain the required reduced pressure.

The Auld Company's "Quitetite" reducing valve may be explained by reference to Fig. 129, and the makers' description. High pressure steam enters valve by branch marked INLET and

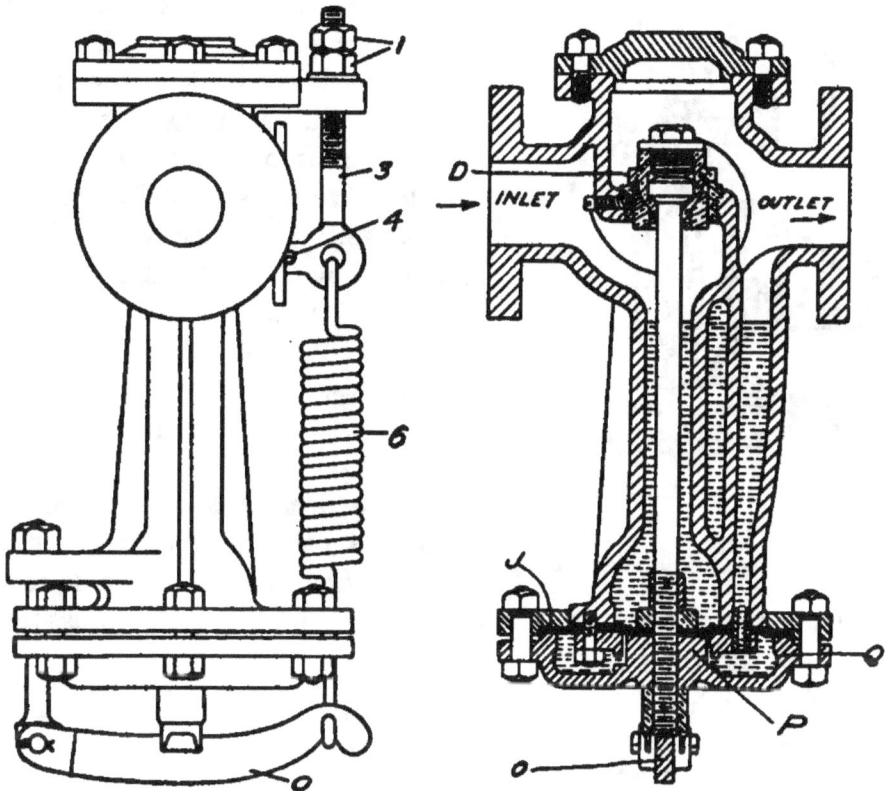

Fig. 129. Auld "Quitetite" Reducing Valve.

acts between valve D and piston P which are of the same area and, therefore, in equilibrium on H.P. side. Reduced pressure is obtained by screwing up adjusting nuts *1* until pointer *4* on Spring Bolt *3* is opposite the figure representing the reduced pressure required. Acting through the lever *0* the extension of spring *6* opens up valve D and passes steam at reduced pressure to outlet side and when the pressure of this reduced steam tends to rise above that required it closes the valve by acting on back of the valve D and chamber Q. When the pressure tends to fall the tension of spring overcomes the force holding valve closed and opens valve, allowing it to admit more steam to the L.P. side, and in this way the reduced pressure is kept constant.

A flexible diaphragm *J* is fitted at lower end of valve body, which makes a frictionless steam-tight packing between the stationary and movable lower parts of the valve. This diaphragm is protected from the action of steam by water of condensation which collects in the lower parts of the valve and keeps the diaphragm cool.

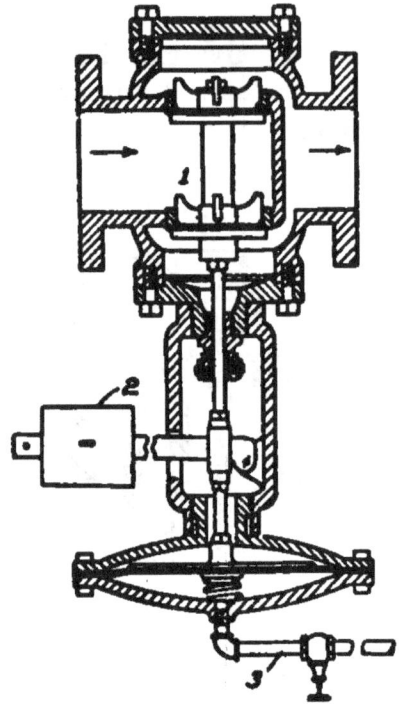

Fig. 130. Fisher Reducing Valve.

The operation of the Fisher reducing valve shown in Fig. 130 is as follows: the inner valve *1* is held open by the lever and weight *2*. The volume of steam which passes through the valve builds up in the low pressure main and enters the diaphragm chamber through the controlling pipe line *3*. When the desired low pressure is reached, a balance is formed with the lever and weight. This action regulates the opening in the valve, and maintains the pressure for which the valve is set.

When a large volume of steam is required at low pressure, such as for heating systems, and it must be reduced from a high pressure, reducing valves may be made with an increased size of outlet. Such valves are used on vacuum systems of steam heating, and for low pressure steam turbines when the supply of exhaust steam is not sufficient and live steam must be reduced from boiler pressure.

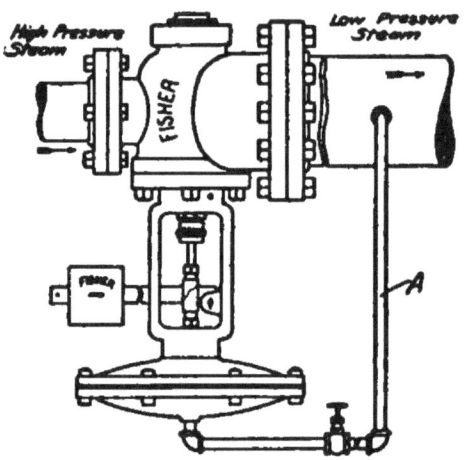

Fig. 131. Increased Outlet Reducing Valve.

The method of piping this type of valve is shown in Fig. 131. The pipe *A* should be tapped into the low pressure main at a distance from the valve so as to get the average low

SPECIAL VALVES

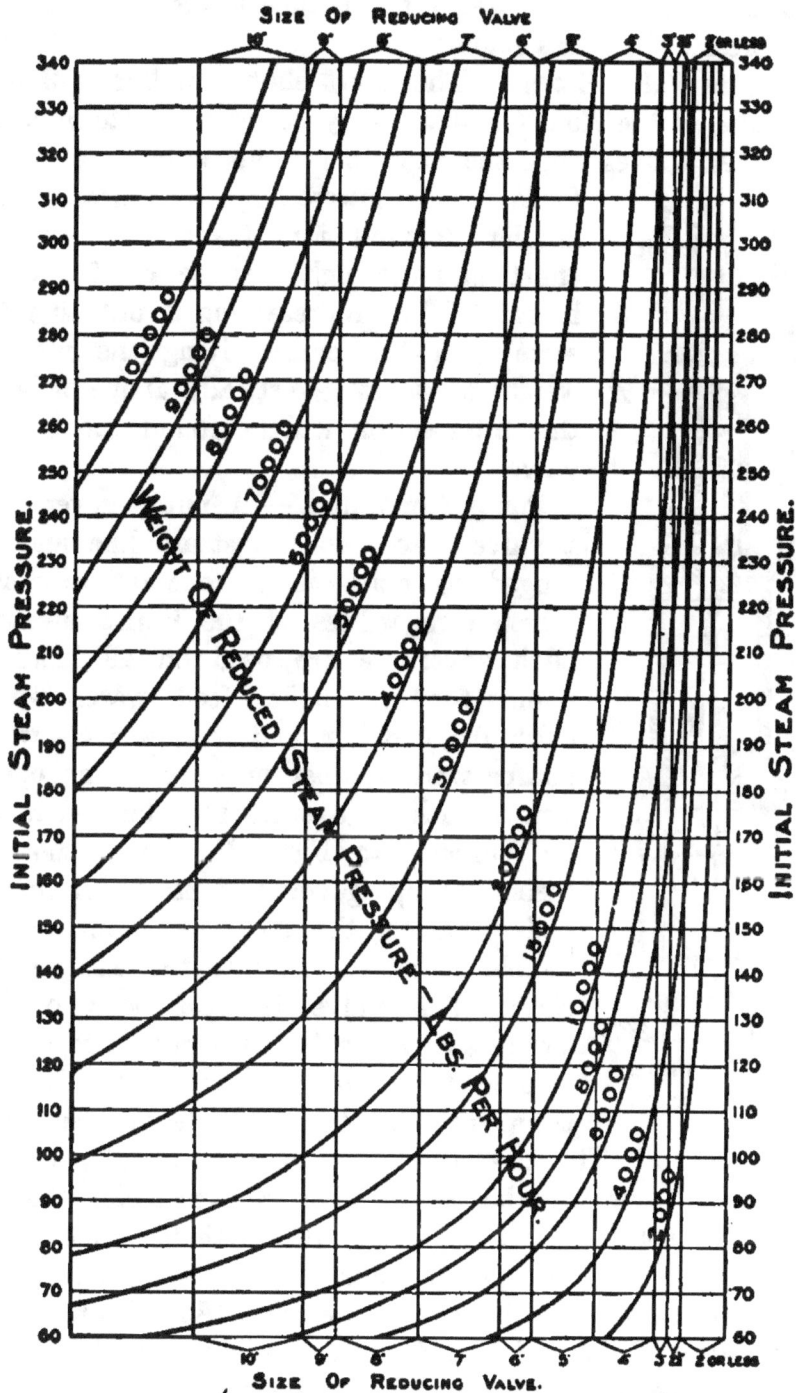

Fig. 132. Pounds of Steam per Hour Delivered by Reducing Valves.

pressure. The outlet is often made double the size of the inlet, thus increasing the area four times.

Reducing Valve Sizes. — The chart shown in Fig. 132 from the catalog of the Auld Company may be used to determine the size of their valves when the reduced pressure is less than three-fifths of the lowest high pressure, with a regular demand for steam. To use the chart, find the high pressure and follow the horizontal line representing it until it intersects with the curve giving the required weight of steam. Vertically above or below this intersection will be found the size of valve.

Pump Governors. — A pump governor is a valve placed in the steam line and arranged to maintain a constant discharge pressure regardless of the initial pressure. Such governors are used on all kinds of pumps for fire, boiler feed, water works, hydraulic, elevator, and other services where pumps work against pressure. The operation of such a governor may be understood by reference to Fig. 133, which shows a Fisher pump governor. Steam from the boiler passes through the semi-balanced double seated valve *1* into the pump steam chest. The valve is held open by the spring shown inside the pressure regulating cylinder *2*. A pipe from the pump discharge is piped to the top of the pressure regulating cylinder at *3*. The discharge pressure acts directly on the piston *4*, and operates the steam valve by overcoming the tension on the spring. In this manner the discharge controls the supply of steam to the pump. For ordinary service the parts are made of cast-iron with bronze trimmings. Superheated steam requires steel bodies and Monel metal or nickel steel trimmings.

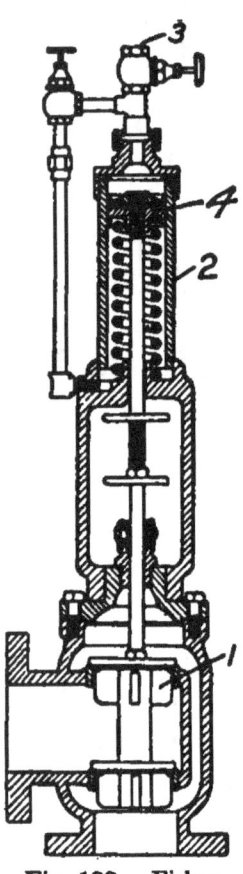

Fig. 133. Fisher Pump Governor.

The arrangement of the piping for a governor used for controlling the discharge pressure from a pump used for boiler feed, water works, and similar service where the pump is operating against pressure is shown in Fig. 134.

SPECIAL VALVES

The method of attaching and operating the governor shown in Fig. 133, as described by the Fisher Governor Company, is as follows:

"*To Attach and Connect.* Place the governor between the steam chest and throttle valve so that governor will stand perpendicular; connect outlet side of governor with the steam pipe on steam chest, then connect the steam pipe to the branch or side inlet, placing throttle valve in most convenient place. Use short nipples and place governor as close to pump as possible.

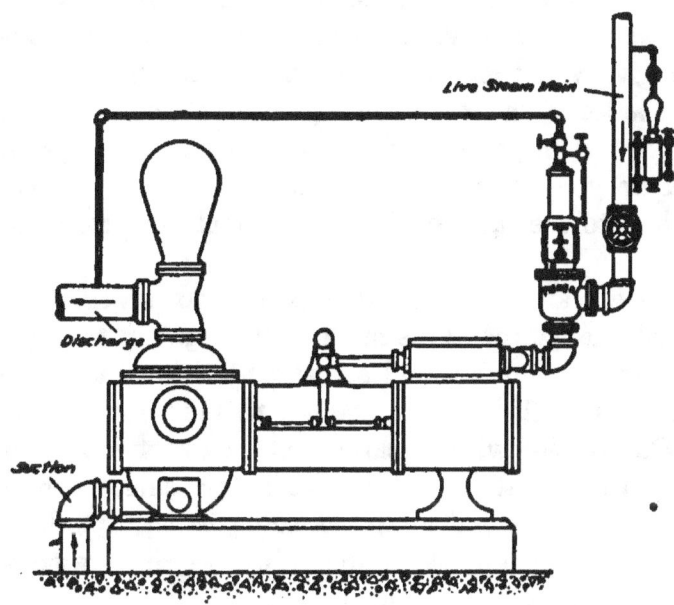

Fig. 134. Piping a Pump Governor.

"For connecting the discharge to governor, tap the discharge main or pipe, if horizontal, on the side, and if for one governor, tap for $3/8$-inch pipe; run pipe up about a foot higher than governor, then over it and down and connect to globe valve on top of pipe work over governor. If for two governors on pump discharging into same main, tap for $1/2$-inch pipe and run up and over until on a line between governors, then put on a "T" and run to right and left until over governor, then connect to globe valve.

If you can tap discharge main or pipe, five or six feet from pump, do so as governor will be less affected by the pulsation of water from pump. However, if you must tap close to pump, this pulsation can be avoided and pump run smoothly by partly closing the upper globe valve. Do not connect close to air chamber. Run piece of 1/8-inch pipe from drip at bottom of brass cylinder to floor or sewer. The drip pipe must never be connected with waste pipe from steam cylinder blow-off cocks or exhaust pipe, as the hot steam will burn out the cup leather piston packing.

"*To Operate.* The upper wheel in yoke is simply for a lock nut. Turn it to the left, then turn lower wheel to the right, which raises and opens the steam valve, when partly open, open your throttle valve and start your steam pump, now close the lower, or angle valve over governor and open the upper globe valve; this will give you the water pressure of the discharge main on piston in water cylinder. Then regulate by screwing up or down on lower wheel in yoke, until your water pressure gauge shows the pressure you desire to carry; then lock in place by turning upper wheel to the right until up tight against bottom end of the piston rod.

"In starting and stopping your pump, do it with the throttle and do not change the adjustment of your governor. Pack valve stem as light as you can and screw stuffing box-nut down lightly with thumb and finger, just enough to hold the steam and no more. Do not use wick packing. Once every month run your engine by the throttle, shut off water pressure, open union in pipe work, take off clyinder cap, take out piston, wipe the cylinder, clean and wipe piston head, and lubricate them with vaseline. Always keep your governor clean."

Back Pressure Valves. — The purpose of this form of valve is to maintain a uniform back pressure in the exhaust pipe from an engine when the steam is used for steam heating, drying, cooking or other purposes.

The Fisher valve shown in Fig. 135 has an inner valve chamber with two accurately machined ports of different areas in which the semi-balanced, double piston type of valve works. This avoids the use of a heavy counterweight and eliminates the tendency to pulsate and hammer. The steam exerts a pressure on both valves, the smaller one tending to close and the larger to open, so that the difference between the two forces tends to keep the valve open. Since the valve stem is connected to the lever arm, the weight tending to keep the valve closed may be moved to a position where the valve will open at the required

SPECIAL VALVES

pressure. The lever and weight control can be adjusted to hold the valve open when no back pressure is wanted.

The Foster back pressure valve shown in Fig. 136 operates with a spring instead of a weight. The valve is made up of two pieces between which the valve seat is clamped. The valve has a piston and guide stem integral with it. A spring and compensating lever hold the valve to its seat. A push rod rests on the bottom of the dash-pot piston and engages with the end of the compensating lever which has its fulcrum at *1*. The spring bears against the lever through a pivot washer *2*, and is adjusted by the screw *3*.

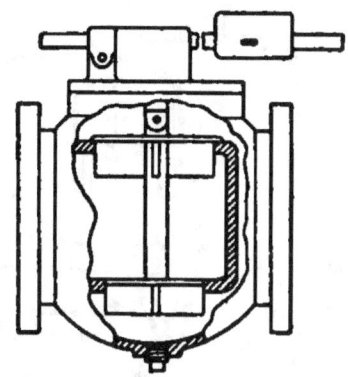

Fig. 135. Fisher Back Pressure Valve.

When the steam pressure lifts the valve, the latter pushes up the compensating lever. As the latter moves, the length of the arm on which the spring acts shortens, so that as the resistance of the spring increases a greater leverage is obtained with the result that the back pressure beneath the valve remains constant regardless of the opening of the valve. When for any reason the flow of steam lessens, the spring forces the valve slowly to its seat, the dash-pot *4* cushioning its movement. Hole *C* is drilled through bottom of the dash-pot to admit of the passage of steam or vapor from or into the dash-pot. A drain pipe is connected to the casing at *D* just above the seat to drain water of condensation. When no back pressure is required, the valve may be thrown out of commission by turning screw *5* to the right to shoulder which carries the valve off its seat.

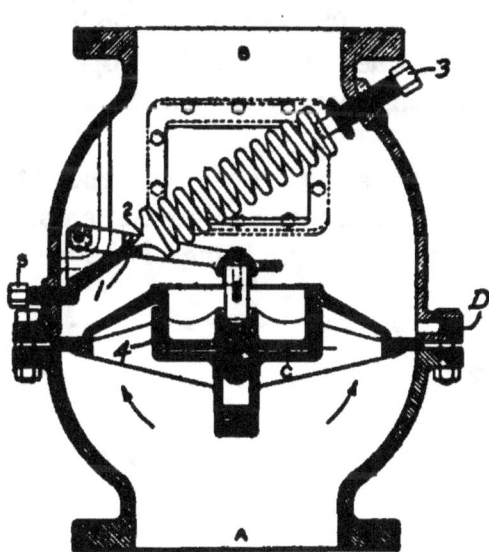

Fig. 136. Foster Back Pressure Valve.

Automatic Exhaust Relief Valves. — With condensing engines and steam turbines it is necessary to use a valve in the exhaust pipe, which will open and allow the steam to exhaust direct to the atmosphere in case pressure accumulates, due to loss of vacuum from any cause. Such valves are designed to remain closed under usual operating conditions, but automatically open to atmosphere as soon as the vacuum is lost. The position of the valve is in a branch leading to the atmosphere and taken from the main exhaust pipe between the engine and condenser.

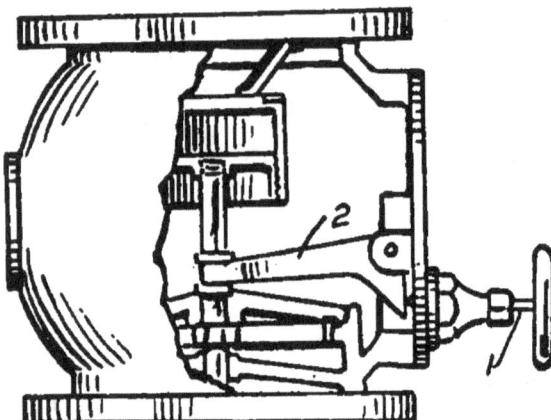

Fig. 137. Fisher Exhaust Relief Valve.

The Fisher exhaust relief valve is shown in Fig. 137. The valve is kept closed by atmospheric pressure. It may be kept open by the screw *1* and lever *2* when desired. The purpose of the internal dash-pot is to prevent hammering when the valve is in operation. A water seal is provided to insure tightness when the valve is used with a high vacuum.

Safety Valves. — The purpose of a safety valve is to relieve the boiler in case the steam pressure rises above the desired amount. There are two general forms, the older form being of the weight lever type, and the modern spring or "pop" type.

The lever type is shown in Fig. 138. The pressure at which the valve will open is regulated by moving

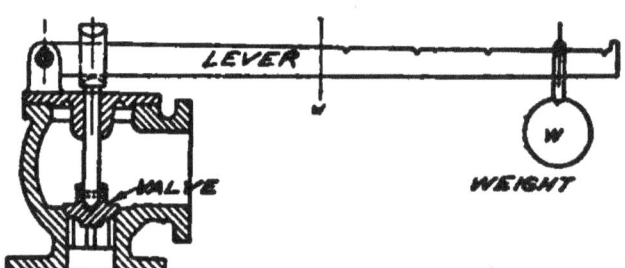

Fig. 138. Lever Safety Valve.

the weight in or out on the lever. This form is open to several objections; the blowing-off pressure is too easily changed, and the action of the valve is likely to be sluggish, both when opening and when closing.

SPECIAL VALVES

A pop safety valve is shown in Fig. 139. Such valves are more certain in their operation, and are almost universally used. The valve operates against a spring which can be set for the pressure at which the boiler is to "blow off." Boiler pressure acting on the under side of the valve raises it slightly, exposing a larger area which causes the valve to "pop" open. The range of operation can be maintained very closely with this type of valve. The lever attachment is for the purpose of operating the valve by hand. The valve shown in the figure is made by Crane Company and is provided with a patented self-adjusting auxiliary disc and spring

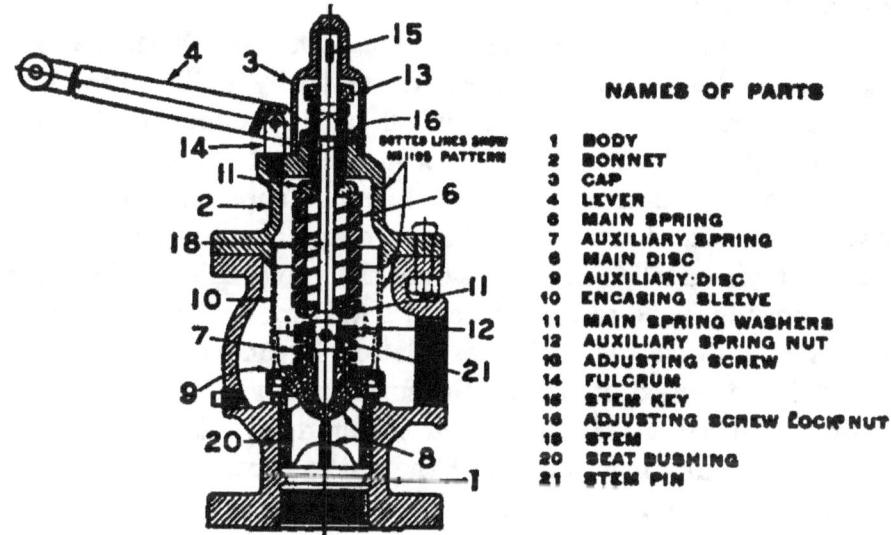

Fig. 139. Crane Pop Safety Valve.

operating independently of the main spring and disc. The device automatically regulates the blow-back of the valve within certain limits and combines the following qualities: high discharging capacity; small blow down of pressure; minimum waste of steam; absence of wiredrawing at the seat and prompt seating without hammering. The dotted lines in the figure indicate a type of valve in which the springs are enclosed in a casing or chamber. This type should be used when the outlets of the valves are piped to the atmosphere and is necessary where a number of valves are connected to one exhaust or discharge pipe. The spring chamber extends over a large portion of the top surface of the valve disc and tends to prevent chattering caused by back-pressure due to long or deflected discharge pipes. It also prevents any tendency of back-pressure from retarding the action of a valve about to pop.

Installation of Pop Safety Valves. — The directions for the installation of iron body pop safety valves are quoted from Crane Company.

"Pop safety valves should be installed on a saddle nozzle if possible. If piping is used between the boiler and the valve, it should be of a larger size than the nominal diameter of the valve. Care should be taken that no chips, scale, red lead or other substances are left in the inlet of the valve or in the boiler connections to it. Where new valves are found to be in a leaky condition, this defect, in most cases, can be traced back to one of the above mentioned causes.

The first time pressure is raised in a boiler on which new pop valves have been installed, open the valve by pulling the lever when the pressure is within about 5 or 10 pounds of the set pressure stamped on the valve, AND KEEP THE VALVE OPEN about one minute or long enough to make sure that all foreign matter has been blown out of the valve and connections.

If piping is installed in the outlet of the valve, this should under no circumstances be reduced in size, and if more than one fitting is used in the line the entire installation beyond the first fitting should be increased in size. Be sure to SUPPORT this piping, as many a perfect valve has been transformed into a leaky one by reason of improper support of the outlet pipe.

"Do not install any pop valve in a horizontal position."

EXTRACTS FROM REPORT OF AMERICAN SOCIETY OF MECHANICAL ENGINEERS BOILER CODE COMMITTEE. (POWER BOILERS)

SAFETY VALVE REQUIREMENTS

269. Each boiler shall have two or more safety valves, except a boiler for which one safety valve 3-in. size or smaller is required by these Rules.

270. The safety valve capacity for each boiler shall be such that the safety valve or valves will discharge all the steam that can be generated by the boiler without allowing the pressure to rise more than 6 per cent. above the maximum allowable working pressure, or more than 6 per cent. above the highest pressure to which any valve is set.

277. The safety valve or valves shall be connected to the boiler independent of any other steam connection, and attached as close as possible to the boiler, without any unnecessary intervening pipe or fitting. Every safety valve shall be connected so as to stand in an upright position, with spindle vertical, when possible.

278. Each safety valve shall have full sized direct connection to the boiler. No valve of any description shall be placed between the safety valve and the boiler, nor on the discharge pipe between the safety valve and the atmo-

SPECIAL VALVES

sphere. When a discharge pipe is used, it shall be not less than the full size of the valve, and shall be fitted with an open drain to prevent water from lodging in the upper part of the safety valve or in the pipe.

280. When a boiler is fitted with two or more safety valves on one connection, this connection to the boiler shall have a cross-sectional area not less than the combined area of all the safety valves with which it connects.

286. A safety valve over 3-in. size, used for pressures greater than 15 pounds per square inch gage, shall have a flanged inlet connection. The dimensions of the flanges shall conform to the American Standard.

SAFETY VALVES FOR HEATING BOILERS

354. No shut-off of any description shall be placed between the safety or water relief valves and boilers, nor on discharge pipes between them and the atmosphere.

355. When a discharge pipe is used, its area shall be not less than the area of the valve or aggregate area of the valves with which it connects, and the discharge pipe shall be fitted with an open drain to prevent water from lodging in the upper part of the valve or in the pipe. When an elbow is placed on a safety or water relief valve discharge pipe, it shall be located close to the valve outlet or the pipe shall be securely anchored and supported. The safety or water relief valves shall be so located and piped that there will be no danger of scalding attendants.

358. The minimum size of safety or water relief valve or valves for each boiler shall be governed by the grate area of the boiler, as shown by Table 74.

TABLE 74
ALLOWABLE SIZES OF SAFETY VALVES FOR HEATING BOILERS

Water evaporated per Square Foot of Grate Surface per Hour Pounds	75	100	160	160	200	240
Maximum Allowable Working Pressure Pounds per Square Inch	Zero to 25 Lbs.	Over 25 to 50 Lbs.	Over 50 to 100 Lbs.	Over 100 to 150 Lbs.	Over 150 to 200 Lbs.	Over 200 Lbs.

Diam. of Valve Inches	Area of Valve Square Inches	Area of Grate, Square Feet				
1	0.7854	2.00	2.50	2.75	3.25	3.5	3.75
1¼	1.2272	3.25	4.00	4.25	5.00	5.5	5.75
1½	1.7671	4.50	5.50	6.00	7.25	8.0	8.50
2	3.1416	8.00	9.75	10.75	13.00	14.0	15.00
2½	4.9087	12.50	15.00	16.50	20.00	22.0	23.00
3	7.0686	17.75	21.50	24.00	29.00	31.5	33.25
3½	9.6211	24.00	29.50	32.50	39.50	43.0	45.25
4	12.5660	31.50	38.25	42.50	51.50	56.0	59.00
4½	15.9040	40.00	48.50	53.50	65.00	71.0	74.25

When the conditions exceed those on which Table 74 is based, the following formula for bevel and flat seated valves shall be used:

$$A = \frac{W \times 70}{P} \times 11 \quad \ldots\ldots\ldots\ldots (18)$$

in which

A = area of direct spring-loaded safety valve per square foot of grate surface, sq. in.

W = weight of water evaporated per square foot of grate surface per second, lb.

P = pressure (absolute) at which the safety valve is set to blow, lb. per sq. in.

CHAPTER VIII

STEAM PIPING

General Considerations. — It is not the purpose of this chapter to deal with pipe lines in an exhaustive way, as there are large books devoted to this one subject, but it is intended to tell something of the general arrangement of pipe lines and some of the things to be considered.

The layout of a piping system is a question of design and ranges from the piping of a single engine and boiler to the complex system of the large power plant. In piping as in all other branches of engineering work "safety first" should be one of the guiding principles. To this end the best of material and workmanship should be called for. These, together with intelligence in design will give both economy and safety in operation and maintenance. The items of general application to any system may be listed as follows:

 A. Reduce the length to the smallest practicable distance.

 B. Have as few fittings and valves as safety and operating conditions will allow.

 C. Make allowances for expansion and contraction.

 D. Make allowances for drainage.

 E. Make allowances for supports.

 F. Eliminate vibration as much as possible.

 G. Make allowances for sectionalizing or shutting off any portion of the system.

 H. Consider the size of pipe from the viewpoints of safety, economy in first cost, economy in operation, radiation losses, loss in pressure, and velocity of flow.

Header System. — There are a number of systems for laying out high pressure steam piping. In every case it is desirable to maintain as uniform a velocity of flow as possible throughout the system. The simplest is the header system. When the engines and boilers are placed back to back as shown in Fig. 140, a small size of header may be used. As the engines and boilers are close together the pipe lines are short and direct. The header may be

138 A HANDBOOK ON PIPING

located either in the boiler room or in the engine room, but preferably in the boiler room. When the engines and boilers are placed end to end as shown in Fig. 141 a larger header is required, as all the steam must pass through the header at a single section. The sections of the header farthest from the engines may be made

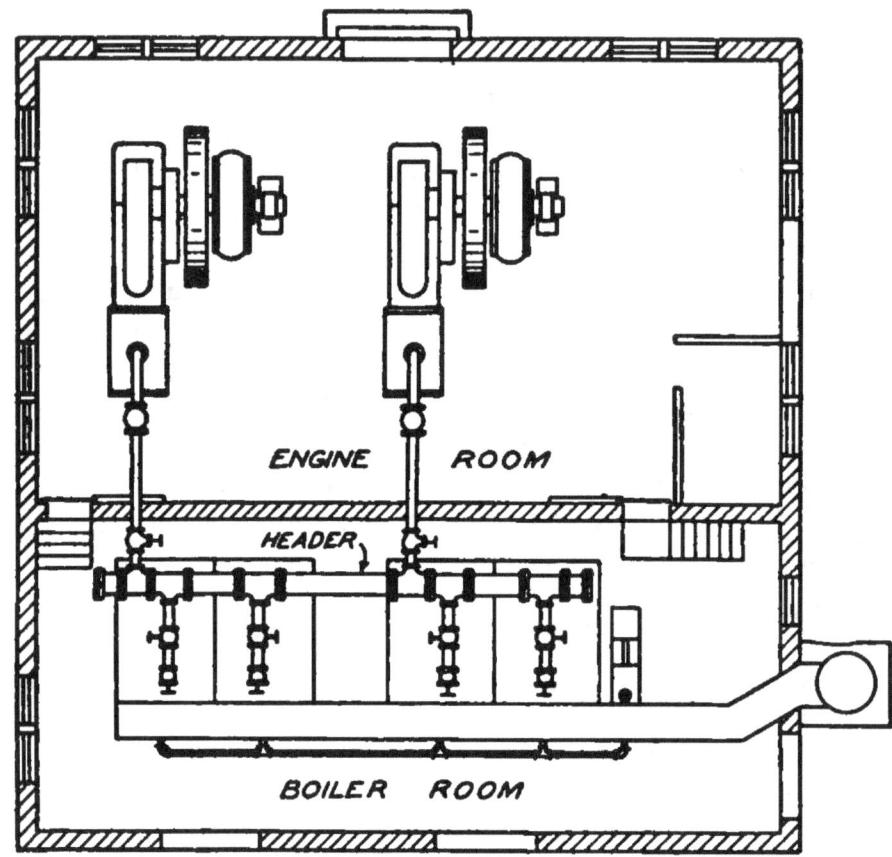

Fig. 140. Header System of Piping.

smaller as they carry only a part of the supply. A separate header may be provided for supplying steam to pumps and other auxiliary apparatus.

Direct System with Cross-over Header. — A direct system of piping with cross-over header used in the Connors Creek Station of the Detroit Edison Company is shown in Fig. 142 from the September, 1915, Journal A. S. M. E., and described by C. F Hirshfield. The live steam piping consists of a run from two boilers to the unit which they serve, all of these runs being cross-connected by a cross-over header. The steam leads from each boiler are of 10-inch pipe and these join together in a Y-fitting, which has a 14-inch discharge. Under full load conditions with two boilers supplying one unit, the steam velocities will be about

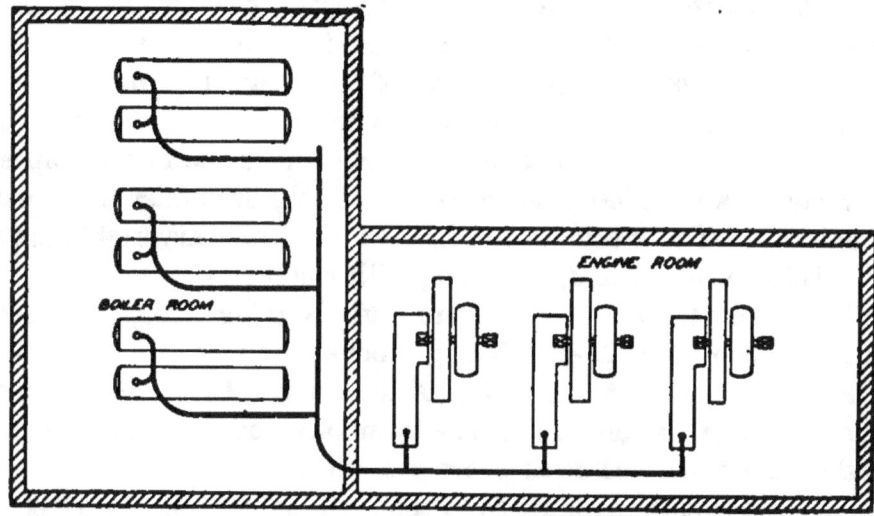

Fig. 141. End to End System of Piping.

10,500 feet per minute in the 10-inch pipe, and 12,000 feet per minute in the 14-inch pipe. With three boilers supplying two units, these velocities will rise to about 14,000 and 16,000 feet per minute, respectively. The cross-over main necessitated a

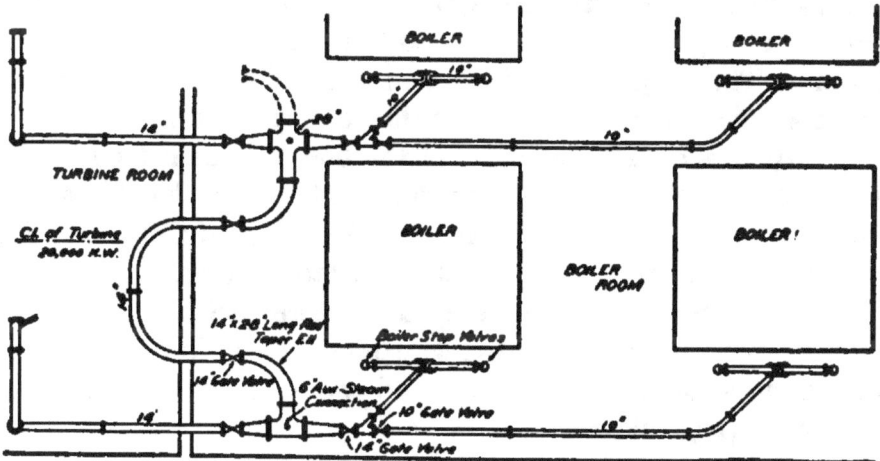

Fig. 142. Connors Creek Station, High Pressure Piping.

design which should permit steam from any two boilers to flow into that main, and steam from the main to flow into any turbine lead, with practically equal facility.

The steam leaving the 10 × 14 × 10 inch Y-branch previously mentioned, passes through a cast steel expanding nozzle which enlarges to a diameter of 28 inches. This in turn leads into a 28-inch cast steel side-outlet T or side-outlet cross. The 28-inch lateral outlets of the latter fittings are the connection points of the cross-over main. The velocity of the steam passing into the cross-over, or from the cross-over main to the turbine lead, is thus reduced to about one quarter of its value in the 14-inch pipes, or roughly, a little less than 4000 feet per minute under the worst conditions. The steam turns through the necessary right angle at this low velocity and, therefore, with small loss.

The steam for the auxiliary turbines is taken from a 6-inch outlet on top of the 28-inch fittings above described.

All superheated steam piping is full weight steel with welded flanges. The flanges are finished smooth and corrugated steel gaskets are used. All fittings are cast steel.

The atmospheric exhaust from the main unit is made of riveted steel pipe and fittings. The auxiliary exhaust piping is lap welded steel with Van Stone joints and fitted with corrugated copper gaskets. All saturated steam piping is extra heavy steel fitted with steel flanges. The fittings are all cast steel and steel valves of American make are used.

Ring System. — The ring system of piping provides a closed ring of piping from the boilers to the engines and back to the boilers. The purpose of this system is to allow operation of the engines from either direction, in order to insure continuous operation. In case of accident parts of the line may be cut out. The extra amount of large pipe, valves and fittings make this sytem heavy, and expensive to install, as well as wasteful in operation due to the large amount of radiating surface and extra valves and joints to keep tight. There are cases where such a system may be desirable, but it is not used so extensively as formerly due to the improvements in materials and workmanship which have lessened piping failures.

The ring main system of piping is shown in Fig. 143, which is a span of the Baltimore high pressure pumping station. This is an instance where reliability outweighs all other considerations. It is described by J. B. Scott in Volume 35 A. S. M. E. Trans. "A 12-inch steam header forms a closed ring around the plant, with long radius expansion bends at all changes in direction. A suffi-

STEAM PIPING 141

cient number of gate valves are placed in the header to sectionalize it, so that any portion may be cut out without disabling more than one boiler or one pump. Pipe is full weight, lap welded, soft open-hearth steel. To provide an independent header for the

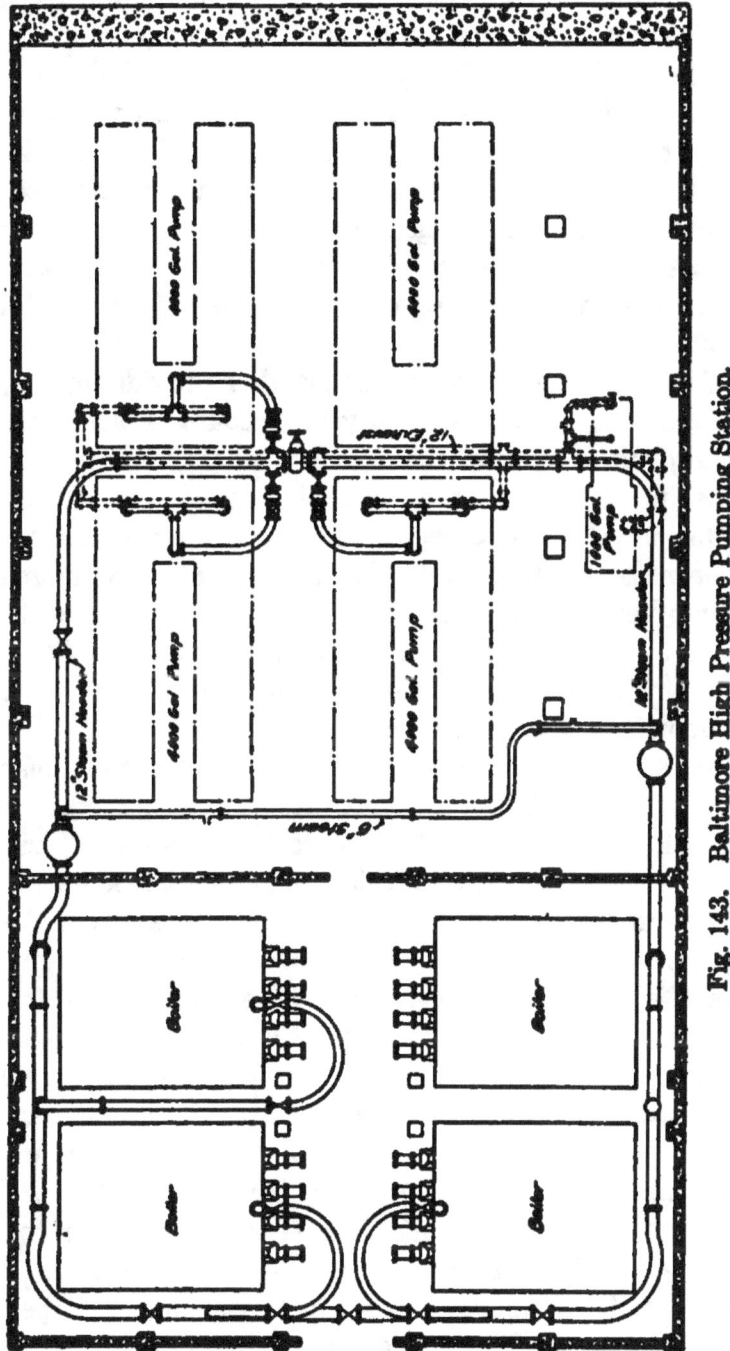

Fig. 143. Baltimore High Pressure Pumping Station.

station auxiliaries, a 6-inch cross connection is made across the centre of the main header, which is capable of being fed from either side of the main header, in case of accident to the other. No fittings whatever are used in the main line, all branches being taken from interlocked welded necks. Boiler branches are provided with non-return valves at the boiler nozzles and gates at the header end. Van Stone flanges are provided for connections to the valves and receivers, which are located so as to avoid as far as possible the necessity for any additional joints in the line. Wrought steel receiver type separators are installed at the low points on each side of the header."

Duplicate System. — The double main or duplicate system provides for two separate sets of piping in any of the following combinations:

A. Two small size mains which together provide for the capacity of the plant. When necessary on account of accidents or repairs the plant can be operated with a single main by increasing the boiler pressure and steam velocities.

B. One large main in regular use and a small idle main for use when necessary to have the large main out of commission.

C. Two large mains, one in use and one idle. The duplicate system is expensive as it requires a large number of fittings and valves. Its purpose is to insure against shut downs, and there may be conditions where its use is desirable.

Steam Velocity. — The velocity of high pressure steam flow in piping is not at all uniform, but ordinarily the average velocity may be taken at from 5000 to 8000 feet per minute. This velocity is often exceeded, especially in large plants. Some values for actual plants are as follows: steam pressures 160 to 210 pounds per square inch, average 175; superheat, 100 to 200 degrees F., average 134; velocity of steam in boiler steam pipe, 3750 to 8700 feet per minute, average 6150; velocity of steam in header, 4200 to 11,400 feet per minute, average 7000; velocity of steam in turbine steam pipe, 3225 to 7900 feet per minute, average 5100.

For turbines smaller pipes may be used than for engines as the flow is constant due to the uniform demand for steam. In the steam plant large piping sometimes acts as a receiver to supply the large amounts of steam required for short periods. Higher velocities result in smaller pipes which are much cheaper to install and maintain. The matter of friction and drop in pressure

STEAM PIPING

is not serious in view of high pressures and superheat commonly employed in large plants. In one large plant operating with 210 pounds steam pressure the average steam velocity is 15,744 feet per minute, and even higher velocities are used. This tendency toward smaller pipes and higher velocities is advantageous in many ways; the first cost is less, the radiation losses are less, smaller repair expenses and provision for expansion is easier.

For exhaust, velocities up to 30,000 feet per minute may be used in estimating sizes of pipe.

Size of Pipe. — The size of pipe may be calculated from the volume of steam and the velocity of flow.

p = absolute pressure, pounds per square inch.
V = velocity of steam, feet per minute.
s = specific volume of steam at given pressure, cubic feet per pound.
a = internal area of pipe, square inches.
w = weight of steam passing through pipe, pounds per minute.

$$w = \frac{aV}{144s} \quad \quad (19)$$

$$a = \frac{144ws}{V} \quad \quad (20)$$

From these formulae the area of the pipe required can be obtained and reference to the pipe tables will give the diameter. When the drop in pressure due to friction is to be considered, Babcock's formula may be used.

L = length of pipe, in feet.
p = drop in pressure, pounds per square inch.
d = inside diameter of pipe in inches.
D = mean density, pounds per cubic foot.
w = weight of steam flowing, pounds per minute.

$$w = 87 \sqrt{\frac{pDd^5}{L\left(1 + \frac{3.6}{d}\right)}} \quad \quad (21)$$

$$p = .0001321 \frac{w^2 L}{Dd^5}\left(1 + \frac{3.6}{d}\right) \quad \quad (22)$$

In addition to the friction of the pipe there is the friction of valves and fittings to be considered. Where long radius pipe bends are used they may be considered as being equal to the same length of straight pipe. Gate valves produce very small losses when fully open. The friction of an ordinary 90 degree elbow, for a globe valve, or for a square end opening may be found from information given in Briggs' paper on "Warming Buildings by Steam." Thus the length of pipe equivalent to a globe valve or square end opening is found from the formula given below.

d = internal diameter, in inches.
E = equivalent length of pipe in feet.

$$E = \frac{9.5d^2}{3.6 + d} \quad \quad (23)$$

For a 90 degree elbow the equivalent length is two-thirds of the above or

$$E = \frac{6.33d^2}{3.6 + d} \quad \quad (24)$$

The curves of Fig. 144 give the values of equation (24) for various sizes.

The allowable drop in pressure varies with conditions and may be from one to ten pounds per square inch. For ordinary plants a drop of five pounds is allowable provided the boiler pressure is high enough to compensate for it so that an economical pressure is maintained at the engine.

Equalization of Pipes. — From formula (21)

$$w = 87\sqrt{\frac{pDd^5}{L\left(1 + \frac{3.6}{d}\right)}}$$

the number of small pipes equivalent to a large one may be found. The variable factor in the formula is

$$\sqrt{\frac{d^5}{d + 3.6}}$$

from which

$$N = \frac{\sqrt{\dfrac{d_2^5}{d_2 + 3.6}}}{\sqrt{\dfrac{d_1^5}{d_1 + 3.6}}} \quad \quad (25)$$

d_1 = diameter of smaller pipe in inches.
d_2 = diameter of larger pipe in inches.
N = number of smaller pipes equivalent to one large one.

STEAM PIPING

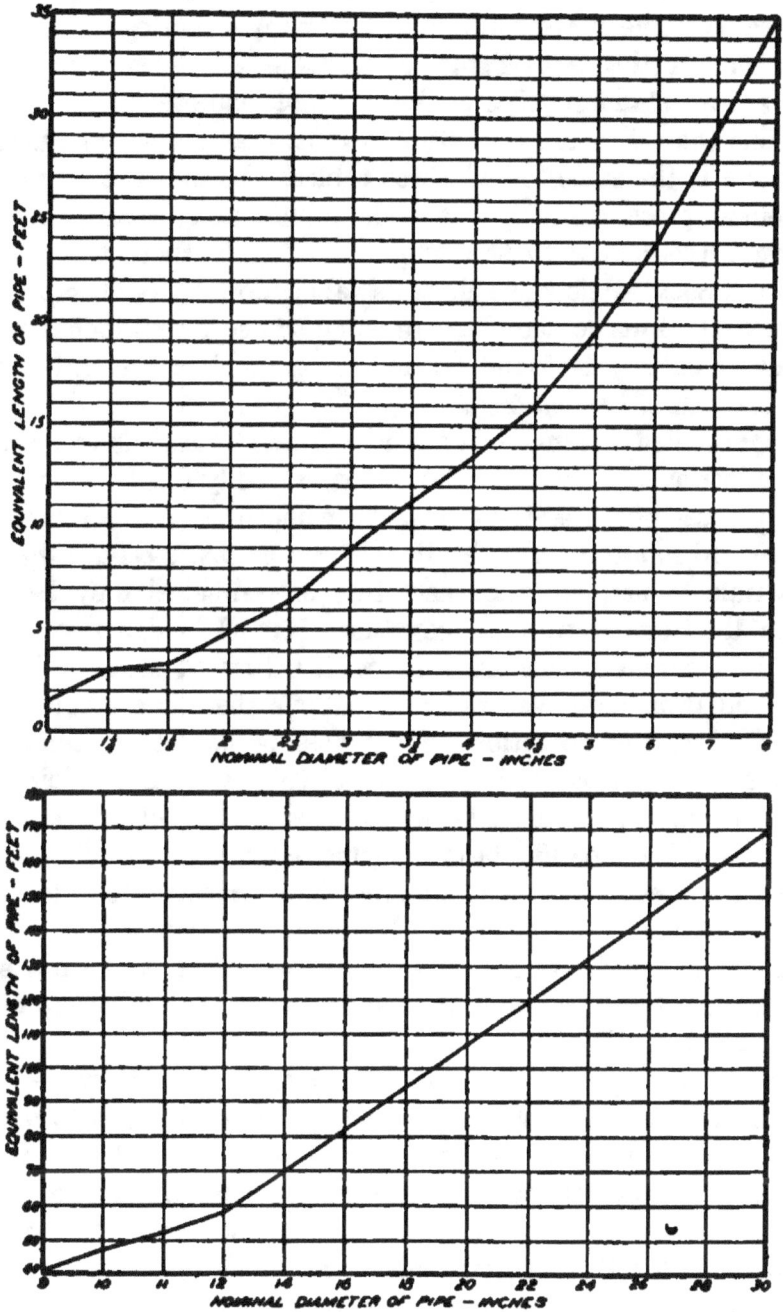

Fig. 144. Length of Pipe in Feet Equivalent to a 90° Elbow.

Values for this formula are given in Table 75. Tables 76 and 77 are from the Watson-Stillman Company's catalog of hydraulic valves and fittings. In Table 75 the values above the heavy black line are for standard pipe of the nominal diameter given. Below the line the values are for actual internal diameters. The method of using is the same for all three tables. To find the number of $1\frac{1}{2}$-inch pipes equivalent to one 6-inch pipe, follow the line marked $1\frac{1}{2}$ across to the column headed 6 where the number given is 39.2. Below the line the table shows that 46 pipes $1\frac{1}{2}$ inches actual inside diameter are equal to one pipe 6 inches actual inside diameter, as found by following the line marked 6 over to the column headed $1\frac{1}{2}$.

Superheated Steam. — When superheated steam is to be used, the selection of materials should be carefully made. Composition or cast iron lose their strength when used with superheated steam and so are unsafe. Malleable iron or cast steel are the best materials to use, although cast iron or semi-steel may be used when the temperature is less than 500° F. Higher velocities are used with superheated than with saturated steam. In this way radiation losses are reduced. While there is a greater drop in pressure, the operation as a whole is generally economical as the heat of friction is given back to the steam. Piping for superheated steam should be well covered as the higher temperatures and low specific heat of superheat make conditions for radiation losses very much greater than for saturated steam at like pressures. Expansion and contraction are much greater with superheated steam and ample provision must be made to care for it. Specifications for superheated steam piping are given in Chapter XIX.

Effect of High Temperature on Metals and Alloys. — The effects of superheated steam due to high temperature is to reduce the tensile strength of metals. An extensive series of tests made in Crane Company's laboratories by I. M. Bregowsky and L. W. Spring are reported in an article read before the International Association for Testing Materials, and published in full by Crane Company. A large number of tests were made upon the materials used by the above company in manufacturing their products and so have an important bearing upon high pressure and superheated steam power plant piping. A number of curves from the report showing the average results of some of the tests are

STEAM PIPING

TABLE 75
Equalization of Standard Wrought Pipe

Dia.	½	¾	1	1½	2	2½	3	4	5	6	7	8	9	10	11	12	13	14	15	16	17	Dia.
½		2.27	4.88	15.8	31.7	52.9	96.9	205	377	620	918	1292	1767	2488	3014	3786	4904	5927	7321	8535	9717	½
¾			2.05	6.97	14.0	23.8	42.5	90.4	166	273	405	569	779	1096	1328	1668	2161	2615	3226	3761	4282	¾
1		2.90		3.46	6.82	11.4	20.9	44.1	81.1	133	198	278	380	536	649	815	1070	1263	1576	1837	2092	1
1½	7.55	9.30	3.20		1.26	3.34	6.13	13.0	23.8	39.2	58.1	81.7	112	157	190	239	310	375	463	539	614	1½
2	24.2	21.0	7.25	2.26		1.67	3.06	6.47	11.9	19.6	29.0	40.8	55.8	78.5	95.1	119	155	187	231	269	307	2
2½	54.8	39.4	13.6	4.23	1.87		1.83	3.87	7.12	11.7	17.4	24.4	33.4	47.0	56.9	71.5	92.6	112	138	161	184	2½
3	102	65.4	22.6	7.03	3.11	1.66		2.12	3.89	6.39	9.48	13.3	20.9	23.7	31.2	39.1	50.6	61.1	75.5	88.0	100	3
4	170	144	49.8	15.5	6.87	3.67	2.21		1.84	3.02	4.48	6.30	8.61	12.1	14.7	18.5	23.9	28.9	35.7	41.6	47.3	4
5	376	263	90.9	28.3	12.5	6.70	4.03	1.83		1.65	2.44	3.43	4.69	6.60	8.00	10.0	13.0	15.7	19.4	22.6	25.8	5
6	686	429	148	46.0	20.4	10.9	6.56	2.97	1.63		1.48	2.09	2.85	4.02	4.86	6.11	7.91	9.56	11.8	13.8	15.6	6
7	1116	656	226	70.5	31.2	16.6	10.0	4.54	2.49	1.51		1.41	1.93	2.71	3.28	4.12	5.34	6.45	7.97	9.31	10.6	7
8	1707	938	322	101	44.5	23.8	14.3	6.48	3.54	2.18	1.43		1.37	1.93	2.33	2.92	3.79	4.57	5.67	6.60	7.52	8
9	2435	1281	440	137	60.8	32.5	19.5	8.85	4.85	2.98	1.95	1.37		1.35	1.71	2.14	2.77	3.35	4.14	4.83	5.50	9
10	3335	1688	582	181	80.4	42.9	25.8	11.7	6.40	3.93	2.57	1.80	1.32		1.26	1.52	1.97	2.38	2.94	3.43	3.91	10
11	4393	2168	747	233	103	55.1	33.1	15.0	8.22	5.05	3.31	2.32	1.70	1.28		1.22	1.63	1.88	2.43	2.83	3.22	11
12	5642	2723	938	293	129	69.2	41.6	18.8	10.3	6.34	4.15	2.91	2.13	1.61	1.28		1.30	1.57	1.93	2.26	2.58	12
13	7087	3326	1146	356	158	84.5	50.7	23.0	12.6	7.75	5.07	3.56	2.60	1.98	1.53	1.22		1.21	1.49	1.74	1.98	13
14	8657	4070	1403	438	193	103	62.2	28.2	15.4	9.48	6.21	4.35	3.18	2.41	1.88	1.50	1.22		1.24	1.44	1.64	14
15	10600	4927	1698	530	234	125	75.3	34.1	18.7	11.5	7.52	5.27	3.85	2.92	2.27	1.81	1.48	1.21		1.17	1.35	15
16	12834	5758	1984	619	274	146	88.0	39.9	21.8	13.4	8.78	6.15	4.51	3.41	2.66	2.12	1.73	1.42	1.18		1.14	16
17	14978	6738	2322	724	320	171	103	46.6	25.6	15.7	10.3	7.20	5.27	3.99	3.11	2.47	2.03	1.66	1.37	1.17		17
18	17537	7810	2691	840	371	198	119	54.1	29.6	18.2	11.9	8.35	6.11	4.63	3.60	2.87	2.35	1.92	1.59	1.36	1.16	18
20	20327	10249	3532	1102	487	260	157	70.9	38.9	23.9	15.6	10.9	8.02	6.07	4.73	3.76	3.08	2.52	2.08	1.78	1.52	20
24	26576	16376	5644	1761	778	416	250	113	62.1	38.2	25.0	17.5	12.8	9.70	7.55	6.01	4.92	4.02	3.32	2.84	2.43	24
30	42624	28990	9990	3117	1378	736	443	201	110	67.6	44.2	31.0	22.7	17.2	13.4	10.7	8.72	7.14	5.88	5.03	4.30	30
36	75453	46143	15902	4961	2193	1172	705	319	175	108	70.4	49.3	36.1	27.3	21.3	16.9	13.9	11.3	9.37	8.01	6.85	36
42	120100	68282	23531	7341	3245	1734	1044	473	259	159	104	73.0	53.4	40.5	31.5	25.1	20.5	16.8	13.9	11.9	10.1	42
48	177724	95818	33020	10301	4654	2434	1465	663	363	223	146	102	75.0	56.8	44.2	35.2	28.8	23.5	19.4	16.6	14.2	48
249351																						
Dia.	½	¾	1	1½	2	2½	3	4	5	6	7	8	9	10	11	12	13	14	15	16	17	Dia.

TABLE 76
EQUALIZATION OF EXTRA STRONG PIPE

Pipe Size	1/8 in.	1/4 in.	3/8 in.	1/2 in.	3/4 in.	1 in.	1 1/4 in.	1 1/2 in.	2 in.	2 1/2 in.	3 in.	3 1/2 in.	4 in.	4 1/2 in.	5 in.	6 in.	Pipe Size
Int. Area	.033	.068	.139	.231	.452	.71	1.271	1.753	2.936	4.209	6.569	8.856	11.45	14.18	18.193	25.97	Int. Area
1/8 in.	1	2.06	4.2	7.	13.7	21.5	38.6	53.2	89.	127.5	199.	268.	347.	430.	550.	786.	1/8 in.
1/4 in.	..	1	2.05	3.4	6.7	10.4	18.8	25.8	43.	62.	96.5	130.	168.5	208.	267.	382.	1/4 in.
3/8 in.	..	..	1	1.66	3.26	5.1	9.2	12.6	21.1	30.3	47.3	63.8	82.5	102.	131.	186.	3/8 in.
1/2 in.	..	..	..	1	1.96	3.08	5.5	7.6	12.7	18.2	28.4	38.4	47.5	61.5	78.8	112.	1/2 in.
3/4 in.	..	..	..	..	1	1.57	2.82	3.9	6.5	9.3	14.5	19.6	25.3	31.3	40.2	57.4	3/4 in.
1 in.	..	..	..	..	..	1	1.79	2.47	4.14	5.93	9.25	12.5	16.1	20.	25.6	36.5	1
1 1/4 in.	..	..	..	..	..	..	1	1.38	2.3	3.31	5.16	6.98	8.4	11.1	14.3	20.4	1 1/4 in.
1 1/2 in.	..	..	..	..	..	..	..	1	1.67	2.4	3.74	5.05	6.54	8.1	10.4	14.8	1 1/2 in.
2 in.	..	..	..	..	..	..	..	..	1	1.43	2.24	3.02	3.91	4.84	6.2	8.9	2 in.
2 1/2 in.	..	..	..	..	..	..	..	..	..	1	1.56	2.1	2.72	3.36	4.31	6.15	2 1/2 in.
3 in.	..	..	..	..	..	..	..	..	..	..	1	1.35	1.75	2.16	2.77	3.95	3 in.
3 1/2 in.	..	..	..	..	..	..	..	..	..	..	..	1	1.29	1.6	2.05	2.92	3 1/2 in.
4 in.	..	..	..	..	..	..	..	..	..	..	..	..	1	1.24	1.59	2.26	4 in.
4 1/2 in.	..	..	..	..	..	..	..	..	..	..	..	..	..	1	1.28	1.83	4 1/2 in.
5 in.	..	..	..	..	..	..	..	..	..	..	..	..	..	..	1	1.37	5 in.
6 in.	..	..	..	..	..	..	..	..	..	..	..	..	..	..	..	1	6 in.
	1/8"	1/4"	3/8"	1/2"	3/4"	1"	1 1/4"	1 1/2"	2"	2 1/2"	3"	3 1/2"	4"	4 1/2"	5"	6"	

TABLE 77
EQUALIZATION OF DOUBLE EXTRA STRONG PIPE

Pipe Size	³/₈ in.	½ in.	¾ in.	1 in.	1¼ in.	1½ in.	2 in.	2½ in.	3 in.	3½ in.	4 in.	4½ in.	5 in.	6 in.	Pipe Size
Int. Area	.042	.047	.139	.271	.515	.93	1.744	2.419	4.097	5.794	7.724	10	12.96	18.66	Int Area
³/₈ in.	1	1.12	3.32	6.45	14.6	22.1	41.5	57.5	95.6	137.	184.	236.	308.	444.	³/₈ in.
½ in.	...	1	2.96	5.77	13.1	19.7	37.2	51.5	87.	123.	164.	213.	276.	398.	½ in.
¾ in.	...	...	1	1.95	4.43	6.7	12.5	17.4	29.4	41.6	55.5	72.	93.5	134.	¾ in.
1 in.	...	...	...	1	2.27	3.42	6.45	8.95	15.1	21.4	28.5	37.	47.9	69.	1
1¼ in.	...	...	...	...	1	1.51	2.84	3.94	6.65	9.4	12.5	16.3	21.1	30.4	1¼ in.
1½ in.	...	...	...	...	...	1	1.88	2.6	4.4	6.21	8.3	10.8	14.	20.	1½ in.
2 in.	...	...	...	...	...	...	1	1.39	2.34	3.31	4.42	5.74	7.45	10.7	2
2½ in.	...	...	...	...	...	...	...	1	1.69	2.39	3.19	4.1	5.33	7.72	2½ in.
3 in.	...	...	...	...	...	...	...	...	1	1.41	1.88	2.44	3.16	4.55	3
3½ in.	...	...	...	...	...	...	...	...	...	1	1.33	1.73	2.24	3.22	3½ in.
4 in.	...	...	...	...	...	...	...	...	...	...	1	1.3	1.68	2.42	4
4½ in.	...	...	...	...	...	...	...	...	...	...	...	1	1.3	1.87	4½ in.
5 in.	...	...	...	...	...	...	...	...	...	...	...	...	1	1.44	5
6 in.	...	...	...	...	...	...	...	...	...	...	...	...	...	1	6
	³/₈"	½"	¾"	1"	1¼"	1½"	2"	2½"	3"	3½"	4"	4½"	5"	6"	

150 A HANDBOOK ON PIPING

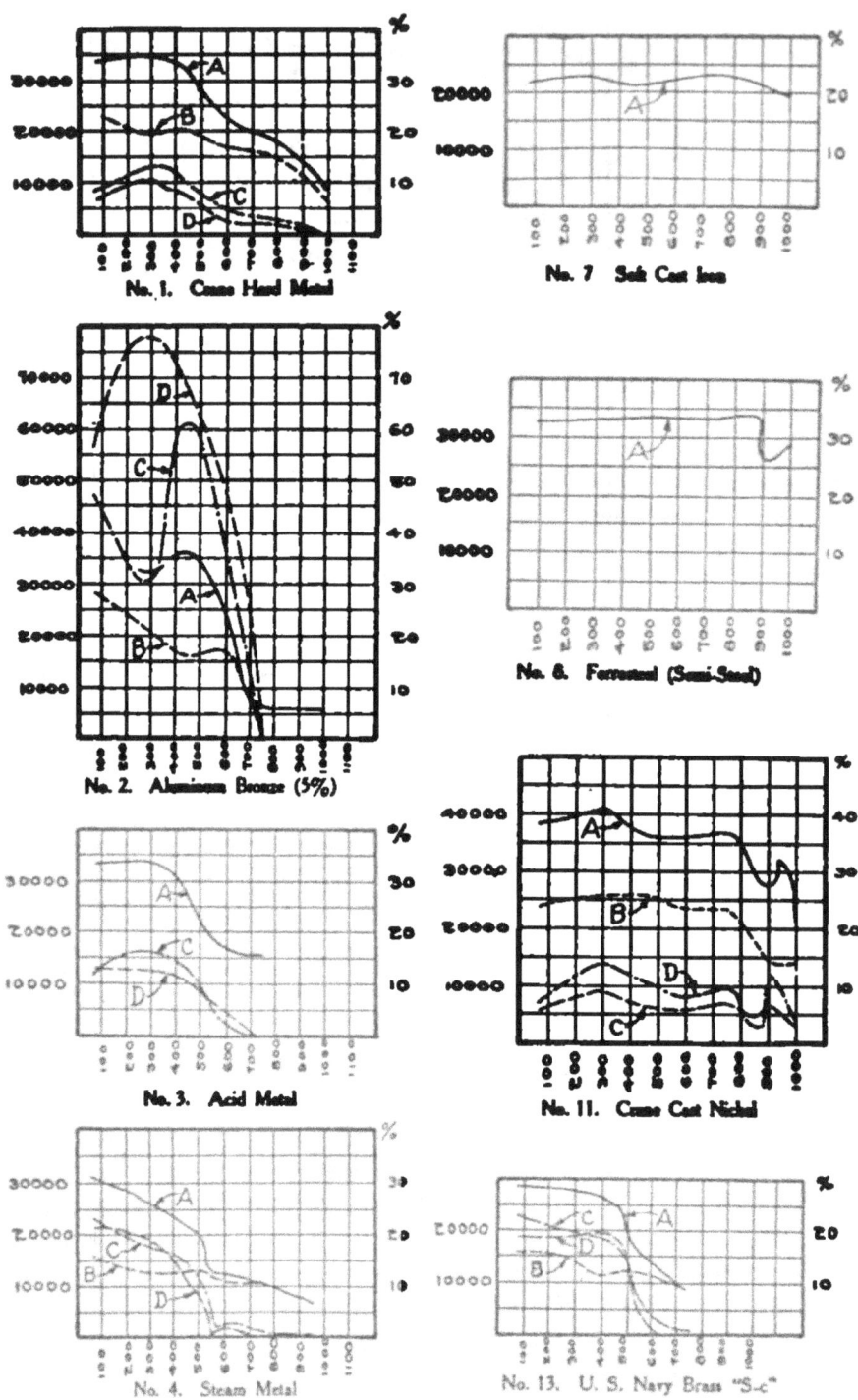

Fig. 145. Effect of Temperature on Strength of Metals and Alloys.

STEAM PIPING

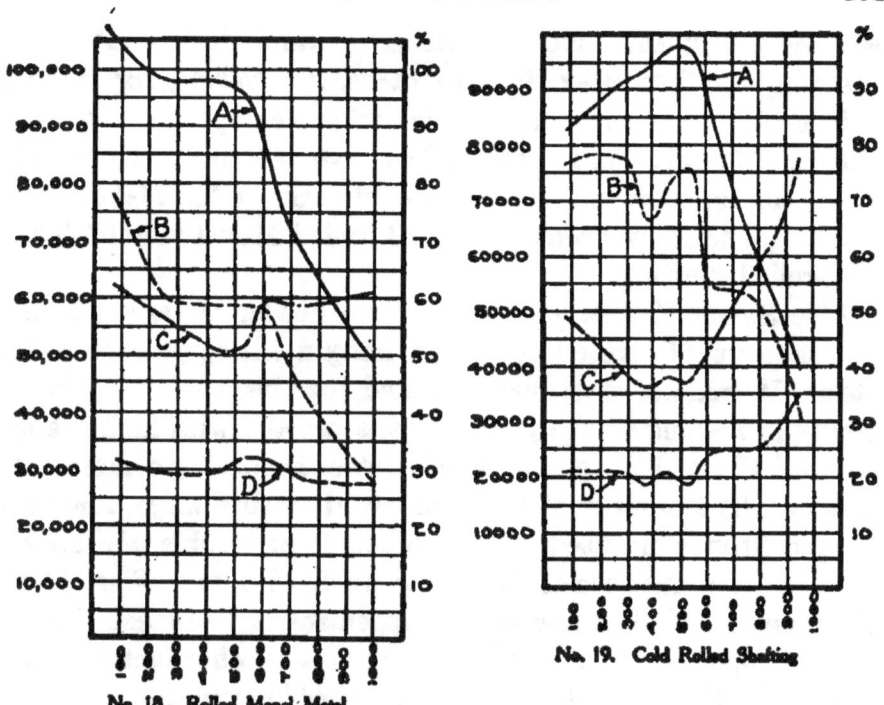

Fig. 145. [cont'd]. Effect of Temperature on Strength of Metals and Alloys.

given in Fig. 145. In each case curve A is ultimate tensile strength, curve B is elastic limit, curve C is per cent. reduction in area and curve D is per cent. elongation. Pounds per square inch are given at the left of the curve, and per cent. at the right.

The materials are: No. 1, Crane "hard metal," a bronze made up of pure copper and tin, alloyed in proportions which give metal of high tensile strength and hardness. No. 2, aluminum bronze (5 per cent. aluminum), a bronze containing 95 per cent. copper and 5 per cent. aluminum. No. 3, acid metal. A phosphorbronze of straight tin and copper. An alloy of high resistance to acids, which is used where ordinary metal would be likely to corrode. Not intended for high temperature purposes. No 4, ordinary steam metal. In general use for all pressures of saturated steam. No. 7, Crane cast iron. A factor which enters into the use of cast iron is the "growth" or "permanent" expansion, which takes place when the metal is alternately heated and cooled a number of times. Crane Company has found that the cast iron of valves used for superheated steam is weaker after a few years of use. Cast steel is considered the best material for use with superheated steam. No. 8, Ferrosteel (semi-steel). Essentially a strong cast iron, used for "extra heavy" valves, for standard valves of sizes over 7 inches and wherever specified for other valves. No. 11, Crane cast nickel. No. 13, U. S. Navy brass "S-c" for government screw pipe fittings. (Cu. 77-80, Sn. 4, Pb. 3, Zn. 13-19 per cent.). No. 18, rolled Monel metal. No. 19 cold-rolled shafting.

Live Steam Header. — A live steam header of large size may be made up of riveted plates, of flanged fittings, or of welded steel. If made of steel plates riveted together there may be difficulty in keeping all the joints tight, especially with high pressure steam. Flanged fittings or welded steel headers are more satisfactory. The number of joints involved when a large number of flanged fittings are used is often a source of trouble and may be avoided by using special fittings or welded headers, Figs. 51 and 52, Chapter V. The size and arrangement of live steam headers depends upon the system of piping used and other factors having to do with the particular design. Further information is given in the articles describing the various systems of piping and the sizes of steam pipes.

Connections Between Boiler and Header. — The pipe between the boiler and header should be arranged so that it will be self-draining, and with provision for expansion. A number of arrangements are shown in Fig. 146. With screwed pipe and fittings, expansion may be taken care of by allowing the pipe to turn on the threads. Bends may be used with either screwed or flanged piping to allow for expansion. Bends are desirable as they offer less resistance to the steam flow and decrease the number of joints to be made and kept tight.

The location of the valves is very important, as it affects the proper draining of the pipe. The valve or valves should be placed at the highest point in the connection to allow condensation to drain from the valves in both directions and so keep the pipe dry. The arrangement when a single boiler is piped with one valve is shown at A, Fig. 146. When more than one boiler is to be used the valve may be placed near the header, as in Fig. 146 at B. Other single-valve arrangements are shown in Fig. 146 at C and D.

Good practice dictates the use of two valves and in many places the law requires two valves on boiler connections. One of

STEAM PIPING

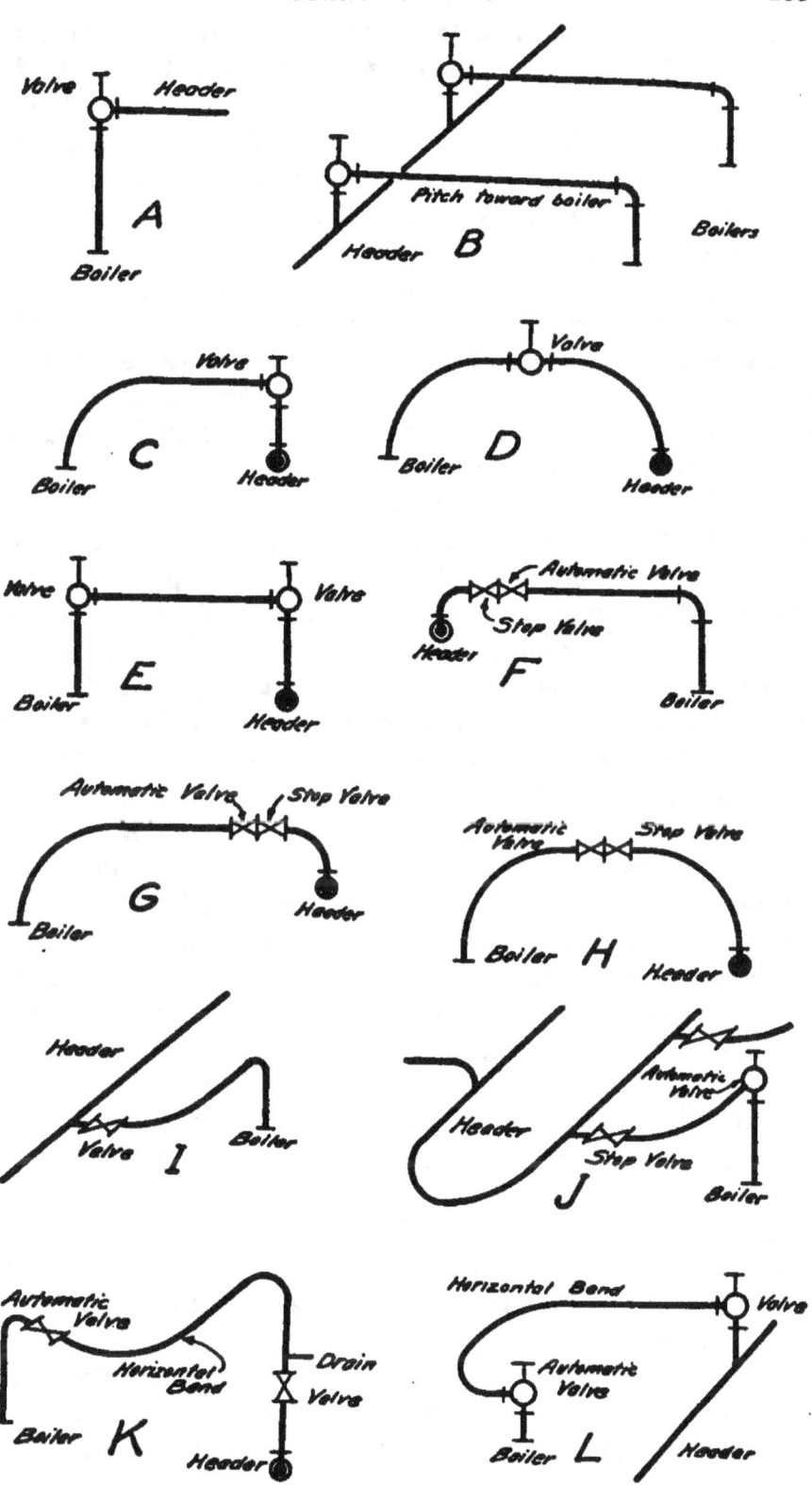

Fig. 146. Boiler to Header Pipes.

these may well be of the automatic stop form described in Chapter VI. With screwed fittings two valves may be arranged as in Fig. 146 at *E*, but some provision should be made for draining the pipe between them, as there is the possibility of condensation accumulating even though the valves are closed. Arrangements are shown with two valves in Fig. 146 at *FG* and *H* in which one of the valves is a non-return valve. Both valves are located at the highest point. Other arrangements of boiler connections with either one or two valves are shown in Fig. 146 at *I*, *J*, *K* and *L*. The necessity for avoiding dangerous water pockets should be kept in mind in all cases and where necessary to place a valve other than at the highest point, provision should be made for draining above the valve before it is opened.

Pipe Lines from Main Header. — Pipe lines from the main header should be designed to allow for expansion and to supply dry steam to the engine or other machine. A separator may be used in the header before the branch is taken off, or if the branch is long the separator may be near the engine. If a receiver separator is employed a smaller pipe may be used between the main and the separator. Several arrangements of engine piping are shown in Fig. 147. Two valves are shown, one a stop valve near the main and the other a throttle valve near the engine. Ordinarily the throttle valve is used, the stop valve being either full open or closed. A drip pipe should be placed just above the throttle to blow out the condensation which collects when the throttle valve is closed. By making the connection from the top of the main there is less danger of water getting into the engine cylinder in case it should come over from the boiler, whereas if the connection is taken from the side or bottom of the main, the engine is almost certain to be wrecked.

Auxiliary and Small Steam Lines for Engines, Pumps, etc. — The same general principles apply to auxiliary steam headers and small steam lines. They should be arranged to provide for expansion and contraction and for ample draining. The expansion can generally be cared for by allowing the pipe to turn on the threads, taking advantage of the necessary changes in direction. For draining, the pipe should slope in the direction of the steam flow and should be provided with a steam trap, drip pipes, or other means of disposing of the condensation. If the branch is taken from the side or bottom of the steam line,

STEAM PIPING

there should be provision for draining, Fig. 148, *A, B, C*. This can be avoided by taking steam from the top of the line, as in Fig. 148, *D*, which also protects the branch while it is in use. Two valves are shown in the illustrations, one a throttle valve

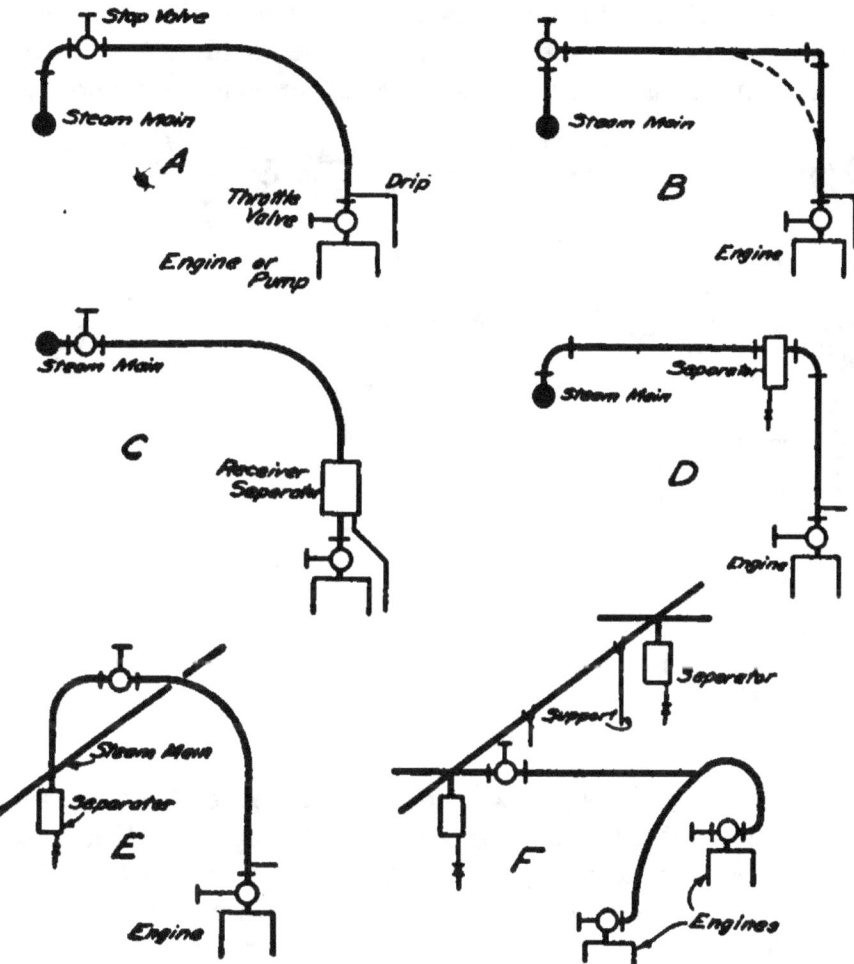

Fig. 147. Header to Engine Pipes.

near the engine or pump, and a stop valve near the steam line. When a throttling governor is used the arrangement may be as at *E*, Fig. 148. Both valves are not always necessary, but they are desirable. The throttle valve can be used to regulate the machine, and the stop valve to close off the branch entirely when necessary. The throttle valve should of course be of the globe or angle pattern. The arrangement of piping when the different

156 A HANDBOOK ON PIPING

forms of valves and regulating devices are used is taken up in connection with the description of the devices.

Steam Loop. — The steam loop is an arrangement of piping for returning condensed steam to a boiler by gravity, as shown in Fig. 149. The water of condensation is carried up the riser along with steam, then into the horizontal pipe where the steam condenses, and flows down the drop leg. When sufficient water has collected in the drop leg, the increase in pressure will open the

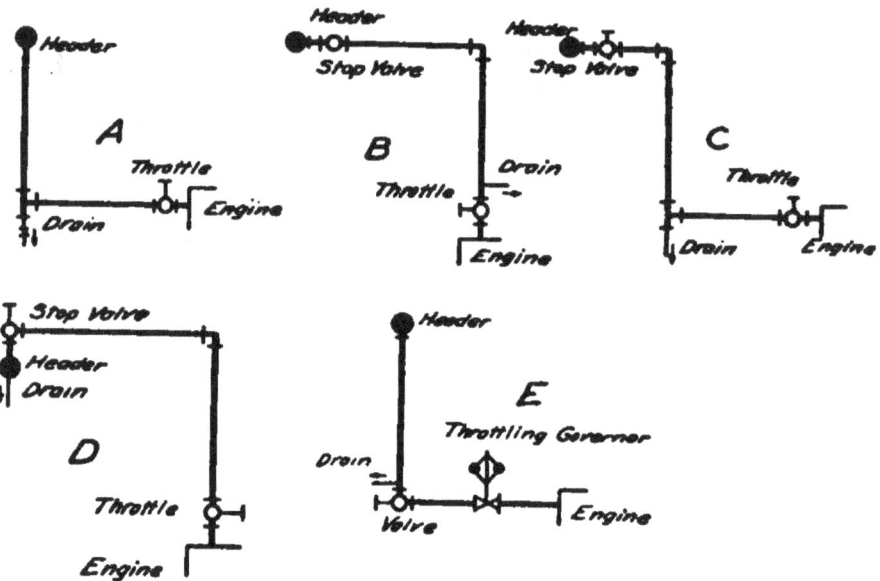

Fig. 148. Branch Pipes.

check valve and the water will flow into the boiler. This operation is repeated automatically as the drop leg fills. The head or pressure in the drop leg must at all times be greater than that in the riser in order to keep the loop in operation. The level of the water when the two pipes are balanced may be about one half way up the drop leg. The drop leg may be from 30 to 50 feet long, depending upon the loss in pressure between the boiler and drop leg and friction of piping and check valve.

Injector Piping. — The general arrangement of piping for an injector is shown in Fig. 150 at *A*. The steam pipe should be taken from as high a point as possible and directly from the boiler. A globe valve should be placed at a convenient point in the steam pipe. The suction pipe should be as short and direct as possible, sometimes a size larger pipe than the injector connection is desirable. A foot valve may be necessary on a long lift. The globe valve is placed close to the injector. The discharge pipe should be the size of the injector outlet or larger, and should contain a check valve, placed at a considerable distance from the injector.

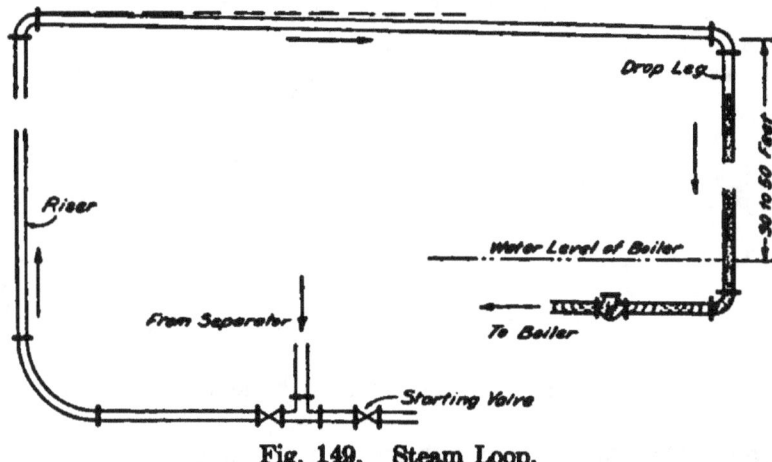

Fig. 149. Steam Loop.

When the water is not lifted the suction pipe should contain two valves, one close to the injector and one far away from it, Fig. 150, at *B*.

Live Steam Feed Water Purifier. — The Hoppes live steam feed water purifier is shown in Fig. 151. It consists of a cylindrical steel shell, within which are located a number of trough-shaped

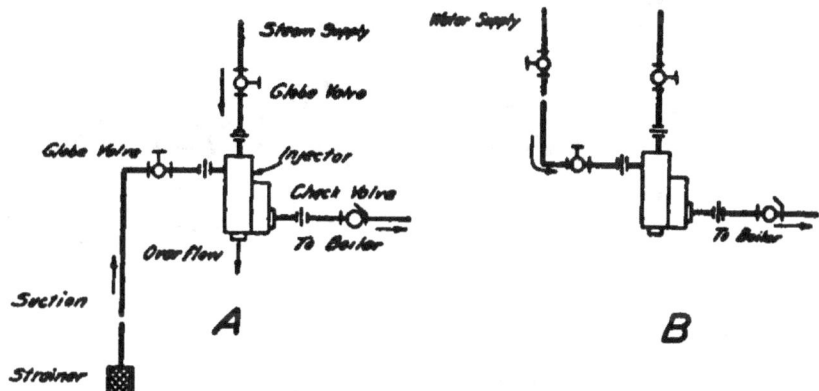

Fig. 150. Injector Piping.

pans. The pans are made of hard sheet steel with malleable iron ends.

The water enters the purifier through pipe *C* and overflows the sides of the pans and follows the under surfaces in a thin film to the lowest point and in direct contact with the steam. The solids in solution are precipitated and adhere to the bottoms of

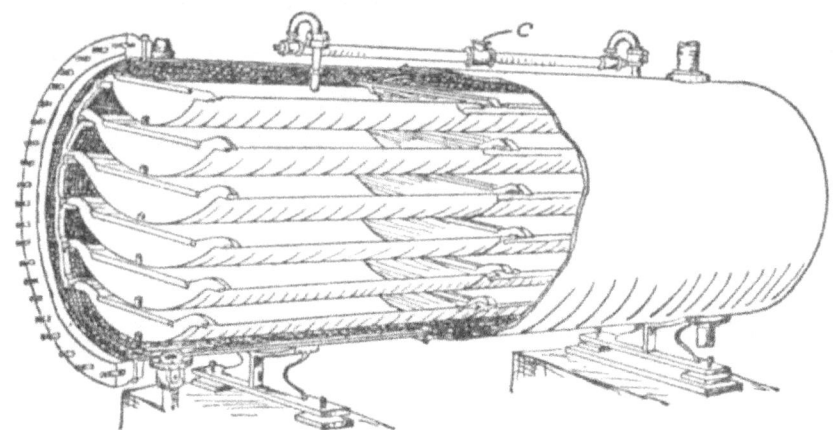

Fig. 151. Live Steam Purifier.

the pans, while those in suspension are retained in the troughs of the pans.

Method of Piping Purifier. — The method of piping a live steam purifier is indicated in Fig. 152. It is generally best to

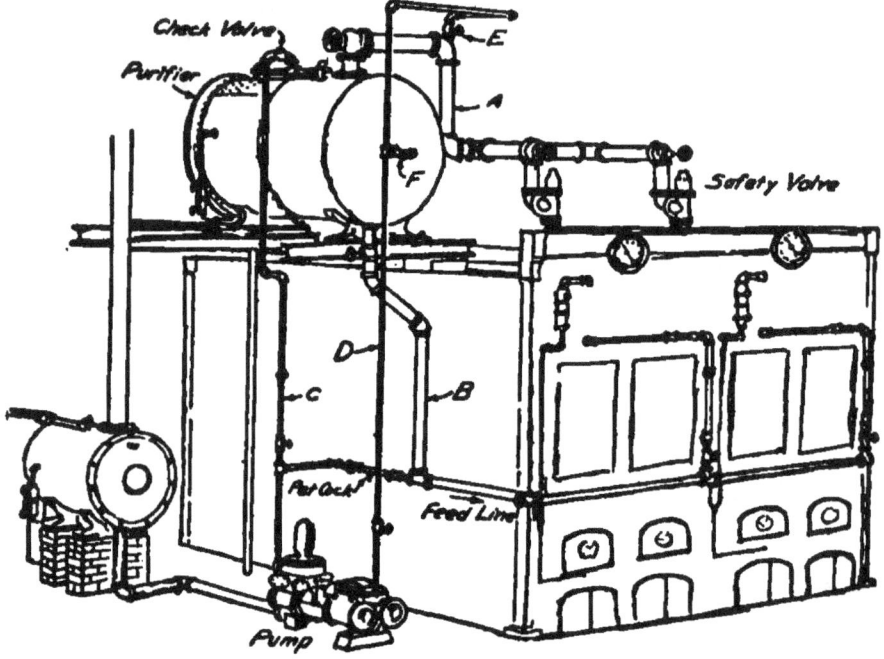

Fig. 152. Live Steam Purifier Piping.

STEAM PIPING

supply live steam to the heater by an independent pipe A in order to be sure of sufficient pressure to allow the water to flow to the boilers by gravity. To cause such a flow, the bottom of the purifier should be placed two or more feet above the water level in the boiler. The feed pipe B from the purifier is connected to the feed line. The pipe C from the pump supplies the feed water to the purifier. This pipe can be used as a direct feed to the boilers by closing the proper valves.

Steam for the pump is supplied by the pipe D. When the purifier is in operation the valve E should be closed and the valve F opened to allow circulation.

Water Column Piping. — A water column is a hollow casting, tapped for three gage cocks, two water gage connections, and for connections to the steam and water spaces of the boiler as shown in Fig. 153. The object of the column is to show the height of water in the boiler. For this

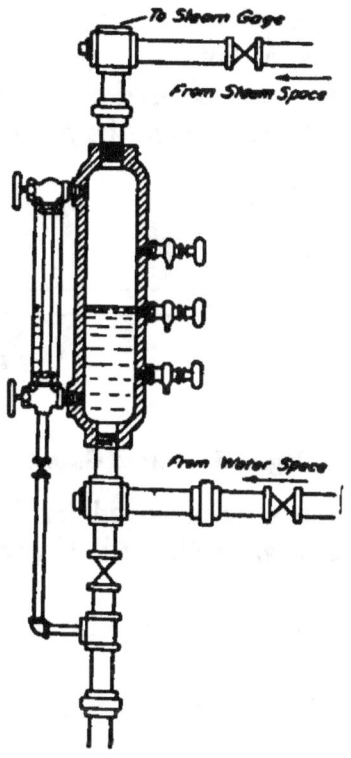

Fig. 153. Water Column Piping.

reason the steam connection should be taken from well above the water level and the water connection well below it. These connections should be made with tees or crosses with plugs instead of elbows. By removing the plugs the connections may be thoroughly cleaned. Extra heavy wrought pipe may be used, but brass pipe of iron pipe size is much better. For small water columns one inch pipe is used, but 1¼ inch is a more usual diameter for all sizes. Valves may be placed in the boiler connections as shown, but should be arranged to indicate plainly when they are closed. In some places such valves are prohibited. The steam

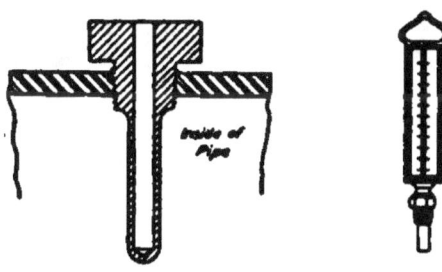

Fig. 154. Thermometer Well.

gage may be piped as indicated, but no other connection should be made from the water column piping.

The Placing of Thermometers in Pipes. — There are many occasions where it is desirable to ascertain the temperature of the medium passing through a pipe. For this purpose thermometers may be used by inserting a thermometer well in the pipe line. The well should be partly filled with oil before inserting the thermometer. The arrangement is indicated in Fig. 154. For permanent locations thermometers are made with a well as part of the casing so that they can be screwed into place. The well should be made of close composition brass and be closed either with a bit of waste or arranged for a screw cap so that water can be kept out when the well is not in use.

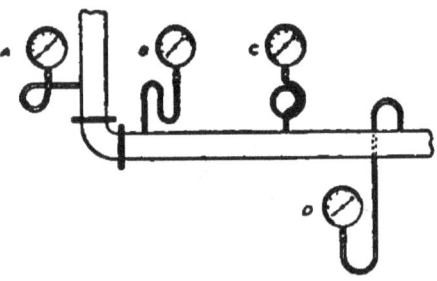

Fig. 155. Steam Gage Locations.

Steam Gages. — The location of gages for steam or water should have careful attention to insure correct readings. With steam gages some arrangement should be made for maintaining water between the gage and the steam, Fig. 155. For this purpose a goose neck may be used, or the gage may be placed below the steam line. When placed as at D a correction should be made for the head of water. The dial hand may be set to make the proper allowance. The location of water pressure gages should receive the same attention in order to avoid erroneous readings.

CHAPTER IX

DRIP AND BLOW-OFF PIPING

Drainage. — Steam piping should be arranged so as to avoid the possibility of condensed steam gathering in pockets and there becoming a source of danger. A slug of water picked up from such a pocket and carried along by a change in velocity of the steam can cause a great deal of damage by its impact with valves, fittings, etc. For this reason efficient drainage must be provided. The slope of horizontal pipes should be at least 1 inch in 10 feet, and in the direction of steam flow. The steam main should be drained from the bottom. The supply pipes should slope from the header to separators, and the engine supply should be taken from the top of the separator. The water from steam main drips can be gathered in a receiver and automatically pumped back to the boilers. The essentials of a properly designed system are provision for drainage when the pipes are full of steam under pressure but not flowing, and the care of all condensation when it is flowing. When a change in size of pipe is necessary, eccentric fittings may be used to keep the bottoms of the lines on the same level. The location of valves should receive careful attention. They should be placed so that the valve body will not form a water pocket. A gate valve with the spindle pointing downward is such a case. Valves should be placed at the high points in the line or ample drain pipe provided to care for the water which collects. When a valve is in a vertical pipe line there must be a drain pipe tapped in immediately above it.

Separators. — Separators are made for the purpose of separating water from steam, or oil from steam. In cases where priming exists in the boilers or where the steam lines are long, water collects in the piping and may be very destructive if carried into the engine cylinder. Further, the presence of moisture in the steam results in loss of economy in the operation of the engine. To remove the water, steam separators of various designs may be used. Baffle plates or changes in direction may be employed in the design of a separator. When steam is to be condensed and returned to the boilers, separators may be placed in the exhaust pipe to remove the oil and water. The principle of operation is the same as for steam separators. A steam separator should be placed as near the engine as possible. The water from the separator may be blown out or may be taken care of automatically by a steam trap.

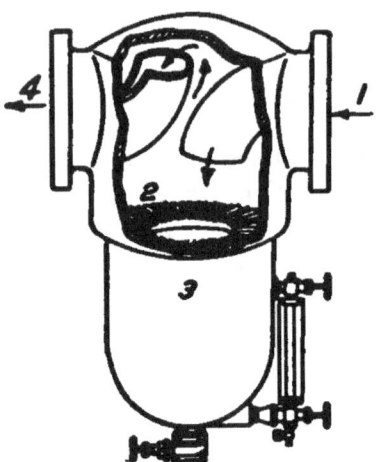

Fig. 156. Pittsburg Separator.

The construction of the Pittsburg separator is shown in Fig. 156. The steam enters at *1* and is turned downward so that it strikes the ribbed annular surface *2* where the oil and water is caught and runs off to the collecting chamber *3*. The steam leaves by the opening *4* near the top.

The construction of the Cochrane separator is shown in Fig. 157, where A is the exterior, B is cross section, and C a longitudinal section. The steam enters at *1* and impinges against a baffle plate *2* having vertical ribs, where the oil or water adheres. This oil or water is directed

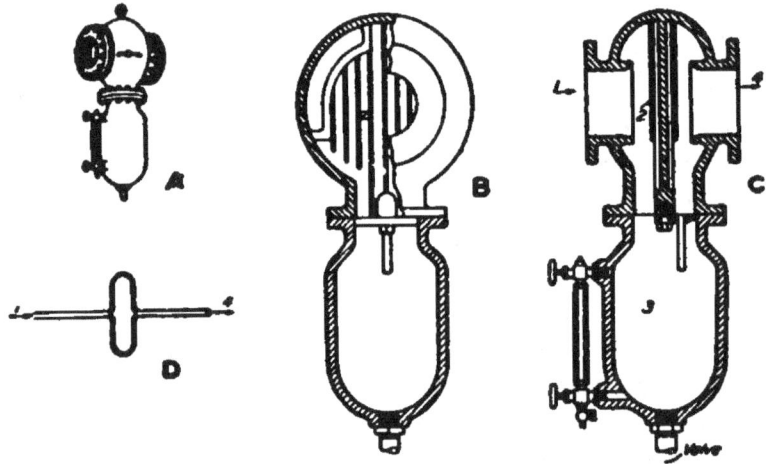

Fig. 157. Cochrane Separator.

to the collecting well *3*. The steam turns to the side of the baffle and leaves the separator at *4*. The path of the steam is indicated at D, Fig. 157.

DRIP AND BLOW-OFF PIPING

The Hoppes steam separator is shown in Fig. 158. Steam enters at the top and plunges downward, the moisture in the steam impinges on the surface of the water in the bottom and is caught and retained. From here it is drawn off through the drain pipe shown. Any entrained moisture creeping along the sides of the separator is intercepted by the troughs, which are partly filled with water and surround both inlet and outlet.

Drip Pockets. — To drain long horizontal pipes properly drip pockets, Fig. 159, should be provided every 75 to 100 feet. In general the drip pocket opening should be the full size of the pipe, as the water is likely to be carried over small openings. From the drip pocket a drain connection is made with a steam trap.

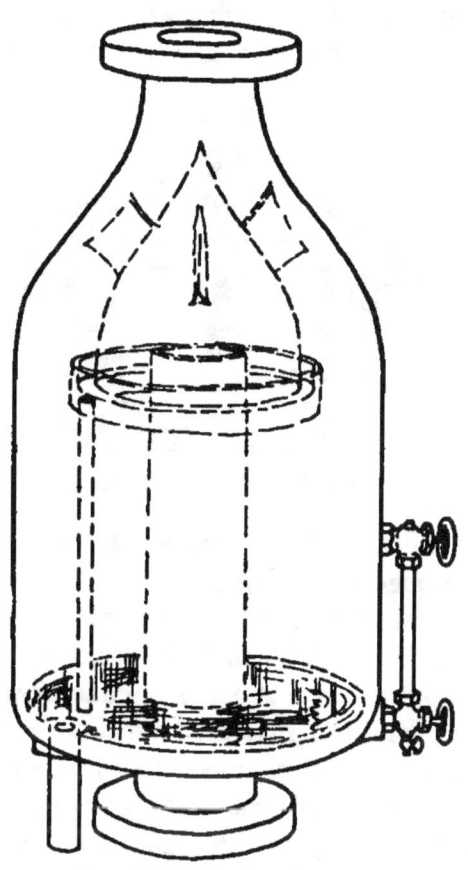

Fig. 158. Hoppes Separator.

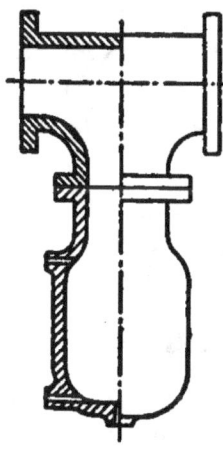

Fig. 159. Drip Pocket.

Steam Traps. — A steam trap is an apparatus made to dispose of the condensed steam from a piping system. The drip pipes from the system are run to the trap which discharges the water without allowing the steam to escape. When this discharge is against atmospheric pressure, as into a hot well or sewer, the trap is called a discharge or NON-RETURN trap. When the hot water is discharged back into the boiler the trap is called a direct RETURN trap. Direct return traps must be located above the boiler. There are numerous forms of steam traps, only a few of which will be described.

The Walworth trap shown in Fig. 160, is operated by a floating bucket. The condensation flows in at *1* around the bucket *2* until it overflows into the bucket and sinks it, uncovering the opening in the spindle at *3*. This allows the water to be driven out at *4*.

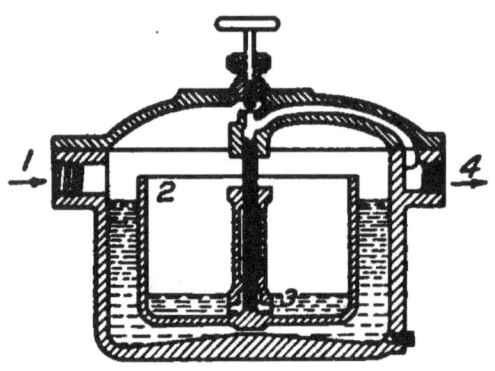

Fig. 160. Bucket Trap.

The McDaniels trap shown in Fig. 161, is operated by a float. The condensation flows in at *1* until it raises the spherical float *2* which opens the valve *3* and allows the water to be forced out at *4*. When the water is drained the float falls and closes the valve *3*. The screw *5* may be used to open the valve *3*.

The Farnsworth trap shown at *A*, Fig. 162, operates by a tilting tank. The tank is composed of a partition and two pipes making two unequal size chambers. The vertical pipe *1* receives condensation into the long chamber *2* until its weight overbalances the full short chamber *3* and opens the valve *4* and the condensation is passed from the bottom of the long chamber through the diagonal pipe *5* into the top of the short chamber, and from the bottom of the short chamber out through the valve, which remains full opened so long as condensation is coming through the vertical pipe, and when the lines or apparatus are finally drained and the long end nearly emptied, the full, short chamber over-balances it and closes the valve against the double seal of water.

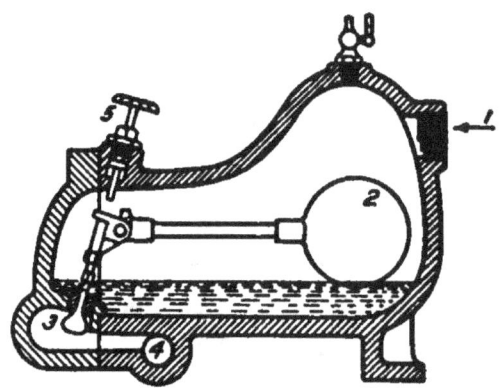

Fig. 161. McDaniels Trap.

Copper flexible hose is used to avoid packed trunnion joints as shown at *B*. This allows the trap to be arranged to operate as a non-return trap or as a return trap.

DRIP AND BLOW-OFF PIPING

The Cranetilt trap is made for a variety of uses, the direct return trap being shown in Fig. 163. Condensation enters the trap through the inlet check valve *1* and passes through the divided trunnion tee into the tank *2*. When the tank fills, the weight of water causes it to drop to the bottom of the yoke *3*. This opens the steam valve *4* and closes the vent valve *5* allowing pressure to enter the steam valve and through the inner pipe

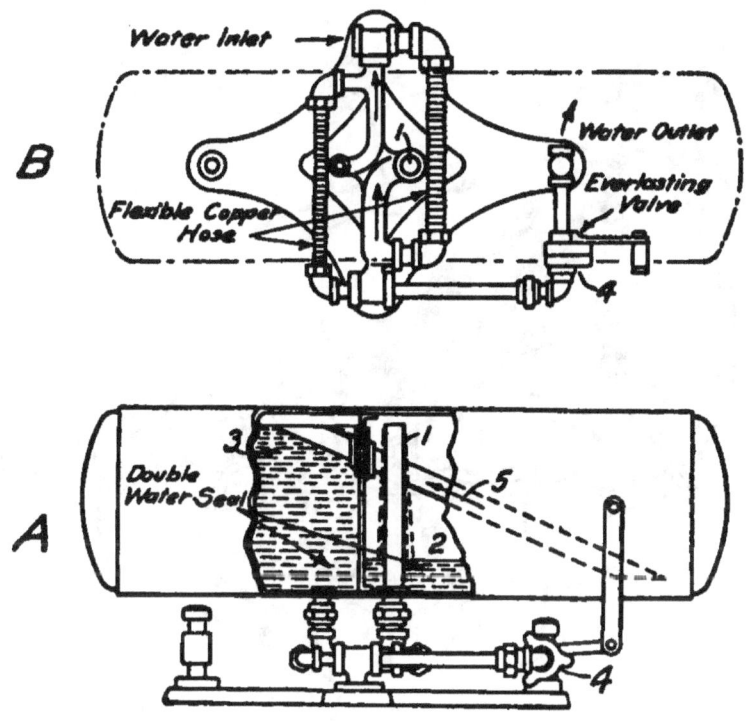

Fig. 162. Farnsworth Trap.

into the space above the water, closing the inlet check valve. The pressure is now the same in the trap as in the boiler and as the trap is above the boiler, the water flows into the boiler by gravity. After sufficient water has left the tank, the counter-weight *6* brings it back into the filling position, this action closing the steam valve and opening the vent valve which allows the pressure in the tank to equal or fall below that in the return lines. The directions given for setting and connecting up a Cranetilt direct return trap are as follows:

"Place the trap at least four feet above the water level of the boiler. The trap does not necessarily have to be directly above the boiler as shown in Fig. 164. In some cases there is not room enough between the top of the boiler and ceiling of the boiler room, in which case the trap can be placed on the floor above, either over or adjoining the boiler house. There are three essential points in connecting up a direct return trap. 1st, the pipe marked 'Discharge to Boiler' in Fig. 164, should have a strong pitch away from the trap along the horizontal line A. 2d, this discharge pipe must not be connected into any pump or injector

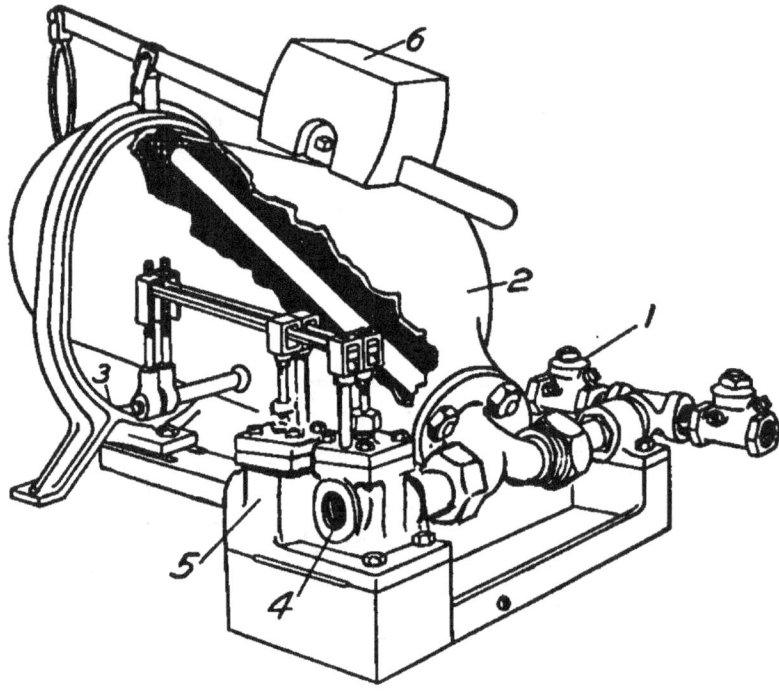

Fig. 163. Cranetilt Trap.

line feeding the boiler but connected independently of other feed lines. 3d, the pipe marked 'steam' must be connected to the boiler at a point where the initial boiler pressure will be secured. Do not connect this line to any steam line connected to an engine, pump or injector. Where the pressure in the receiver is not sufficient to elevate the condensation to the trap a Cranetilt lifting trap should be located below the receiver and connected to it. The lifting trap will elevate the condensation through the pipe marked 'discharge to trap' to the direct return trap. As the amount of water which the trap handles at each operation will vary only slightly, the attachment of a revolution counter, recording each operation, will give a close average."

DRIP AND BLOW-OFF PIPING

Drips from Steam Cylinders. — Steam cylinders should be provided with drain connections at both ends, and may have

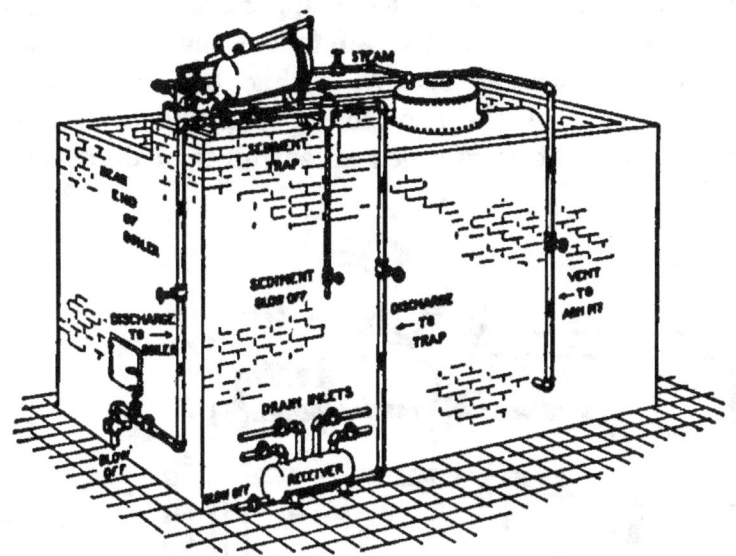

Fig. 164. Setting for Direct Return Trap.

automatic relief valves or hand operated valves. If hand operated the valves should always be opened before starting the engine or pump. For small engines and pumps pet cocks screwed directly into the cylinder are frequently used. In most cases, however, it is preferable to pipe the drips to a drain. The size of pipe should be sufficiently large to care for condensation and not be easily stopped up. The arrangement of drips for steam and exhaust pipes from cylinders is treated in connection with the piping of engines.

Drainage Fittings. — Condensed steam should be drained by gravity whenever possible. When conditions are such that this cannot be

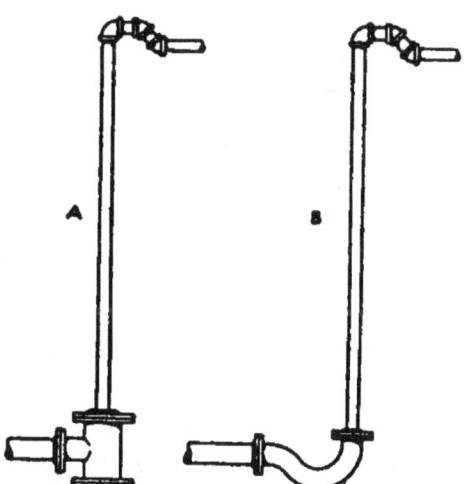

Fig. 165. Drainage Fittings.

done, lifts as shown in Fig. 165 at A and B may be employed. The principle of operation is the same. The water of condensation gathers in a pocket until it closes the pipe and the steam pressure forces it up the riser in slugs. Condensation may be lifted by high vacuum by the same apparatus. The diameter of the riser should be about one half to one third that of the horizontal pipe. The arrangement at *A*, Fig. 165, is composed of a tee with the ends of the riser lower than the horizontal pipe. The fitting shown at *B* is called an entrainer or drainage fitting.

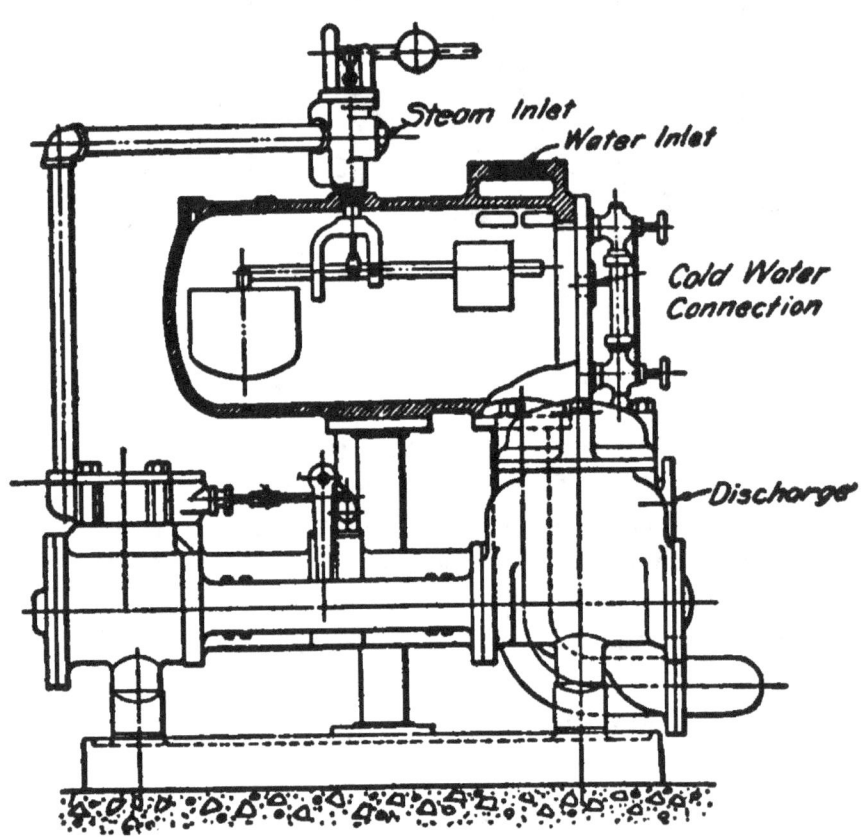

Fig. 166. Automatic Pump and Receiver.

Automatic Pump and Receiver. — A combination of receiver and pump, Fig. 166 provides an effective arrangement for draining radiators, steam jackets, steam coils and heaters. The water of condensation enters at the top of the receiver. A float in the receiver maintains a constant water level and regulates the pump. When used for boiler feed, cold water may be admitted directly to the receiver to make up for losses or in case of excessively high temperature. As with other apparatus the piping should be

DRIP AND BLOW-OFF PIPING

arranged with a by-pass so that the receiver may be cut out when repairs are necessary.

Blow-off Piping. — The size of the blow-off pipe from a boiler may be from one to $2\frac{1}{2}$ inches in diameter. The wrought-pipe

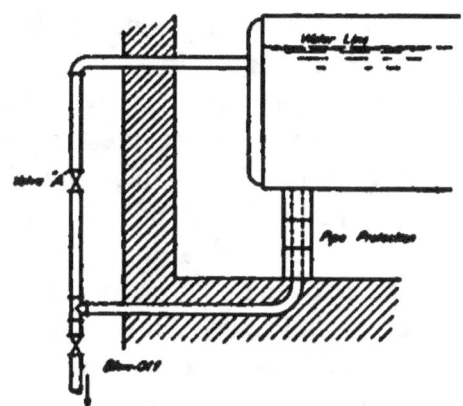

Fig. 167. Blow-off Piping.

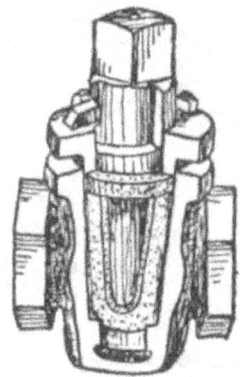

Fig. 168. Asbestos Packed Cock.

fittings and valves should be extra heavy as made for 250 pounds pressure. When the pipe passes through the combustion chamber it should be protected from the hot gases by magnesia, asbestos, or fire brick, or it may be enclosed in a larger pipe of either tile or cast iron. Further protection may be had by arranging the piping as shown in Fig. 167, which allows a continuous circulation to be maintained. The valve A is closed before the blow-off valve is opened. Ample provision should be made for the movement of the pipe due to expansion.

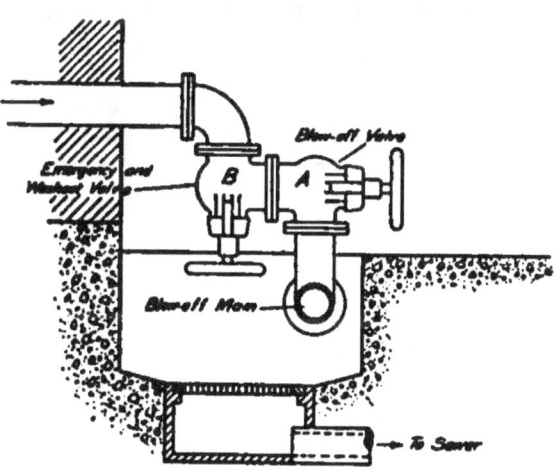

Fig. 169. Arrangement of Blow-off Valves.

The blow-off pipe should be arranged so that the discharge is visible, otherwise failure to close the valves may not be noticed until the boiler is damaged. Any leaks can be seen and attended to. It is well also to have the blow-offs from different boilers independent of each other.

170 A HANDBOOK ON PIPING

In general, two valves should be used in each blow-off pipe, one a valve and the other a cock. The asbestos-packed cock, shown in Fig. 168, is very commonly used. The valve should be placed nearest to the boiler. The cock should be opened before the valve and closed after the valve. In this way it will be possible to keep it tight for a longer time, as it will not be under pressure when operated. Blow-off valves are often made up in pairs, Fig. 169. When cleaning the boiler the valve A is kept closed and the valve B opened and its bonnet removed, allowing the wash water and scale to fall upon the floor which is connected to a drain. The blow-off valve being closed the boiler cleaner is safe from any back blow from the pipe.

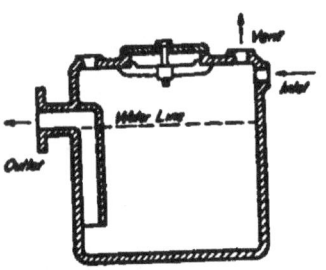

Fig. 170. Cast Iron Blow-off Tank.

Blow-off water and steam can sometimes be discharged into the open. When it must be cared for by a sewer it should first be allowed to cool in some form of sump or tank, Figs. 170, 171 and 172, as the heat from the blow-off water will crack drain tile, allowing it to be crushed and so become closed. Aside from this, the escape of steam through street openings from the sewers is objectionable. Blow-off tanks are made of cast-iron, steel or wrought-iron plate, and brick or concrete. A blow-off tank should have a vapor pipe carried up through the roof to carry off

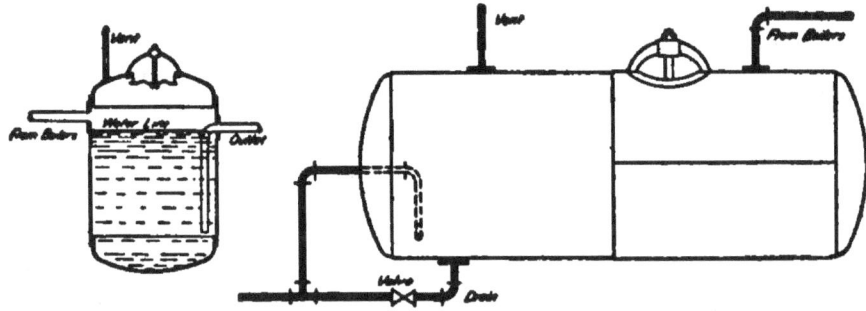

Fig. 171. Steel Blow-off Tanks.

the steam and vapor, a manhole for cleaning, and if there is a chance for the accumulation of pressure, a safety valve should be added. The outlet of the blow-off pipe should be above the water line, as otherwise condensation in the pipe will create a vacuum

DRIP AND BLOW-OFF PIPING

and draw water from the tank or sump back into the pipe, often with injurious results. It is well to have a partition between the inlet and outlet parts of a sump or tank, or other arrangements to form a trap and so prevent steam from entering the tile drain.

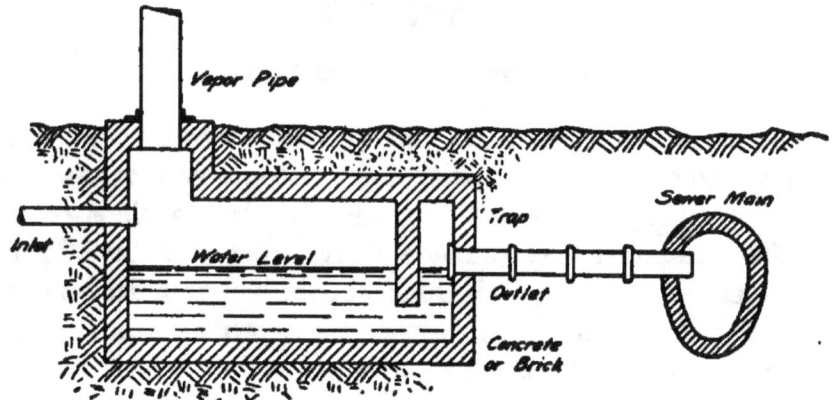

Fig. 172. Concrete Sump.

A small cast-iron blow-off tank is shown in Fig. 170. Riveted steel-plate tanks may be of cylindrical form with bumped heads, Fig. 171. For a common blow-off from a number of boilers a concrete sump may be constructed, similar to Fig. 172.

CHAPTER X

EXHAUST PIPING AND CONDENSERS

Exhaust Piping. — Exhaust piping to the atmosphere can be made of light-weight pipe, with light fittings and valves. For small sizes wrought pipe or tubing may be used, while sizes 24 to 30 inches and larger may be made of riveted steel plates. Riveted pipe, when less than $1/4$ inch thick, should be galvanized to assist in keeping the joints tight; thicker plate can be calked. Large fittings may be made of steel plates riveted together, and with

Fig. 173. Riveted Steel Plate Fittings.

cast-iron flanges, Fig. 173. Where flat surfaces occur they should be braced to withstand pressure from the outside, as there may be a vacuum due to condensation.

Exhaust lines should be designed carefully as to drainage. They should pitch in the same direction as the flow. An exhaust-steam separator may be used to separate the oil and water from the steam, if it is to be used for heating and other purposes. The drip from the oil separator or from a drip pocket may be discharged through a loop, as shown in Figs. 174 and 175. The drop leg should be long enough so that a possible slight vacuum in the exhaust pipe will not raise the water from it. This will require

EXHAUST PIPING AND CONDENSERS

from three to six feet. Fig. 175 shows a loop applied to a tee at the bottom of a vertical exhaust pipe.

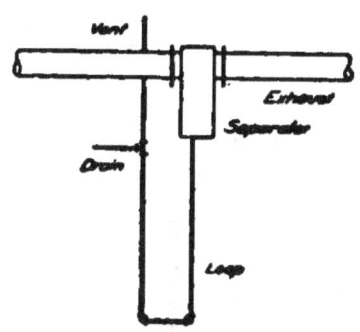

Fig. 174. Method of Draining Separator in Exhaust Pipe.

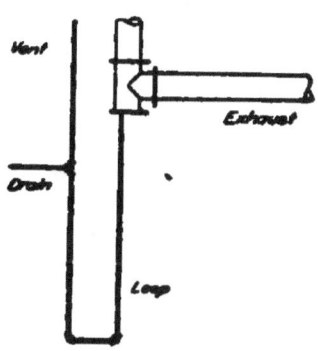

Fig. 175. Method of Draining Vertical Exhaust Pipe.

When exhaust steam is used for heating purposes, a single vertical pipe may be used for atmospheric exhaust and as a heating riser for an overhead system. This arrangement is shown in Fig. 176, where a heating main connection is made well up the pipe. A smaller heating connection is shown near the base of the riser. The back-pressure valve is placed just above the upper heating main. A drip pipe is taken from the exhaust head, and another from just above the back-pressure valve.

Exhaust from Small Engines, Pumps, etc. — The arrangement of small exhaust pipes is indicated in Figs. 177 and 178. The piping as shown enters an exhaust main. The valve near the main serves to close the branch for repairs, or when working on the machine, and also to keep the branch from filling with condensation when not in use. The exhaust pipe should be drained from its lowest point, and should slope from the highest point to the main.

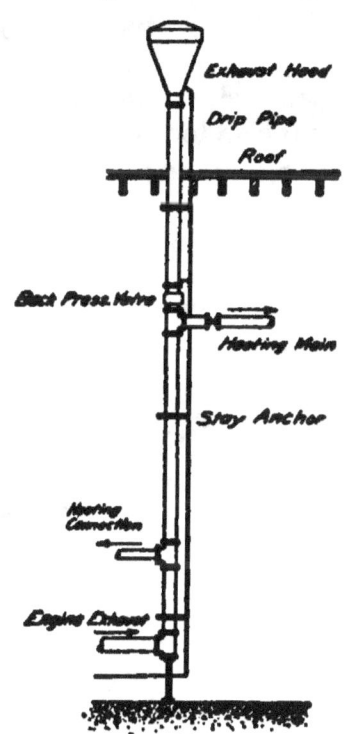

Fig. 176. Combination Exhaust Pipe and Heating Riser.

If changes in direction occur, it may be necessary to provide more than one drip pipe. As in all steam lines, pockets where water may collect should be avoided. Either an angle stop valve or a gate valve should be used, as the passage of the

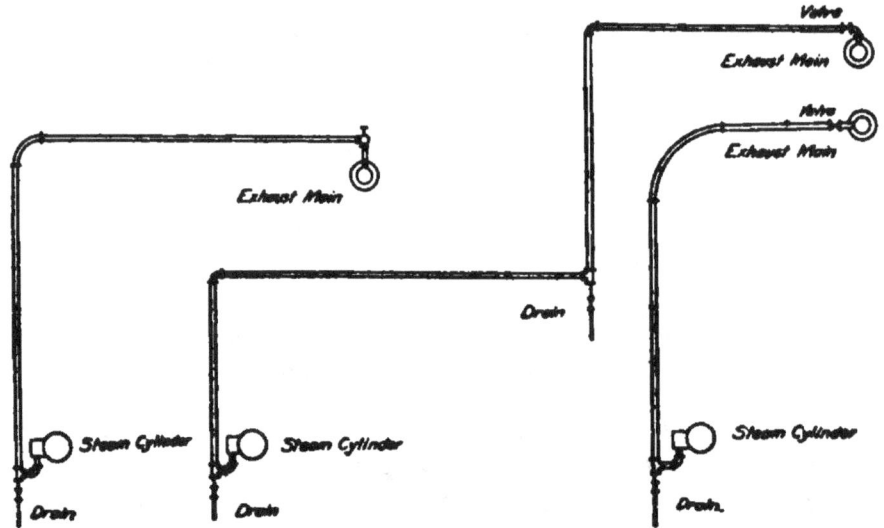

Fig. 177. Connections to Exhaust Main.

exhaust steam should not be restricted. Care should be taken to provide for movement due to expansion, and also to allow for making up the pipe, lack of exact alignment, etc.

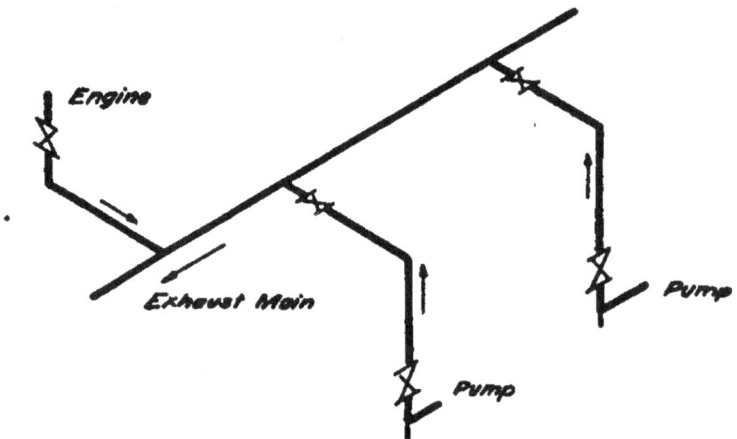

Fig. 178. Connections to Exhaust Main.

Exhaust Heads. — When steam is exhausted from an engine to the atmosphere, some form of exhaust head should be used to catch and return the oil and condensation. Such heads may be made of galvanized iron or cast iron, and should be so designed as not to cause back pressure. The Swartwout cast-iron exhaust head is shown in Fig. 179. The steam passes through a long helix, from which it emerges with a whirling motion. The particles of water which have been thrown into the outer surface of the tube are flung forward. The extension of the tube forms an annular chamber in which the water collects, and from which it is removed through the drip.

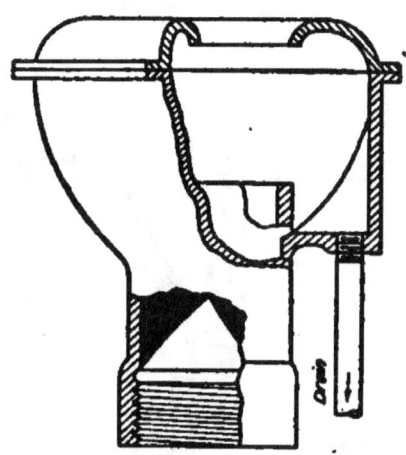

Fig. 179. Swartwout Exhaust Head.

The Hoppes cast iron exhaust head is shown in Fig. 180. When the steam enters the head it expands gradually into a large chamber several times the area of the pipe, while the particles of oil and water in the centre of the current are separated by impinging on the cone, and those on the outer edges strike against and adhere to the side of the separating chamber. A trough partly filled with water surrounds the outlet and prevents creeping. This trough is connected with the drain by the pipe shown.

Vacuum Exhaust Pipes. — Vacuum exhaust pipes should be as short and direct as possible, but with ample provision for expansion. Various forms of expansion joints for exhaust lines are used, three of which are shown in Figs. 181, 182 and 183. The first is of corrugated copper, the second is a steel plate or diaphragm, and the third is the Badger copper expansion joint. Because of the range in temperature, a considerable movement should be allowed for. The increase in volume of steam at low pressures, makes it desirable to have such pipes of as large a diameter as possible. If long pipes must be used, the diameter should be increased. The material of which the pipes are made may be cast iron, wrought

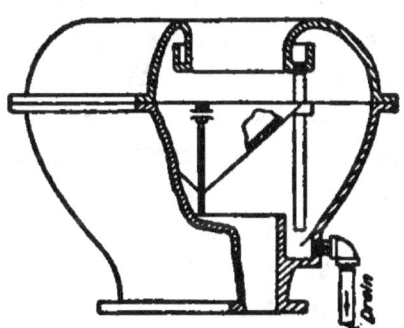

Fig. 180. Hoppes Exhaust Head.

iron or steel, or riveted steel. It is of course essential that the pipe and its joints be tight, as a very small leak will seriously affect the vacuum. Gate valves should be used where valves are required in vacuum lines in order to keep the full opening of

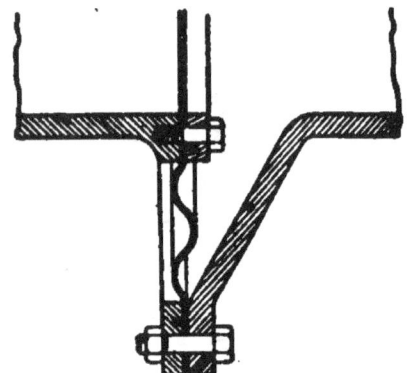

Fig. 181. Corrugated Copper Expansion Joint.

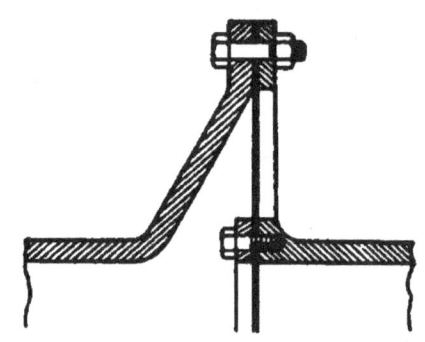

Fig. 182. Steel Plate Expansion Joint.

the pipe. Light-weight valves should be avoided if tightness is to be maintained. Automatic relief valves should be provided in the vacuum exhaust pipe from engines or turbines to condensers. In case of an accident to the condenser, the pressure will build up in the exhaust pipe and open the relief valve, thus allowing the steam to exhaust to the atmosphere.

Classes of Condensers. — Condensers are used to reduce the back pressure in steam cylinders and turbines by condensing the steam and producing a vacuum. The different classes of condensers are the surface condenser, jet condenser, and barometric or siphon condenser. These may be further subdivided, as the surface condenser may be either vertical or horizontal, and with the steam either inside or outside the tubes; the jet condenser is made in a variety of forms, and the barometric condenser may be either the nozzle or spray type.

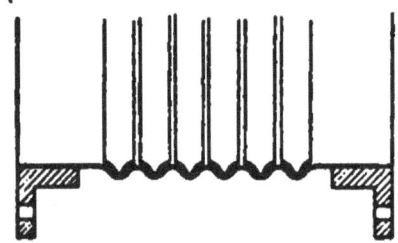

Fig. 183. Badger Copper Expansion Joint.

Surface Condensers. — Essentially a surface condenser, Fig. 184 comprises a shell or casing containing tubes through which cooling water is circulated. The tubes range in size from $5/8$ inch to one inch in diameter, and generally are made of brass or composition. An air pump is connected to the condensing chamber to remove the condensed steam and air. Often the condenser is mounted above the air and circulating pumps. The exhaust steam from the engine, upon entering the condenser, comes into contact with the external surface of the tubes which are kept cool by the water circulated through them. This condenses the steam which falls to the bottom of the casing, and is removed by the air pump and may be used over again in the boilers. The air pump should always be placed on a lower level than the condenser so that the condensation can flow to it by gravity. The shell of

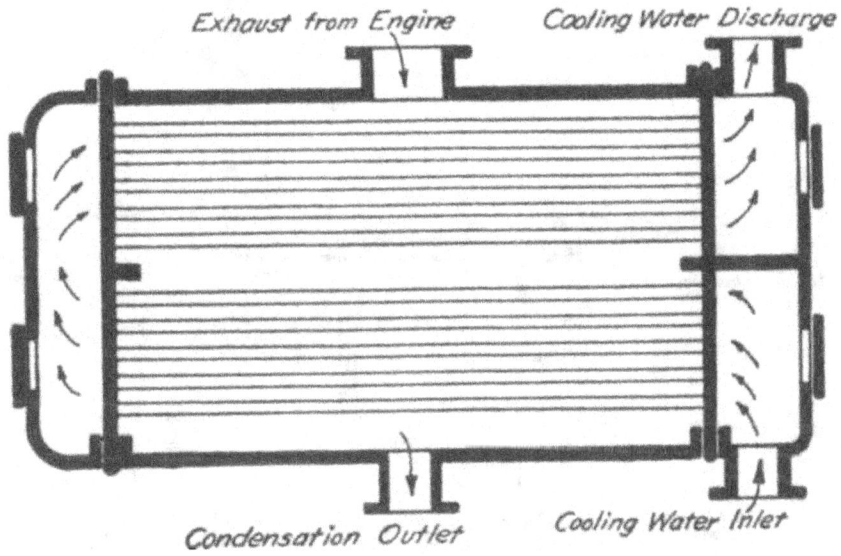

Fig. 184. Surface Condenser.

the condenser may be either circular or rectangular in cross section and may be set with the tubes either vertical or horizontal.

Piping for Surface Condenser. — Arrangements of piping for surface condensers are shown in Figs. 185, 186 and 187. The condenser should be placed near the engine or turbine in order to make short piping and so avoid joints with possible leaks tending to destroy the vacuum. As shown in Fig. 185, the condenser is mounted above the air and circulating pumps, which are placed in the basement below the engine. The valve in the pipe to the condenser is for the purpose of cutting out the condenser when exhausting to the atmosphere. The atmospheric relief valve is placed in the atmospheric exhaust pipe and automatically opens should the condenser lose its vacuum due to failure of either of the pumps. The branch containing the relief valve may be one size smaller than the main exhaust pipe.

A steam turbine is sometimes mounted directly on the condenser with the air and circulating pumps separate. Any form of pump may be used for circulating the cold water. The arrangement of a steam turbine in connection with a surface condenser and dry vacuum pump is shown in Fig. 186. The higher the vacuum is, the larger will be the volume of steam and air to be handled, and larger pipes should be provided. In order to main-

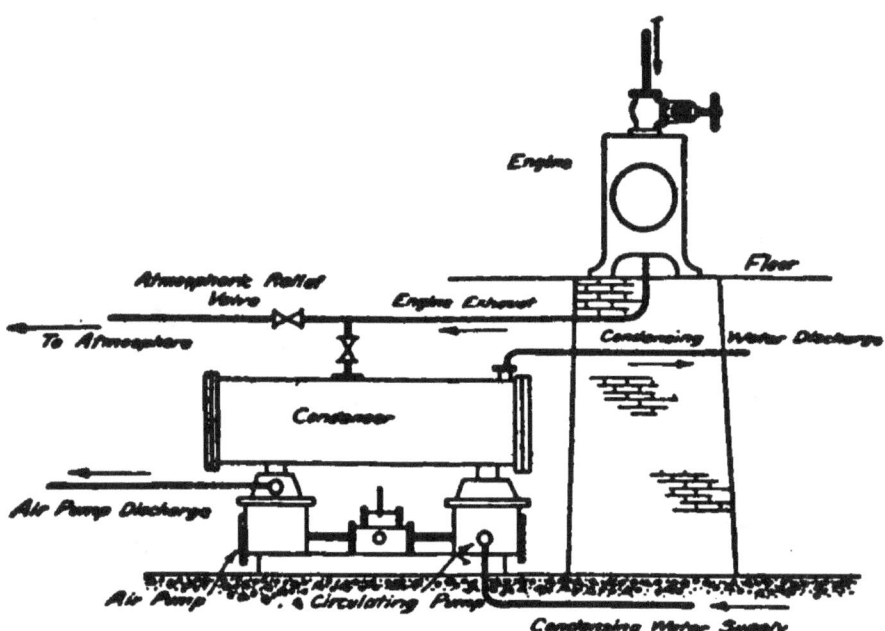

Fig. 185. Steam Engine and Surface Condenser.

tain a high vacuum without an excessively large condensing surface and air pump, it is usual to provide a separate pump for removing the air, called a dry vacuum pump, which is piped from the air space of the condenser. Such pumps generally run at a high speed, and have small clearance spaces, and so should not be expected to handle water without disastrous results. To this end the piping from the condenser should slope toward the pump and should not rise at any point or have any places for condensation to collect. By this arrangement, such condensation as occurs will pass to the pump in a vaporous condition and be safely handled. Where several condensers or pumps are used in connection with a vacuum main, the pipes from the condenser should

EXHAUST PIPING AND CONDENSERS

enter at the top of the main, and those to the pumps should be taken from the bottom of the main. The air discharge from the

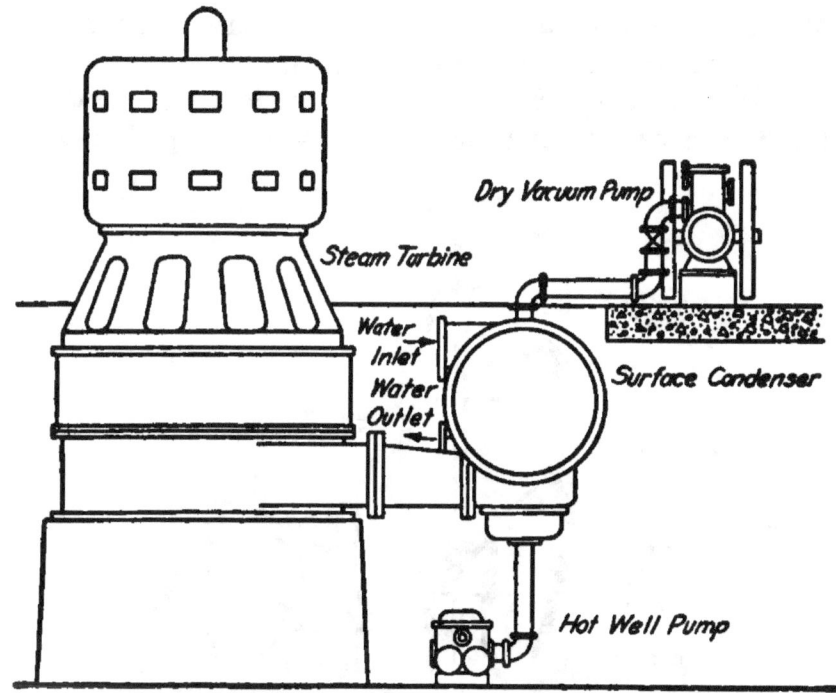

Fig. 186. Steam Turbine and Surface Condenser.

vacuum pump may be discharged through a pipe to the atmosphere, or into the atmospheric exhaust pipe of the engine.

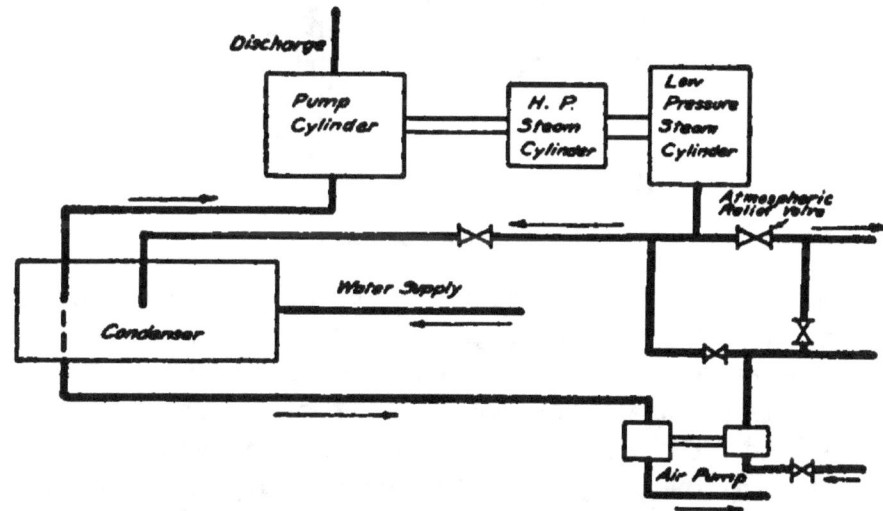

Fig. 187. Steam Pump and Surface Condenser.

A compound pumping engine may be piped with a surface condenser as shown in Fig. 187. The water supply to the pump is taken through the condenser. Sometimes a surface condenser is placed in the discharge pipe. In either case a separate air pump is necessary to remove the condensate.

Jet Condensers. — The form of jet condenser illustrated in Fig. 188 is made by the Blake and Knowles Pump Works. As

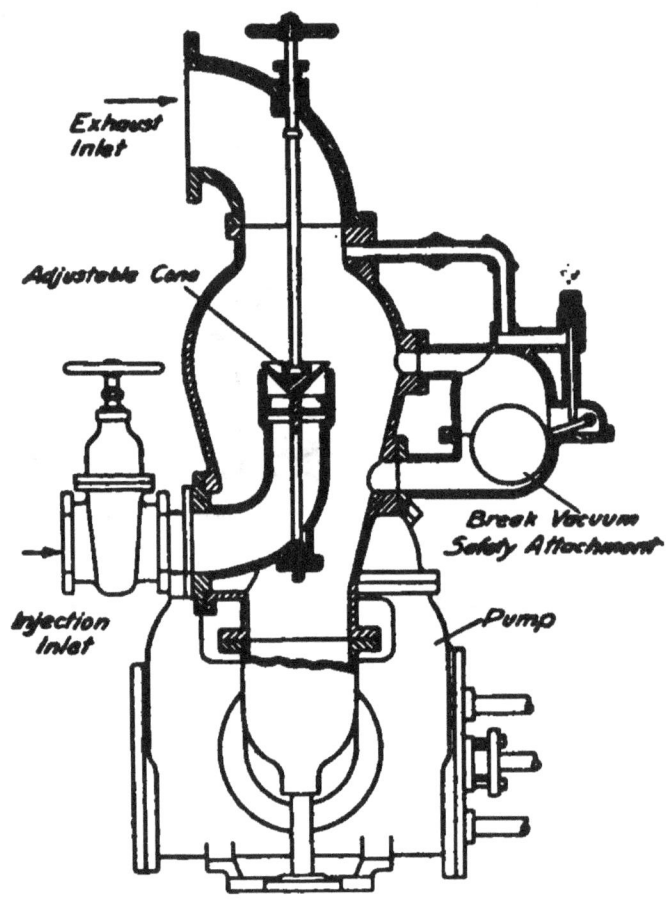

Fig. 188. Jet Condenser.

shown, it consists of a condensing cone in which the exhaust steam and cooling water mingle, and a pump for removing the resulting air and water. The exhaust steam enters at the top and meets the injection water which enters through a cone or spray head. The cooling water enters due to the partial vacuum produced by the pump. The vacuum breaking device is automatic in its operation. Its purpose is to prevent the water rising above the proper level in the condenser. In case the pump should stop for any reason, the water will continue to rise until

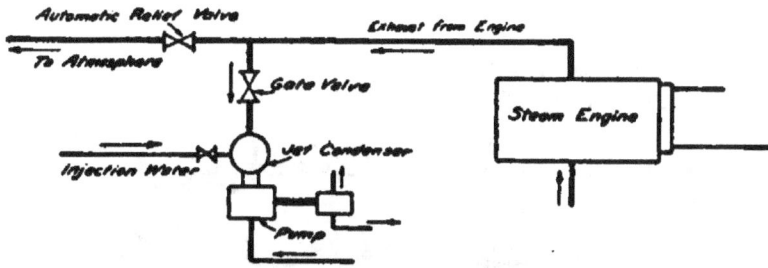

Fig. 189. Steam Engine and Jet Condenser.

it lifts the float. This float will then open a relief valve which admits air, and so destroys the vacuum. In this way the water is prevented from rising in the exhaust pipe, and possibly wrecking the engine. When the pump is again started, the float falls to its normal position, and the condenser is put into operation.

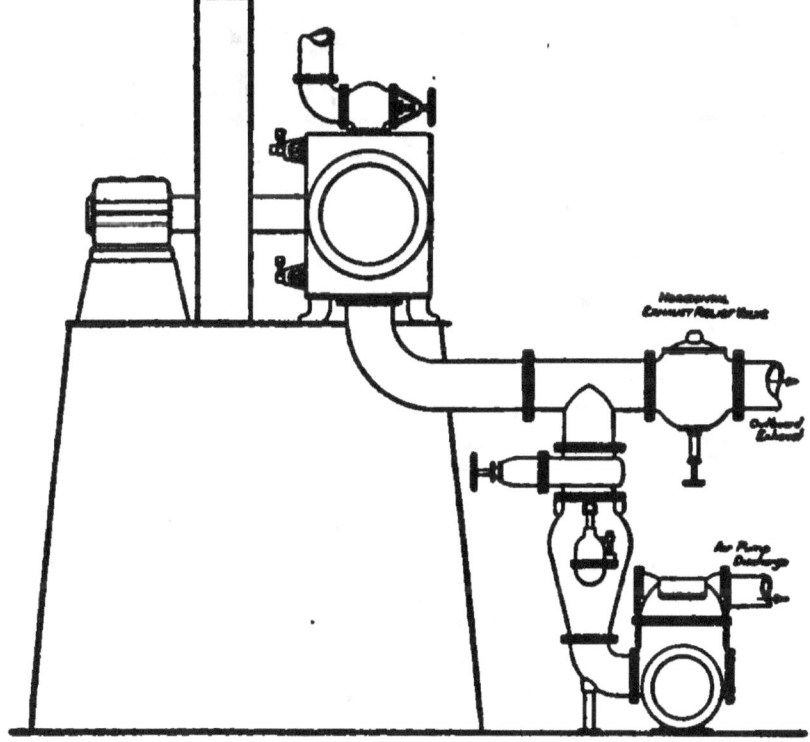

Fig. 190. Steam Engine and Jet Condenser — Elevation.

Jet Condenser Piping. — Arrangements of piping for jet condensers are shown in Figs. 189, 190, 191 and 192. A steam engine

182 A HANDBOOK ON PIPING

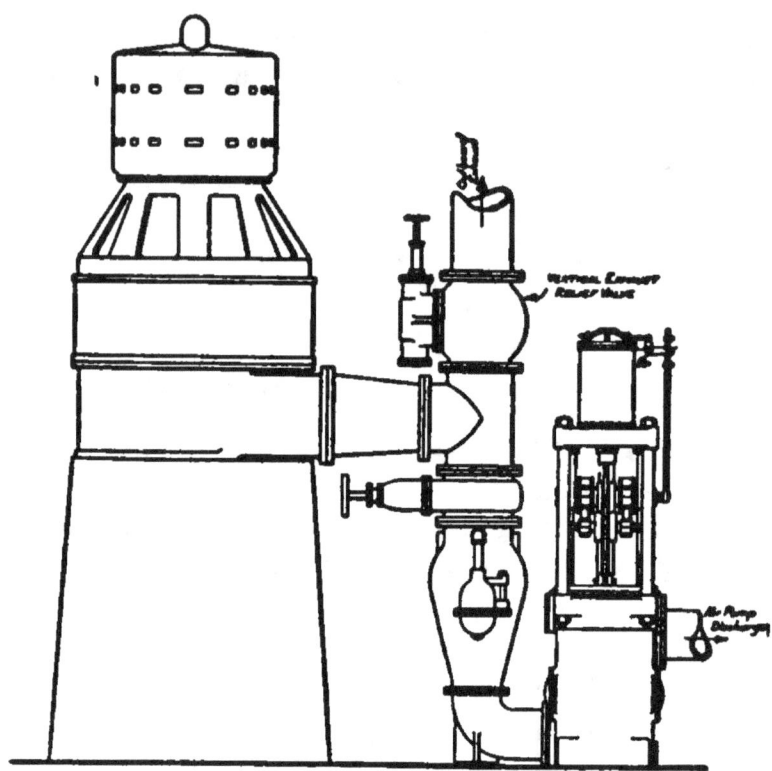

Fig. 191. Steam Turbine, Jet Condenser Single Acting Air Pump.

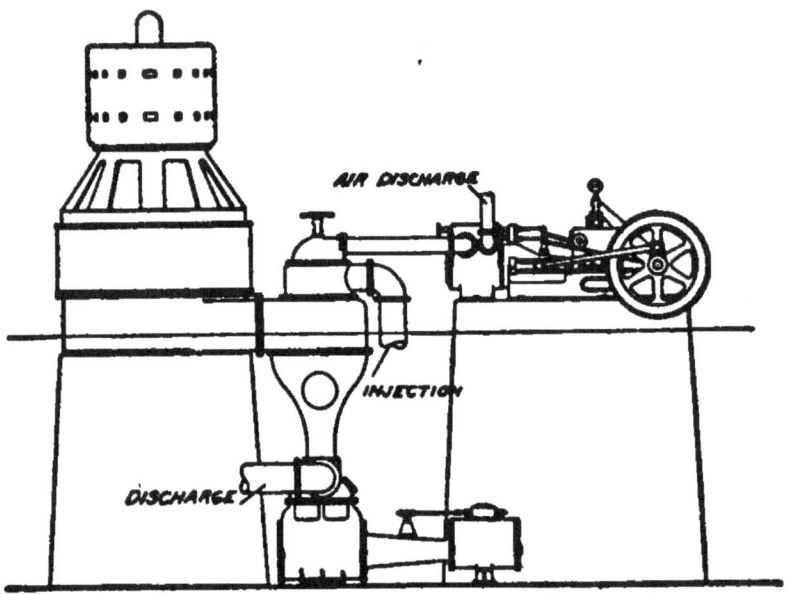

Fig. 192. Steam Turbine, Jet Condenser and Dry Vacuum Pump.

EXHAUST PIPING AND CONDENSERS

and jet condenser are shown in Fig. 189. An automatic relief valve is provided in the atmospheric exhaust pipe, and a gate valve in the pipe to the condenser. The exhaust from the pump may be arranged to connect into the condenser, to a feed water heater, or to an atmospheric exhaust pipe. Several arrangements of Blake condensing apparatus are given in Figs. 190, 191 and 192. A steam engine piped to a jet condenser and double acting vacuum

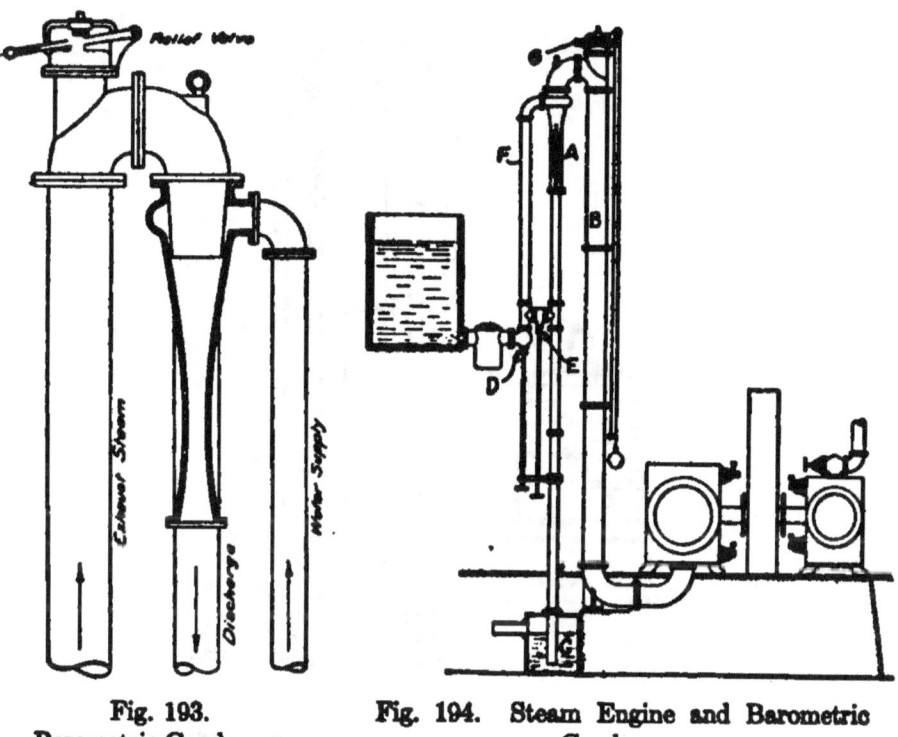

Fig. 193. Barometric Condenser.

Fig. 194. Steam Engine and Barometric Condenser.

pump is indicated in Fig. 190, a steam turbine, jet condenser, and single acting twin beam air pump in Fig. 191, and a steam turbine arranged with a jet condenser, air pump, and rotative dry vacuum pump in Fig. 192.

Barometric Condenser. — One form of barometric condenser is shown in Fig. 193. The exhaust steam enters through a conical nozzle, and passes down into a combining tube. The cooling water enters at the side and around the steam nozzle, then passes downward in a thin film or sheet. The steam meeting this water is condensed and is carried down the discharge or tail pipe with the water, thus creating a vacuum in the pipe above. The tapering form of the condenser is such that the water acquires a high velocity in passing the contraction and is enabled to carry the entrained air and vapors along with the condensed steam. This apparatus requires no pumps if the water supply pipe has less than 20 feet lift. If over 20 feet, a pump must be used to supply the cooling water. It is necessary, however, to have the condenser at a height of about 34 feet above the hot well in which

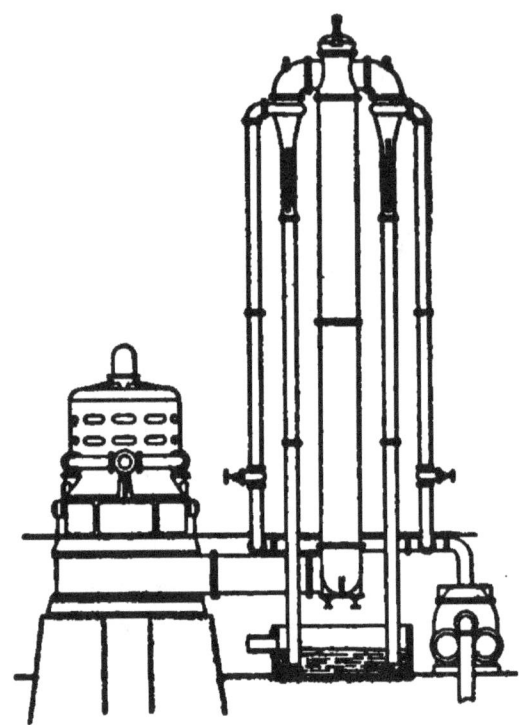

Fig. 195. Steam Turbine and Barometric Condenser.

the lower end of the discharge pipe is immersed. As the atmosphere will not support a column of water at such a height, the cooling water supplied will fall through the condenser and discharge pipe.

Piping for Barometric Condenser. — When the source of cooling water is a tank, or is otherwise located not more than 20 feet below the condenser, the water may be siphoned by the condenser. If the water must be raised it may be pumped direct to the condenser, or to a supply tank. Both methods are indicated in Fig. 194, which shows the arrangement of piping, with the parts lettered as follows: *A* is the condenser; *B* is the exhaust pipe from the engine; *C* is the hot well; *D* is the injection water valve;

E is the starting valve; F is the water supply pipe; and G is the atmospheric relief valve. Either an open or closed relief valve may be used, according as to whether the exhaust pipe is outside or inside of a building. An arrangement of twin spirojector con-

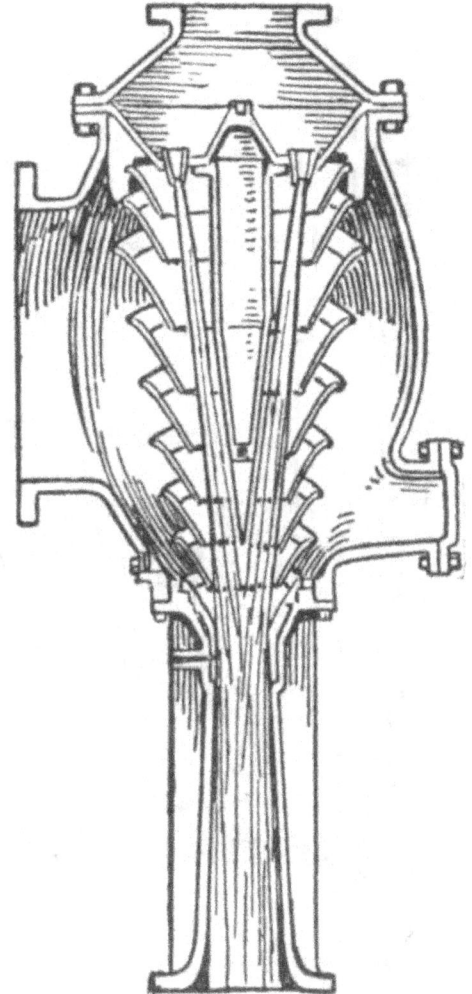

Fig. 196. Eductor Condenser.

densers is shown in Fig. 195, as recommended for units larger than 500 K.W. by the Blake-Knowles Pump Works. It is advisable as being more economical and flexible. When running under light loads, or with low temperature cooling water, one condenser may be cut out.

Multi-jet Educator Condenser. — This form of condenser is made by Schutte & Koerting Co., and is shown in Fig. 196. With this condenser no air pump is required. The cooling water enters through a number of converging jets which meet and form a single jet in the lower part of the condensing tube. Exhaust steam enters through the side connection and flows through the

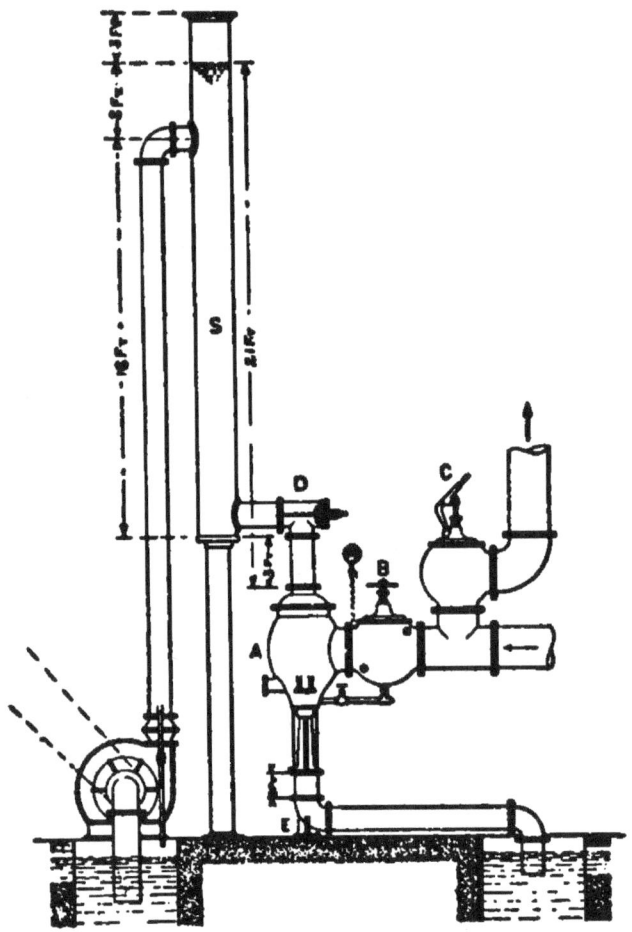

Fig. 197. Piping for Eductor Condenser.

annular passages which guide it so that it impinges on the condensing jet. This steam is condensed and the particles of water into which it is changed are united with the water jet with which it is discharged, together with the entrained air against atmospheric pressure.

The method of piping is shown in Figs. 197 and 198, the first of these being the preferred one. Here a standpipe is used. By pumping the water up into the standpipe it is possible to get rid of the air contained in the water. If water is available with a head of 21 feet, or 9 pounds per square inch at the inlet flanges of the condenser, no pump is necessary. Instead of a standpipe the water may be delivered direct to the condenser by a pump, as shown in Fig. 198. A water check valve in the exhaust pipe

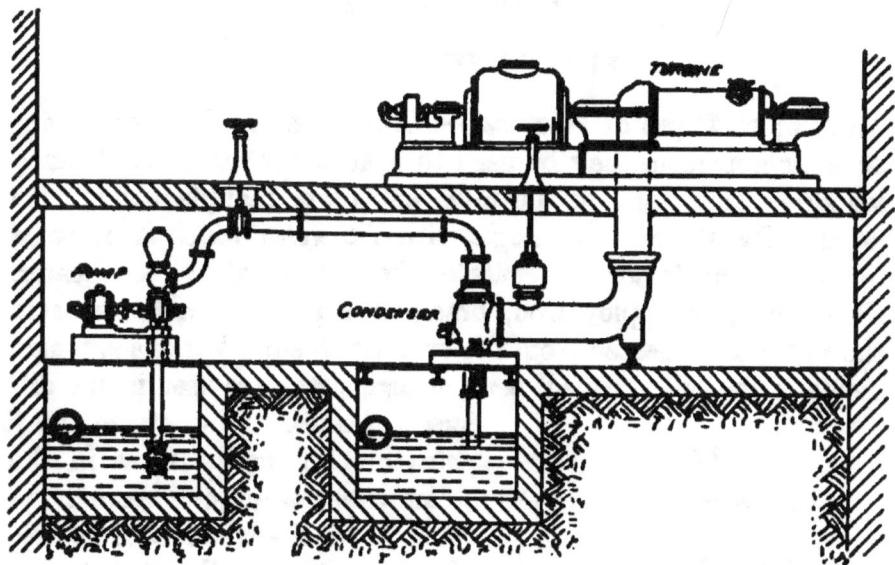

Fig. 198. Steam Turbine and Eductor Condenser.

prevents water from flowing back from the condenser to the engine, but allows the exhaust steam to pass to the condenser. A steam turbine in connection with a multi-jet condenser is illustrated in Fig. 198. In this case the water is supplied by a centrifugal pump.

CHAPTER XI

FEED WATER HEATERS

Uses and Types of Heaters. — Exhaust steam from an engine or other apparatus may be used to heat water for boiler feeding laundries, paper and textile mills and other manufacturing purposes. The steam may mingle with the water which it heats as in an open heater or be separated from it as in a closed heater. Closed heaters employ iron, brass, or copper tubes to separate the water to be heated from the exhaust steam. Various arrangements of the tubes, coiled, bent, straight, etc., are used in the different makes. The steam may pass through the tubes as in the steam tube heater, or surround the tubes, as in the water tube heater. The advantage of the closed type is that the steam does not come into contact with the feed water and so keeps oil from entering the boiler. However, if a scale forming water is used the open type is to be preferred as the scale can be formed in the heater and removed from time to time. The closed heater is under pressure and tight joints must be maintained as well as provision for expansion. All the exhaust steam may be passed through the heater or only a part, if all is not required to heat the water. Sometimes the exhaust steam is not sufficient, and provision must be made to supply live steam. In the open heater the steam mingles with the water which it heats and an oil separator should be used, either separate or as a part of the heater.

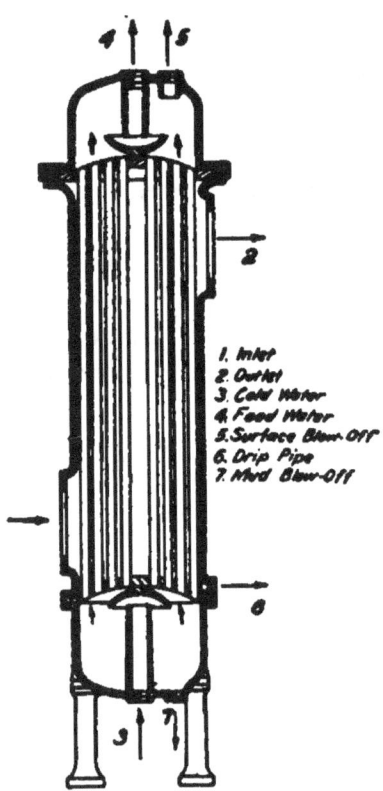

Fig. 199. Goubert Closed Heater.

1. Inlet
2. Outlet
3. Cold Water
4. Feed Water
5. Surface Blow-Off
6. Drip Pipe
7. Mud Blow-Off

FEED WATER HEATERS

Closed Feed Water Heaters. — The Goubert closed feed water heater is shown in Fig. 199, where the various connections are indicated. Most of the oil in the steam is removed and passes off with the water of condensation through the drip pipe. The cold feed water enters at the bottom and meets the deflector, which spreads it out, allowing the mud or sediment to settle before the

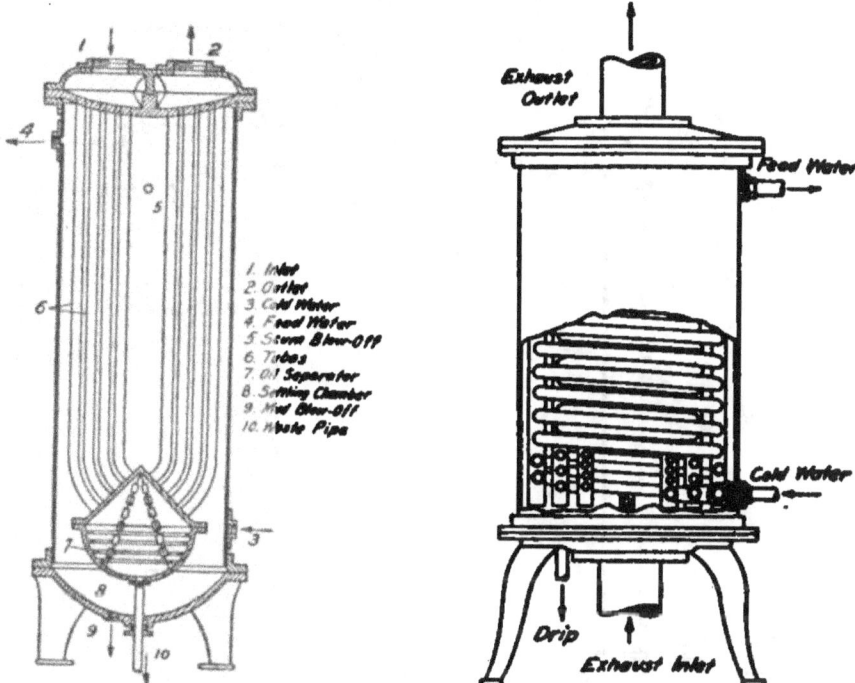

Fig. 200. Otis Closed Heater.   Fig. 201. National Closed Heater.

water passes upward through the tubes. The surface blow at the top permits the removal of scum.

The Otis heater is shown in Fig. 200. As shown by the openings, the exhaust steam enters at the top, passes down one section of tubes to an oil and water separator, and then up the other section of tubes to the outlet from which it is exhausted or used for other purposes. The cold water enters near the bottom and passes out near the top when heated.

The National heater shown in Fig. 201 consists of coils through which the water to be heated passes. The exhaust steam enters at the bottom of the shell and leaves at the top. In some forms both exhaust and live steam coils are used to maintain the required temperature.

190   A HANDBOOK ON PIPING

Closed Heater Piping. — The arrangement of piping for a closed feed water heater may be such as to allow all of the exhaust to pass through the heater, or only a part of it. This will depend upon the source of supply. If the main exhaust is used, and is

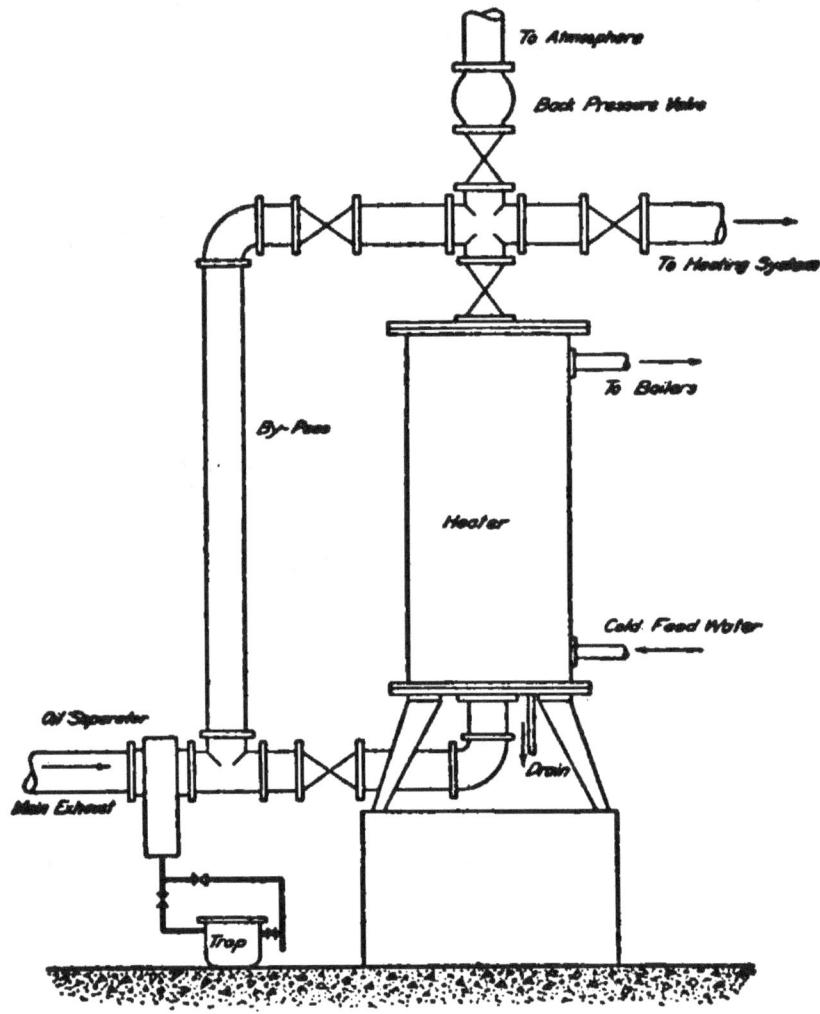

Fig. 202. Piping for Closed Heater.

more than sufficient to heat the feed water, a branch may be used to supply the heater and the extra steam used for heating or other purposes. When the main exhaust is condensed and only the exhaust from the pumps and other auxiliaries is passed into the heater, the entire amount of steam can be passed through the heater. A method of piping for a closed heater is shown in

FEED WATER HEATERS

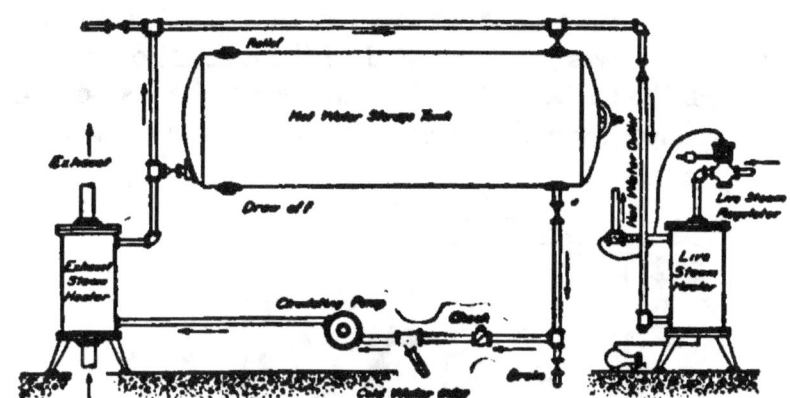

Fig. 203. Piping for Combination Exhaust and Live Steam Heaters.

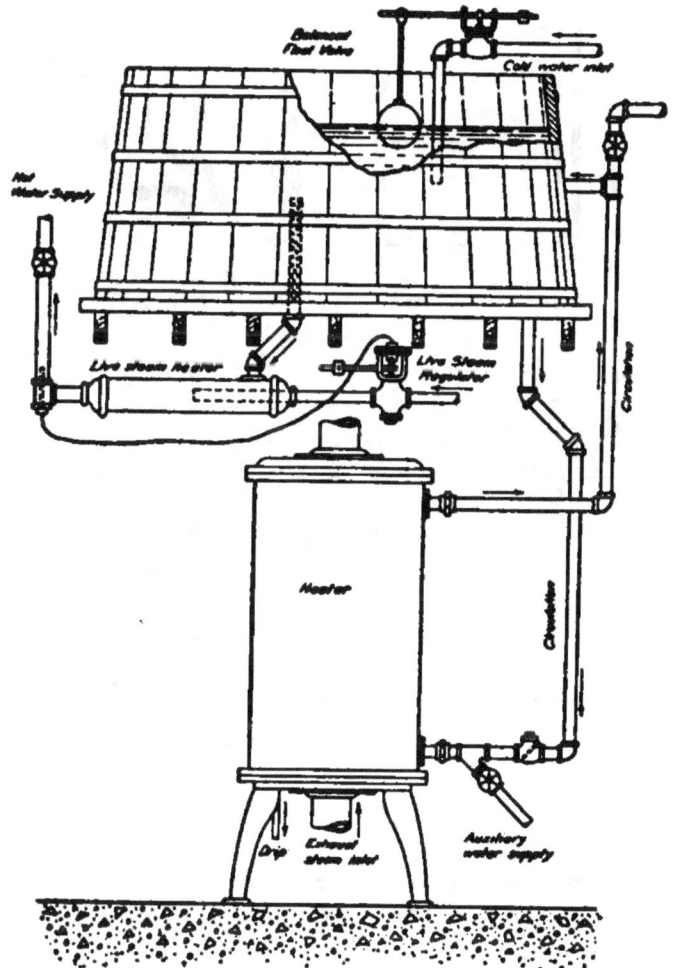

Fig. 204. Piping for Heater and Storage Tank.

Fig. 202. The by-pass is arranged so that the heater may be cut out when necessary, or to regulate the amount of steam passing through the heater. The oil separator may be placed near the heater as shown, or if the steam is from the main exhaust it may

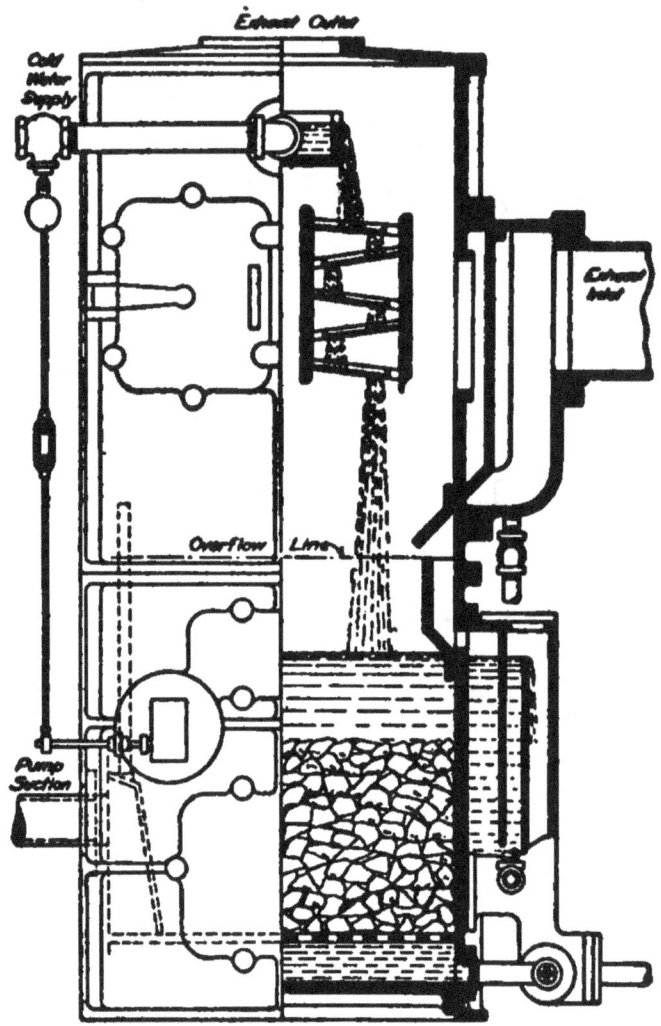

Fig. 205. The Cochrane Open Heater.

be near the engine. As shown, the trap is arranged with a by-pass for use if necessary.

The arrangements shown in Figs. 203 and 204 are from the National Pipe Bending Company's book of plans. In Fig. 203 the piping is given for using a live steam heater in connection with an exhaust heater where more or hotter water is wanted.

FEED WATER HEATERS

The piping in Fig. 204 is for a closed heater in connection with a live steam re-heater and wooden storage tank.

Open Feed Water Heaters. — As stated before, the water is heated in an open heater by direct contact with the exhaust steam. Such heaters are usually designed to combine the functions of heater, purifier, receiver, and filter. The water enters at the top of a chamber and drips down over trays while being heated by the steam. The water then passes through filtering material contained in the lower part of the chamber to the pump suction. The cold water supply is regulated by a valve with a float control. One of the advantages of an open heater is that its efficiency as a heater is not affected by conditions as to cleanliness of surfaces. The details of several forms are shown in the following figures.

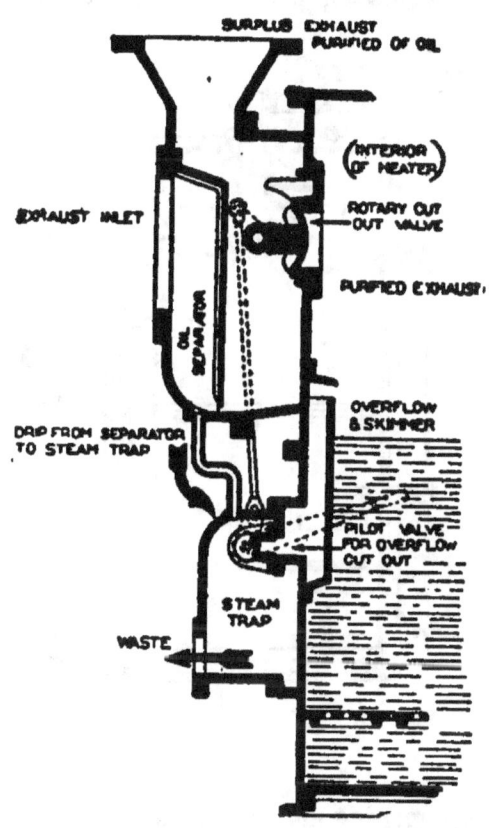

Fig. 206. Cochrane Steam-stack and Cut-out Valve.

The Cochrane heater is shown in Fig. 205. The steam enters through an oil separator forming a part of the heater, while the water enters at the top and overflows from a trough over and through a series of perforated trays, inclined first one way and then the other, each tray catching the drips from the one above. From the last tray the water falls into a settling chamber. This has a perforated false bottom for carrying a filter bed. The boiler feed pump receives its supply from the space underneath the filter bed. The body of the heater is made of cast iron and the fittings of copper and brass.

A partial section of the Cochrane steam-stack and cut-out heater is shown in Fig. 206. The steam enters through an oil separator, near the top of which is a flanged outlet for passing through the surplus exhaust steam to the heating system or atmosphere. The opening to the heater is controlled by a special valve. When this valve is open it occupies such a position that the heater has the "preference" for the steam. That is, in its open position the valve diverts a portion of the steam from the

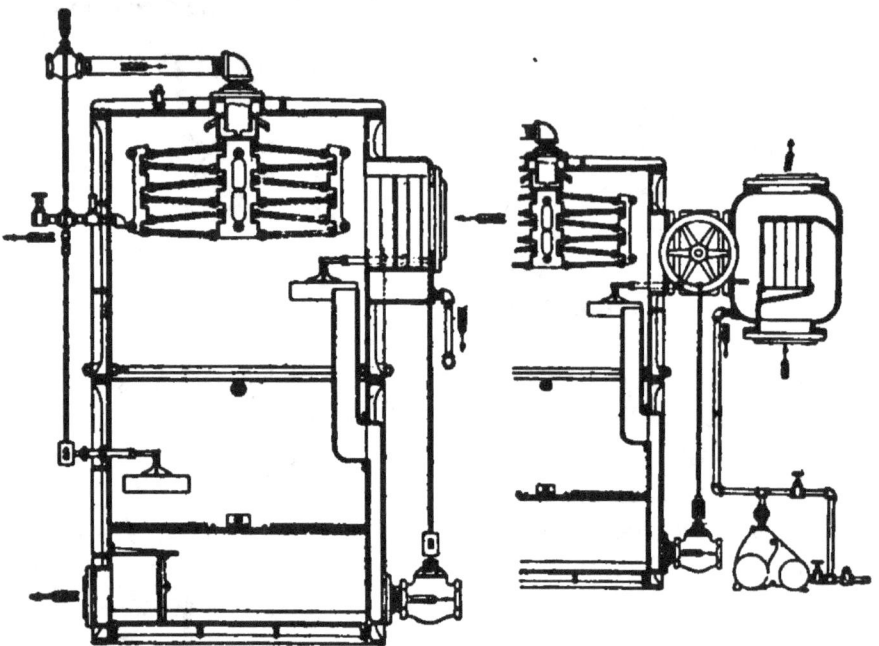

Fig. 207. Webster Feed Water Heater.

Fig. 208. Webster Feed Water Heater.

top opening and directs it into the heater, at the same time allowing surplus steam to escape through the upper opening. The valve may be closed and so cut out the heater without the necessity for extra valves and fittings for a by-pass. A vent pipe provides a means for the escape of air and gases.

The Webster feed water heater and purifier is shown in Figs. 207 and 208. Water is admitted through an automatically controlled valve and is discharged into a trough which forms a water seal. From this trough the water overflows to oppositely inclined and perforated copper trays. In this manner it mingles with the steam and becomes thoroughly heated. It then flows downward through a filter bed and to the pump suction chamber. Fig. 207 is the Standard type built on the induction principle,

FEED WATER HEATERS

with the oil separator attached to the heater shell. Fig. 208 is the preference type which is a cut-out heater using a gate valve in connection with an oil separator of sufficient size to purify all steam passing through the exhaust main to both the feed water heater and to a heating or drying system, or to low pressure turbines.

A typical installation of a Webster feed water heater for power service is shown in Fig. 209 and for a gravity return heating system in Fig. 210.

The Hoppes feed water heater and purifier is shown in Fig. 211. The steam enters through an oil separator, passes through the

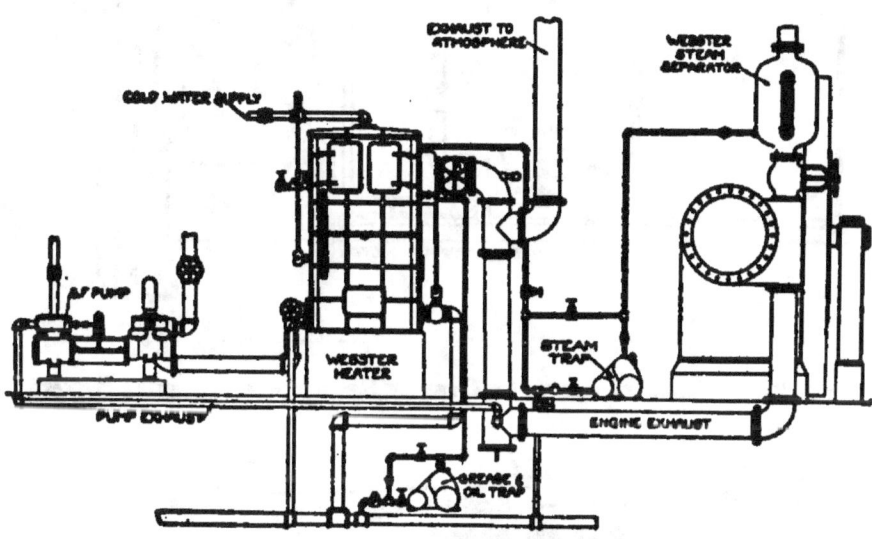

Fig. 209. Piping of Heater for Power Service.

heater and escapes by the outlet near the front end. Water is admitted through a balanced regulating valve and evenly distributed to the top pans by inside feed pipes. The water overflows the edges of the pans and follows the under side to the lowest point and drops into the next pan below until it reaches the bottom of the chamber and passes to the main pump suction through a hooded opening. The troughs of the pans provide settling chambers and so eliminate the necessity for a filter. Solids precipitated from solution are deposited and retained on the under side of the pans.

The Hoppes induction chamber shown in Fig. 212 takes the place of by-pass piping, and is described as follows:

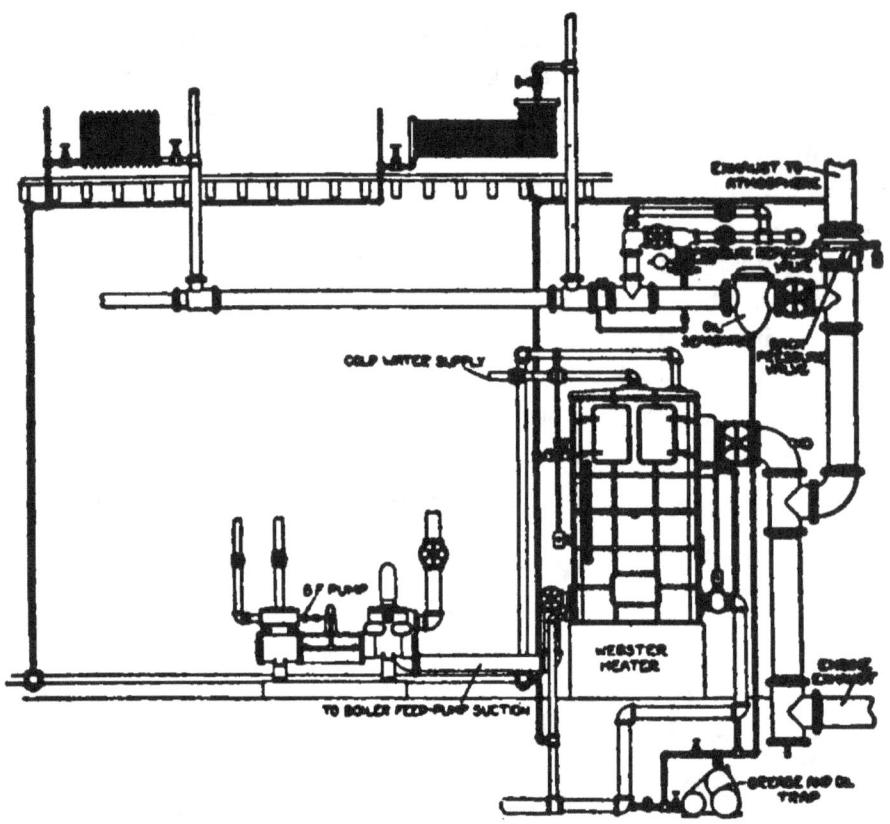

Fig. 210. Piping of Heater for Gravity Return Steam Heating System.

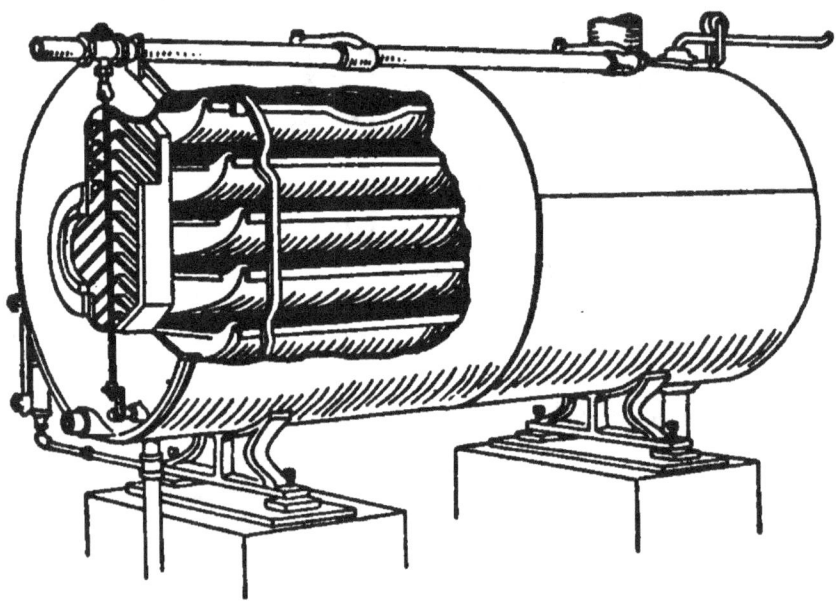

Fig. 211. Hoppes Feed Water Heater.

FEED WATER HEATERS

"This device may be used for any size of exhaust pipe in connecting Hoppes heaters of any type to the exhaust line, effecting a saving proportional to the size of the exhaust pipe by doing away with large and expensive valves and fittings.

"The steam enters the chamber at the bottom, and flowing upward, part of the current enters into the mouth of a downwardly curved pipe, supplying the heater with an ample amount of exhaust steam to heat the water to 210 degrees, even though the heater is worked considerably beyond its rated capacity. The remainder of the steam passes out at the top, either to atmosphere or heating system as the case may be."

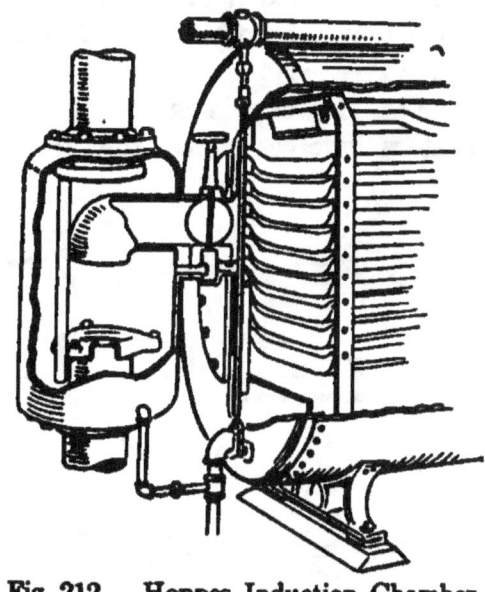

Fig. 212. Hoppes Induction Chamber.

A good way of connecting a Hoppes heater to an exhaust steam heating system is shown in Fig. 213. The piping is arranged so

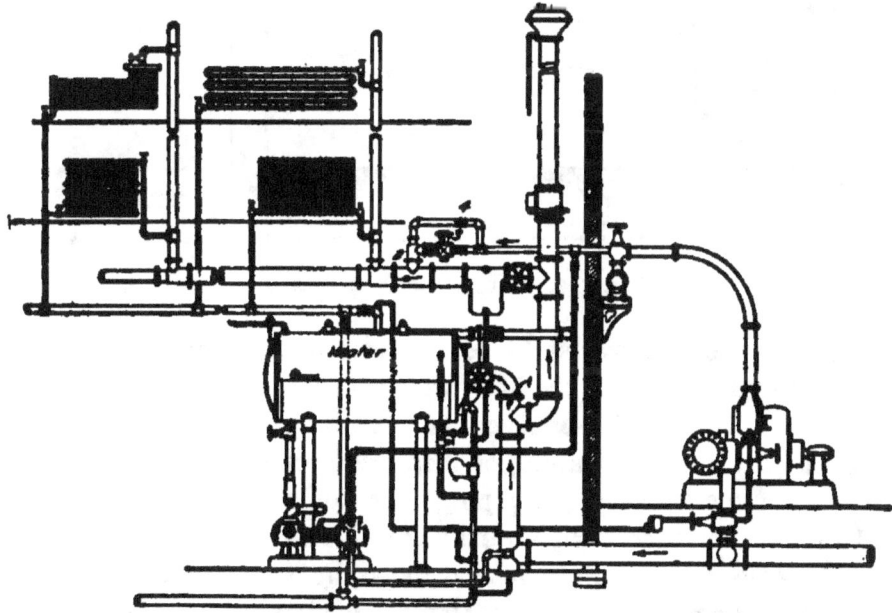

Fig. 213. Piping Arrangement for Hoppes Heater and Exhaust, Heating System.

that the heater has preference, the surplus steam passing out at the side of the tee *1* to the heating system. A live steam connection is provided at *2* through reducing valve *3*. The reducing valve should be provided with a by-pass *4* so that it can be cut out if necessary.

Open Heater Piping. — The arrangement of piping for open heaters involves much the same considerations as for closed

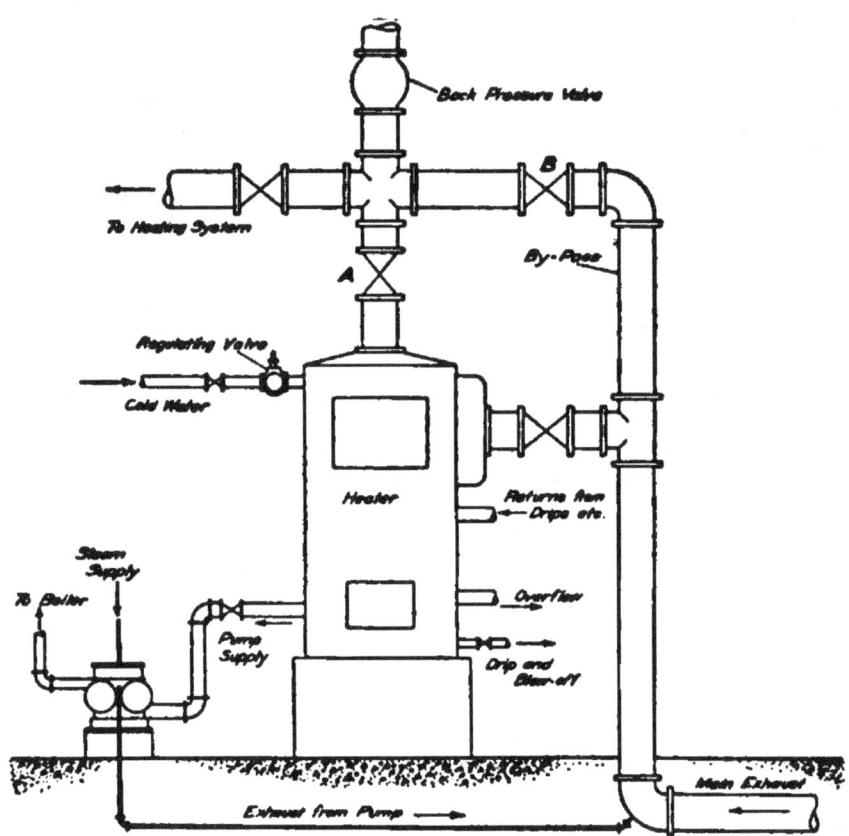

Fig. 214. Piping for Open Heater.

heaters. The heater should be placed two or three feet higher than the pump so that the hot water will flow into the pump suction by gravity. A by-pass should be arranged so that the heater may be cut out for cleaning or inspection, or when all the steam is needed for heating. A piping arrangement is shown in Fig. 214 for heater used with an exhaust heating system.

The cold water supply is controlled by a float inside of the heater. As noted, the returns from drips, etc., or from the heating

FEED WATER HEATERS

system are connected directly to the heater. Should the valves A and B both be closed at the same time, the starting of the

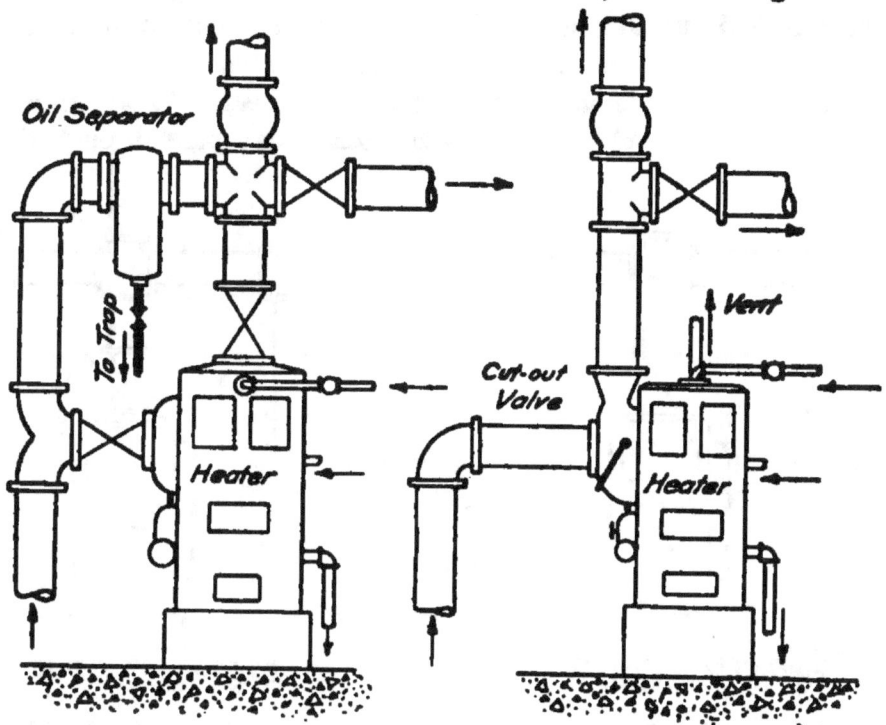

Fig. 215. By-pass Piping. Fig. 216. Cochrane Cut-out Valve in Place of By-pass.

engine can produce a sufficient pressure to rupture the heater unless the valve A is arranged to open under such conditions or a relief valve provided with direct connection to the heater. Some

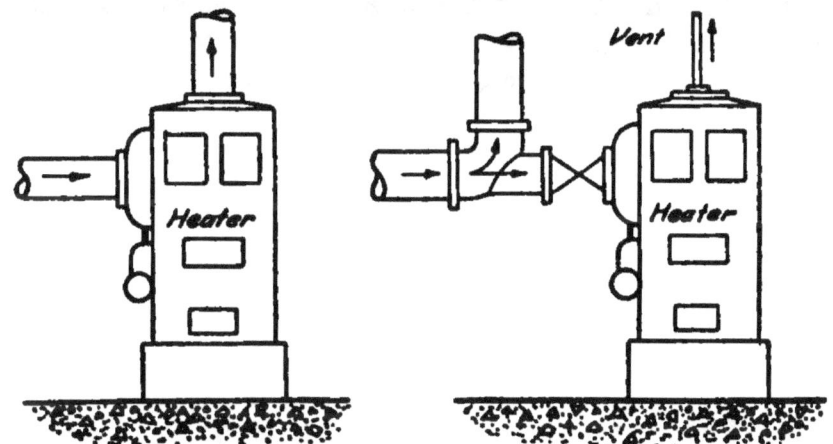

Fig 217. Thoroughfare Heater. Fig. 218. Preference Heater.

heaters have the by-pass made as part of the main casting, thus effecting a considerable saving in valves and piping, as indicated in Figs. 215 and 216, where Fig. 215 shows a piping by-pass and Fig. 216 a by-pass contained in the cut-out valve.

All of the steam may pass through the heater to the atmosphere, as in Fig. 217. Part of the steam may pass through the

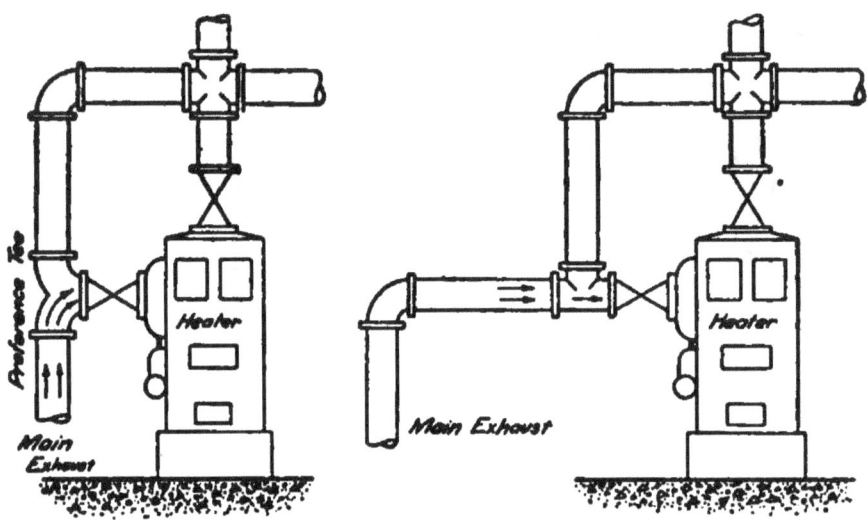

Figs. 219 and 220. Preference Connections for Heaters Used with Exhaust Heating Systems.

heater and part to the atmosphere, as in Fig. 218, where only sufficient steam is admitted to the heater to heat the water. Other forms of "preference connections" may be used, as shown in Figs. 219 and 220, where the tendency of the steam is to enter the heater, the excess passing on. A preference tee or a plain tee arranged as shown may be used for this purpose.

CHAPTER XII

PIPING FOR HEATING SYSTEMS

Piping for Heating Systems. — The purpose of this chapter is to illustrate the general arrangement of piping and connections as used for heating systems. No attempt is made at completeness, but it is hoped that sufficient material is included to be of value to those who wish to learn something about the different systems of piping.

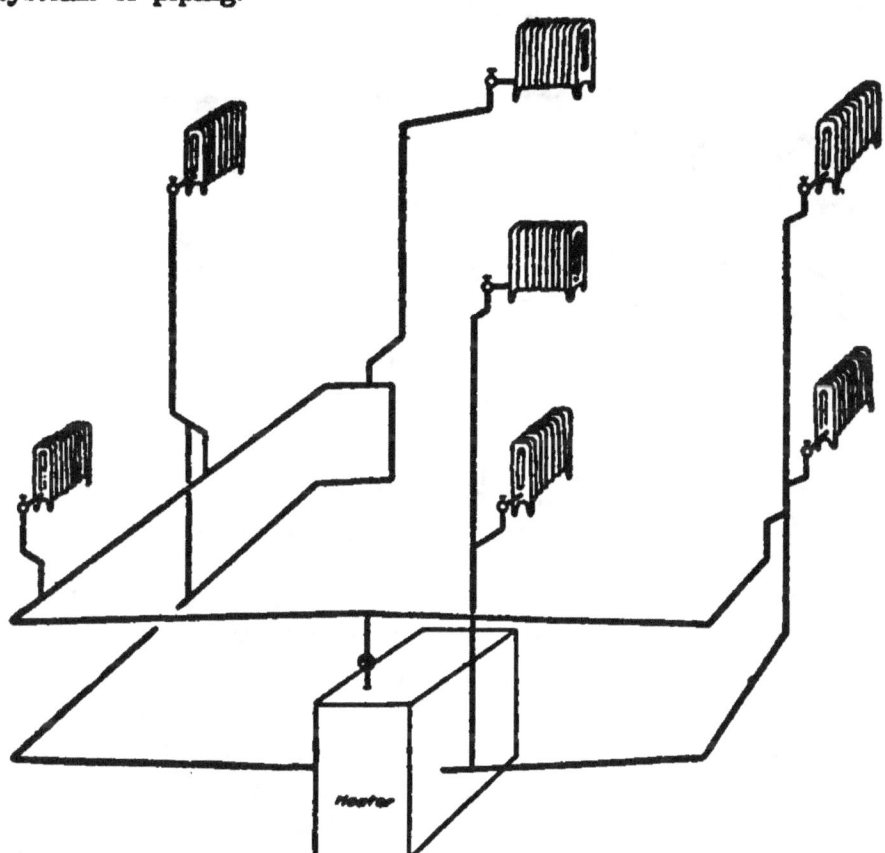

Fig. 221. One-pipe Wet System.

Steam Heating Piping Systems. — For supplying steam to radiating surfaces and removing the condensed steam, there are two general arrangements of piping "one-pipe" systems and "two-pipe" systems. A one-pipe wet system is shown in Fig. 221. As the same pipes are used to supply steam and to return the condensation from the radiators, they must be large. The main steam pipe is sloped away from the boiler. A return main is run under the supply main and is pitched toward the boiler, entering it below the water line. The risers are taken from the top of the steam main to supply the radiators, and these same

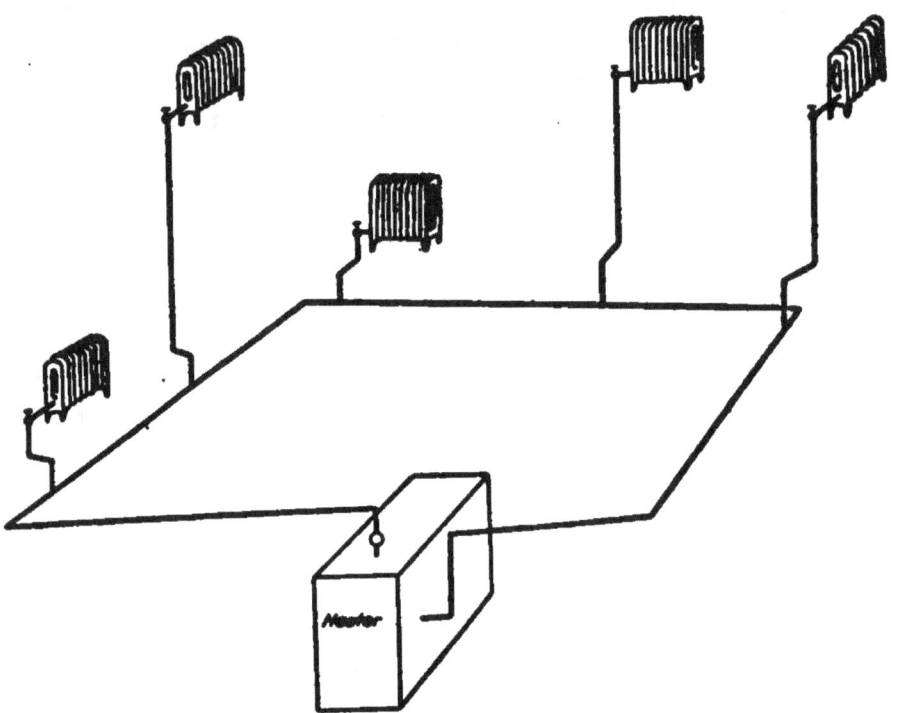

Fig. 222. One-pipe Circuit System.

risers are used by the condensation which drains into the return pipe. A single radiator on the first floor may be used without connecting to the return pipe. A one-pipe circuit system is shown in Fig. 222. In this system the main steam pipe makes a complete circuit of the basement, at the same time pitching away from the boiler, and on returning enters it below the water line. The radiators are supplied with steam and are drained by the same riser which is made large enough for this purpose. The condensation, after reaching the circuit pipe, is forced along in the same direction as the steam and completing the circuit is returned to the boiler. The steam main should be of one size and large enough so that there will be plenty of room for both the steam and water of condensation. With tall buildings, the use of the same pipe for supply and drain is objectionable, due to the interference. In such cases the one-pipe system shown in Fig. 223, may be used. The supply main in this case is run to the top of the building, and then the radiator branches are taken off from drop pipes. In this way the steam and condensation both flow downward except in the short connections between the drop and the radiator. The drop pipe connects into a drain pipe which returns the condensation to the boiler. A few radiators may be connected into the main riser.

The arrangement of a two-pipe system is shown in Fig. 224. As shown, steam is supplied at one end of the radiator and drained from the other, the steam and drain pipes being entirely separate. The radiators are supplied by risers from a steam main located near the basement ceiling. The drain pipes drop to a return main located near the floor of the basement or below the water line in the boiler.

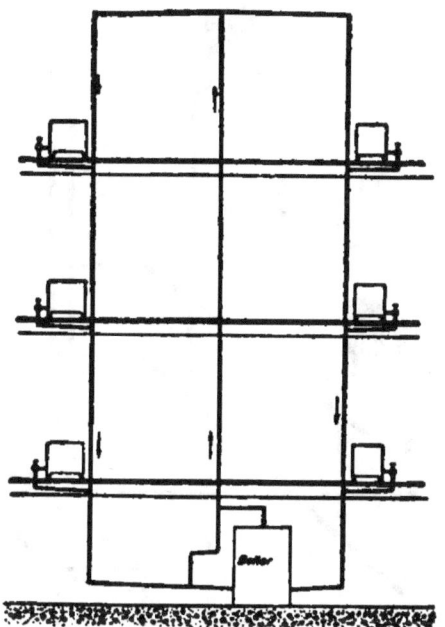

Fig. 223. One-pipe Down Flow System.

Steam Radiator Pipe Connections. — Several methods of making radiator connections are shown in Fig. 225. There should always be provision for expansion and contraction. The connection at A is for a radiator and main, unconcealed, while B shows a similar connection but using a 45 degree branch because of limited room above the main. At C and D are shown methods of connection between radiators and risers, one-pipe system. At E and F are shown methods of connection for the two-pipe system.

The sizes of pipe for which radiators are tapped as used by the American Radiator Company are given in Table 78, which is for one and two-pipe direct steam radiators. If the connection between the radiator and the riser is short these same sizes may be used.

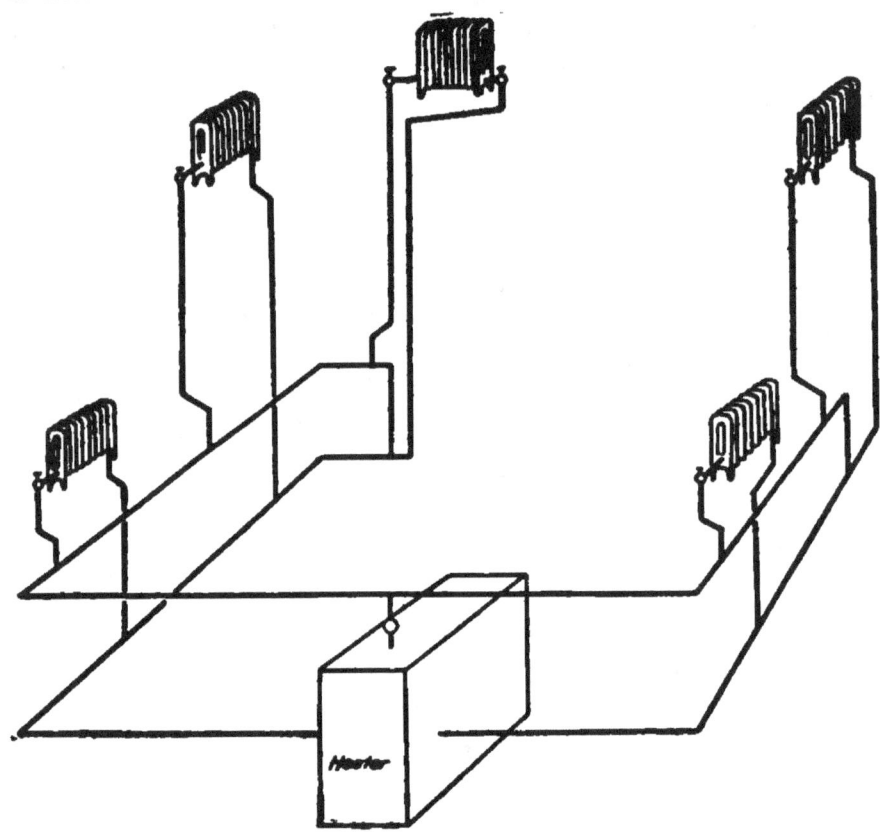

Fig. 224. Two-pipe System.

TABLE 78

PIPE SIZES FOR STEAM RADIATORS

Square Feet of Radiation	One-pipe System	Two-pipe System	
		Supply	Return
Up to 24	1	...	...
24 to 60	1¼	...	...
60 to 100	1½	...	...
Above 100	2	...	...
Up to 48	...	1	¾
48 to 96	...	1¼	1
Above 96	...	1½	1¼

Sizes of Steam Heating Pipes. — Steam pipes for heating should always be of ample size and carefully drained. The steam main should never be less than 1½ inches in diameter, and should

PIPING FOR HEATING SYSTEM 205

be larger if more than 30 feet in length. Risers should be at least one inch in diameter. All branches should be taken from the top of the main or at an angle, but never from the side so as to avoid getting water with the steam. To insure good drainage

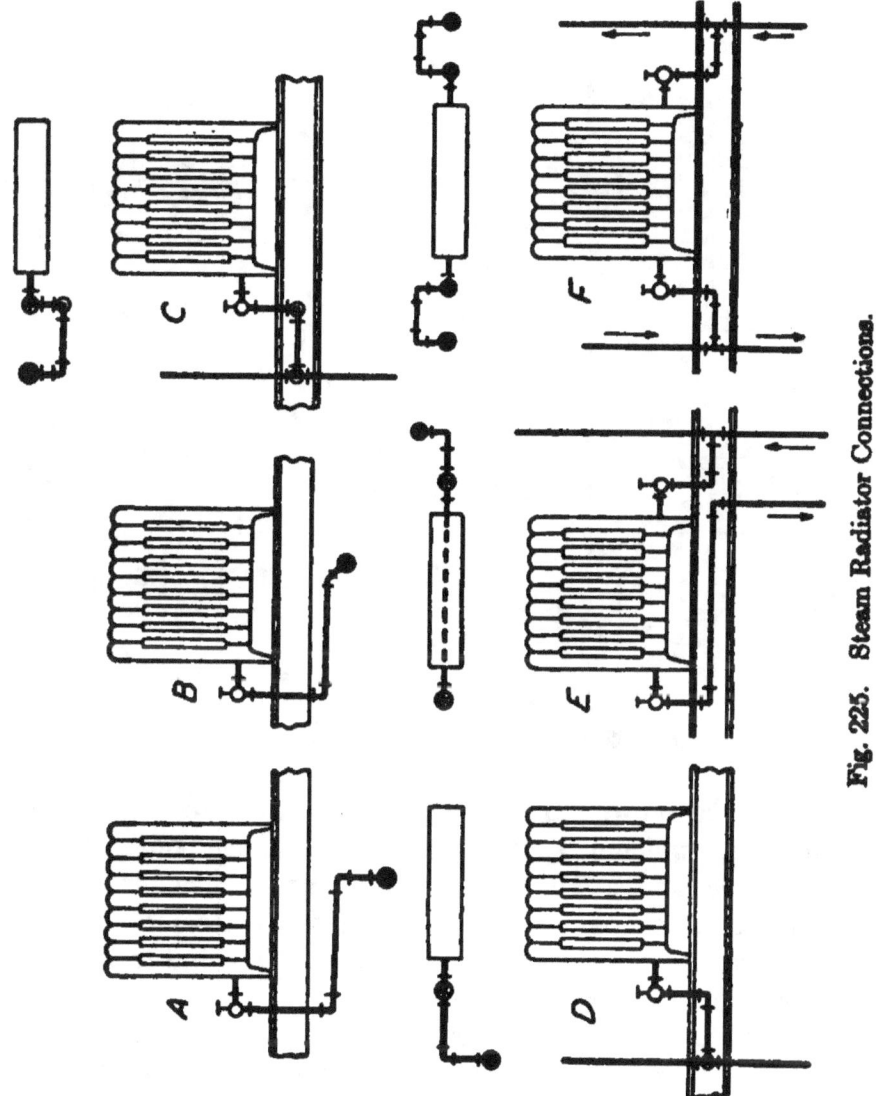

Fig. 225. Steam Radiator Connections.

the steam main should have a slope of at least one inch in ten feet and branches should have twice this slope.

The sizes of steam mains and risers may be obtained from Fig. 226, where average values are plotted, the sizes being proportioned to the radiating surface. The sizes of returns for the two-pipe system are not given as they should be determined from the conditions in connection with each installation. For small supply pipes they may be one size smaller. For large supply pipes the returns may be very much smaller, but dry returns should be larger than wet returns. A dry return is one

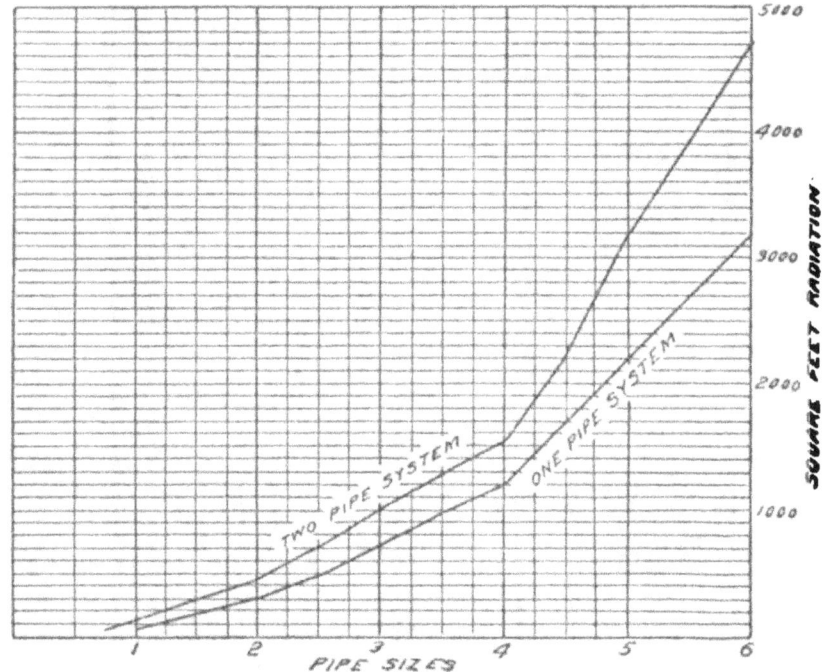

Fig. 226. Sizes of Steam Mains and Risers.

in which the return pipe is above the water level, and a wet return is one which is below the water level of the boiler, and consequently is always full of water. The dry return pipe exposes the surface of the water flowing along the bottom of the pipe, and is likely to cause water hammer and noises due to the rapid condensation of the steam.

Hot Water Heating Systems. — There are two systems of hot-water heating, the open tank system shown in Fig. 227, and the closed tank system. The arrangement of piping is the same for both systems, but the piping may be somewhat smaller for the closed system, and a safety valve must be provided. This safety valve is usually set for ten pounds pressure. The system illustrated in Fig. 227 shows the supply mains rising from the heater and the return mains sloping toward the heater and entering it at as low a point as possible. The risers to the radiators are taken from the top of the supply mains. The mains and risers may be reduced as the radiator branches are taken off.

For large buildings a single supply pipe may be carried to the expansion tank, and from there the branch down-feed pipes run

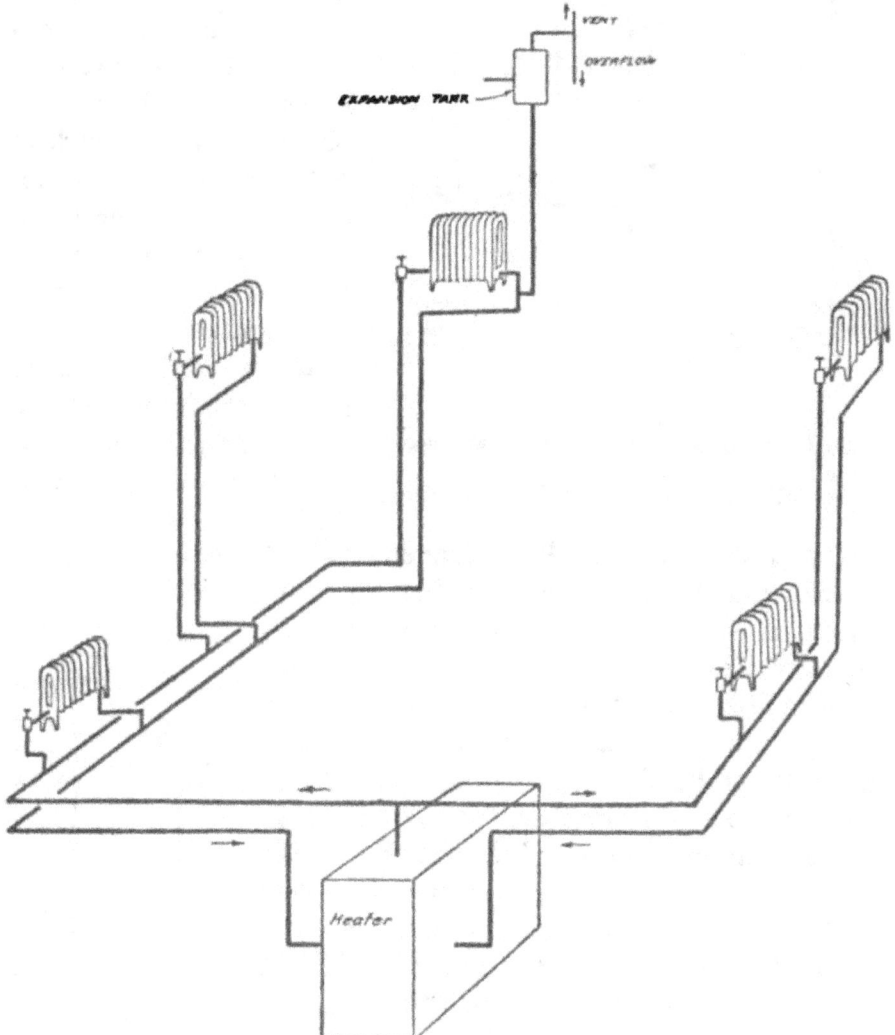

Fig. 227. Open Tank System.

to the radiators. This system is shown in Fig. 228. Circulation is caused by the fact that water expands when it is heated, therefore it becomes lighter than cold water and rises through the system, allowing the cold water to flow downward to the heater. This method of operation is known as a gravity system. For large buildings it is necessary to use a pump to circulate the water and it is then called a forced circulation system.

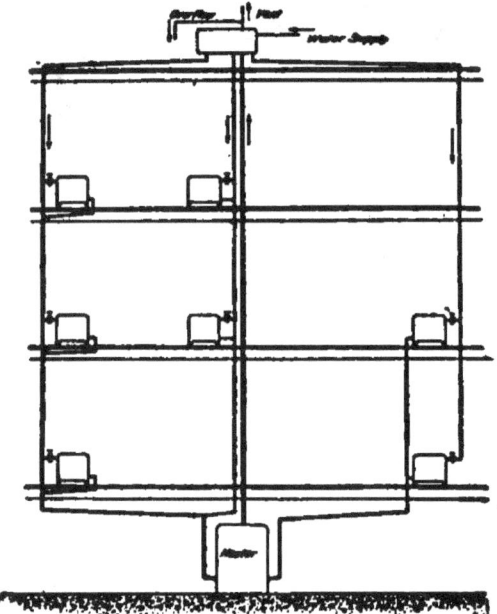

Fig. 228. Down-feed Hot Water System.

Expansion Tanks.— The purpose of the expansion tank is to care for the changes in volume of the water as it is heated. It should be placed above the highest radiator in the system, and should be provided with a vent pipe, and an overflow pipe connected to a drain. The ordinary form of tank made of galvanized iron is shown in Fig. 229, together with the necessary piping connections.

Hot Water Radiator Pipe Connections.— Several methods of making radiator connections for hot water are shown in Fig. 230. The connection for horizontal mains is shown at D, and for vertical pipes or risers at A and C. The supply pipe may be connected at the top of the radiator, as shown at B, which makes the valve handy. Two methods of connection for overhead supply systems are shown at E and F. In the method shown at F the water passes through each radiator separately, entering all of them at practically the same temperature, while in the method shown at E it passes through each of the radiators in succession, necessitating larger radiators on the lower floors. The sizes of pipe for which hot water radiators are tapped as used by the American Radiator Company are as follows. The same sizes are used for both supply and return pipes. The size of pipe refers to the nominal diameter of standard wrought pipe.

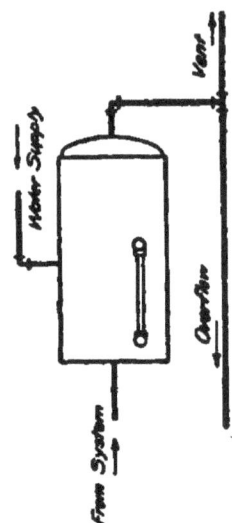

Fig. 229. Expansion Tank Connections

PIPING FOR HEATING SYSTEM

Radiators containing 40 square feet and under.................	1 inch
Above 40, but not exceeding 72 square feet....................	1¼ "
Above 72 square feet...	1½ "

Vapor tappings, top and bottom opposite ends, supply ¾ inches, return ½ inch.

Unless otherwise ordered, all openings of Direct Radiators will have right-hand threads (except that of Wall Radiators where tapped 1½ inch, in which case tapping at one end is right-hand and left-hand on other end).

All air-valve tappings of Direct Radiators are regularly made ⅛ inch.

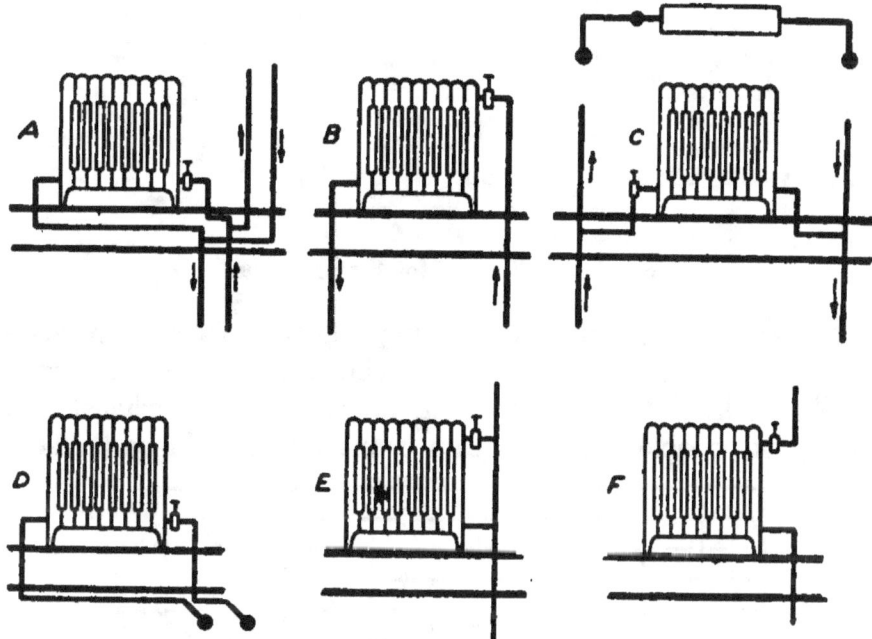

Fig. 230. Hot Water Radiator Connections.

Sizes of Hot Water Pipes. — The factors involved in determination of sizes of hot water piping are: the amount of radiating surface; the location of the radiating surface, both elevation above and distance from the heater; and the difference in temperature.

The sizes of hot water mains may be obtained from Fig. 231, where average values are plotted, the sizes being approximately proportional to the radiating surface. In a similar manner average values are plotted in Fig. 232, for sizes of pipes to supply various amounts of radiating surface on the different floors of a building.

Exhaust Steam Heating. — Steam that has been used in a steam engine or other power apparatus may be exhausted to a heating system. Any system of heating may be used in connection with exhaust steam by installing the proper apparatus.

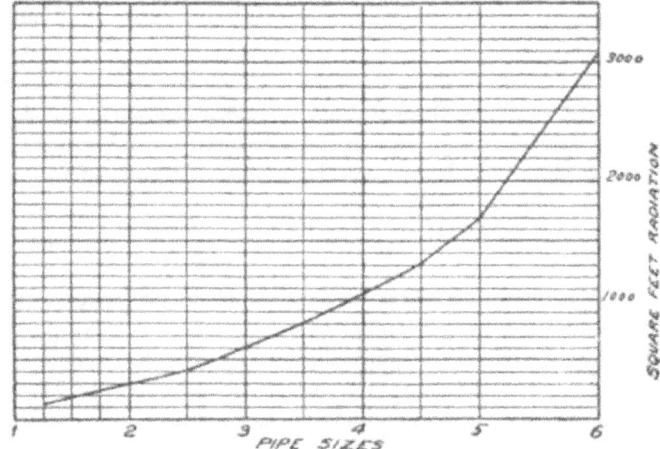

Fig. 231. Sizes of Hot Water Mains.

Factories and large buildings having a power plant often make use of exhaust steam in this way. The piping for such a system should be large to keep the back pressure in the exhaust pipe as low as possible. A live steam connection should be made to

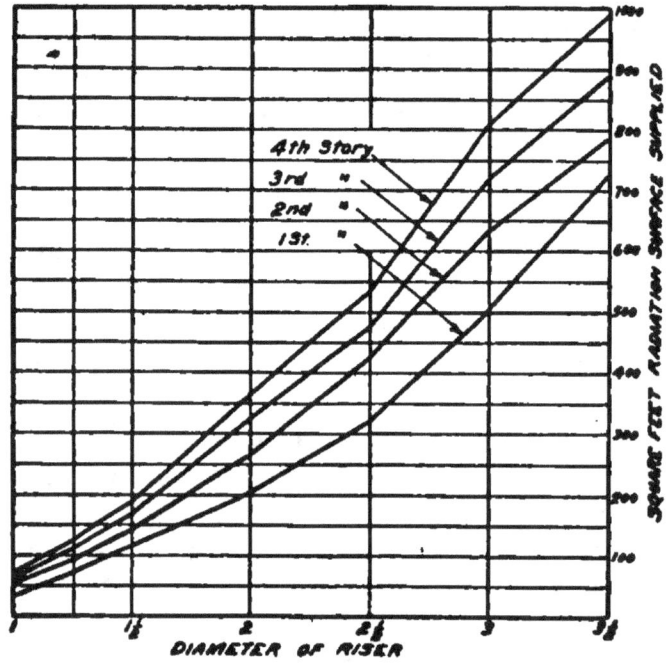

Fig. 232. Sizes of Hot Water Risers.

PIPING FOR HEATING SYSTEM

the heating pipe, using a reducing valve to lower the pressure, and a relief or back pressure valve should be placed in the exhaust pipe to prevent excessive back pressure. If the condensation is to be returned to the boiler, an oil separator should be placed in the exhaust pipe before the connection is made with the heating system. Steam traps, automatic pump and receiver, and other devices used in connection with exhaust heating are described in other parts of this book, and may be located by reference to the index. The piping for feed water heaters is shown in Chapter XI.

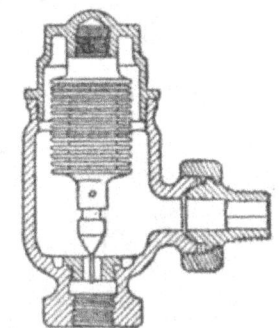

Fig. 233.
Webster Sylphon Trap.

The Webster Vacuum System of Steam Heating. — The Webster system is used here to illustrate a method of heating with a pressure lower than atmospheric. A vacuum system necessitates the removal of air from the system by means of a pump. This establishes a lower pressure in the returns, after which the pump removes the condensation and entrained air. The steam condenses in the radiators and so induces a further supply of steam. This removal of air and condensation makes a positive circulation, and insures complete filling of the radiators with steam.

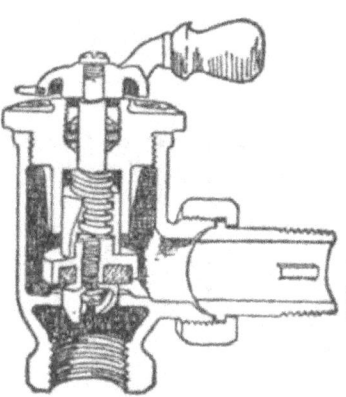

Fig. 234.
Webster Modulation Valve.

If exhaust steam is used there will be very little back pressure upon the engines.

One of the essential features of the Webster system is the outlet valve used on radiators and coils. The form shown in Fig. 233 is the Webster sylphon trap. It is operated by a sylphon bellows. The sum of the small movement of each of the folds gives the necessary lift to the valve. This trap will close quickly and positively when steam reaches the bellows, but at a slightly lower temperature the water and air will be withdrawn or discharged. Since the valve is wide open when cold, the radiator is sure to be drained.

The circulation of steam may be controlled and modulation of temperature secured by throttling the inlet valve on any radiator. The Webster modulation valve shown in Fig. 234 is made so that less than a full turn is required from shut to full opening, the area of the opening increases in proportionate progression, and a pointer and dial are used to indicate the degree of opening.

Radiator Pipe Connections. — The size of radiator tappings as given by Warren Webster Company are shown in Table 79.

TABLE 79

CAST IRON RADIATOR TAPPINGS

Square Feet of Direct Radiating Surface Condensing Normally not to exceed 1/4 Pound per Square Foot per Hour	Normal Maximum Pounds of Condensation per Hour	Pipe Size of Supply Tapping. Customary Practice followed by Engineers when Ordinary Radiator Valves are Used	Supply Tapping when the Webster Modulation Valve is Used	Pipe Size of Return Tapping
1–25	7	1/2	3/4	1/2
26–50	13	3/4	3/4	1/2
51–100	25	1	3/4	1/2
101–175	44	1 1/4	3/4–1	1/2
176 and over	75	1 1/2	1	3/4

NOTE. 3/4" Webster Modulation Valve is used for radiators up to 150 square feet; 1" above 150 square feet, with interchangeable "Modulation" sleeves to secure throttling control.

PIPE COIL TAPPINGS

Square Feet of Direct Radiating Surface Condensing Normally not to exceed 1/4 Pound per Square Foot per Hour	Normal Maximum Pounds of Condensation per Hour	Pipe Size of Supply Tapping	Pipe Size of Return Tapping
42	13	3/4	1/2
84	25	1	1/2
146	44	1 1/4	1/2
250	75	1 1/2	3/4
528	158	2	3/4
924	277	2 1/2	1

The figures refer to vacuum systems only. If the condensation is greater than that given for the radiating surface the pipes should be based upon the condensation rate. The run-outs from supply risers to radiators should be one size larger if more than four feet long.

PIPING FOR HEATING SYSTEM

Typical Arrangement Webster Systems. — A typical arrangement of the Webster vacuum system is shown in Figs. 235 and 236, taken from the Warren Webster Company's catalog and described by them as follows (the numbered parts are all of Webster manufacture):

"The engine *A* is protected by a steam separator *2* dripped through a high pressure trap *3*. The exhaust steam from the engine passes through an oil separator *8*, dripped through grease

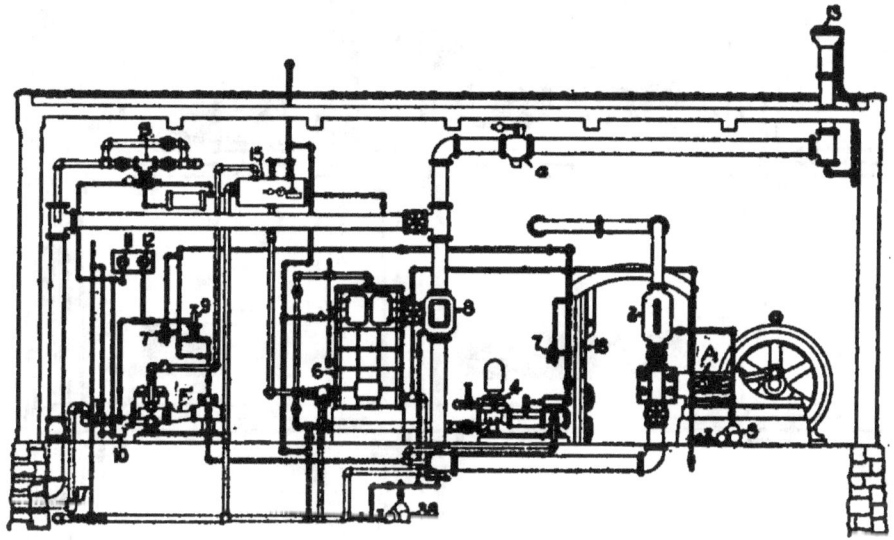

Fig. 235. Webster Vacuum System.

trap *38*, thence to the heating system. A pressure reducing valve *B* with by-pass is provided to make up any deficiency in the volume of exhaust steam or for heating when the main engine is shut down.

"The supply main is dripped as it enters the building through a heavy-duty thermostatic trap *22*, protected by a dirt strainer *19*. The steam supply risers in larger buildings may require to be dripped through traps of the proper size and type.

"Steam is supplied to the various types of heating units through Webster modulation valves *21*, although the system will work in harmony with automatic temperature control. We have shown ordinary radiator supply valves on some of the units. A particular type of Webster modulation valve, with chain attachment, is shown for the overhead radiator *C*.

"Each heating unit is drained through a Webster sylphon

trap *20* into the return risers, the larger heating coils being protected by dirt strainers *19*.

"Steam is also supplied to tempering and re-heating coils,

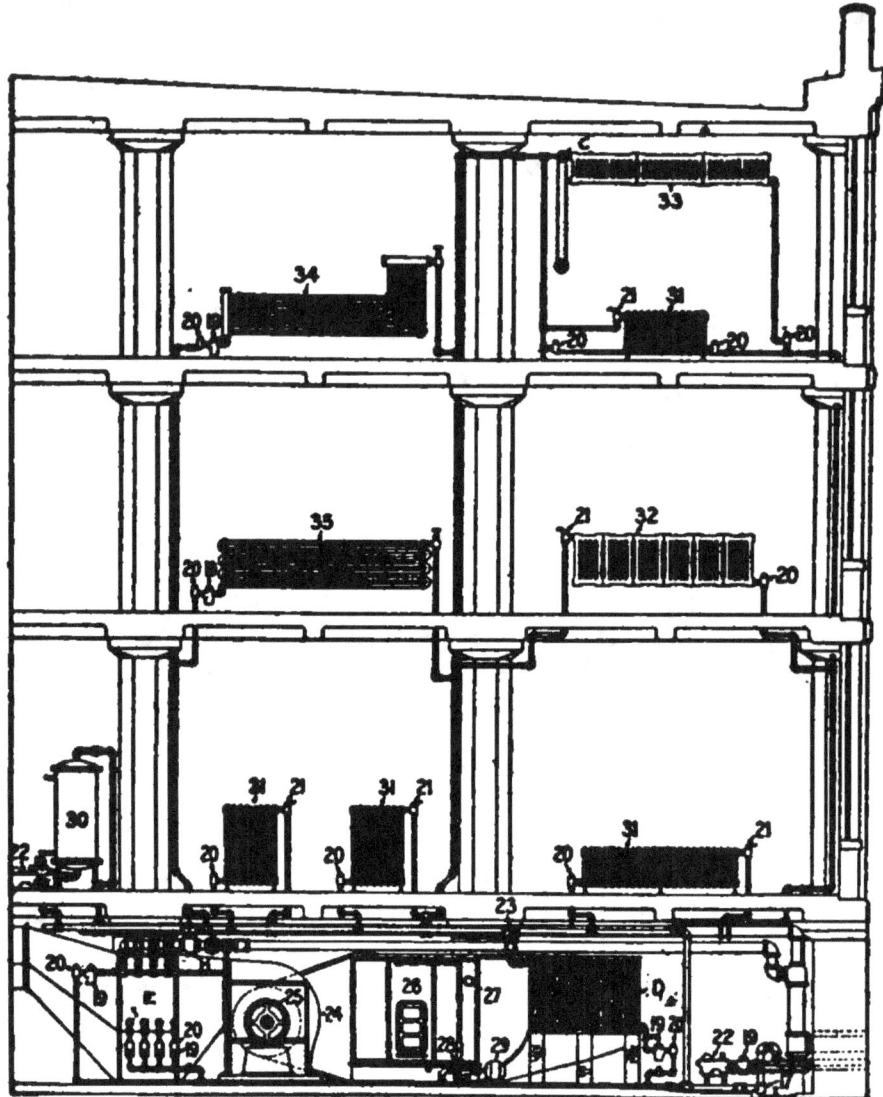

Fig. 236. Webster Vacuum System.

D–E which are also drained at the return ends of each group through traps *20*, protected by dirt strainers *19*.

"All the returns join and lead to a vacuum pump, *F* protected by a suction strainer *10*, the steam supply to the pump being automatically controlled by the vacuum pump governor *9*. Gauges on slate board *11-12* are shown with connections taken from the heating main and the vacuum return line.

"The vacuum pump discharges through an air separating tank *15*, to a feed water heater *6*. The illustration shows the preference type heater, the oil separator *8* being so constructed that a sufficient quantity of exhaust steam is directed toward the heater, the balance is available for the heating system, while any excess

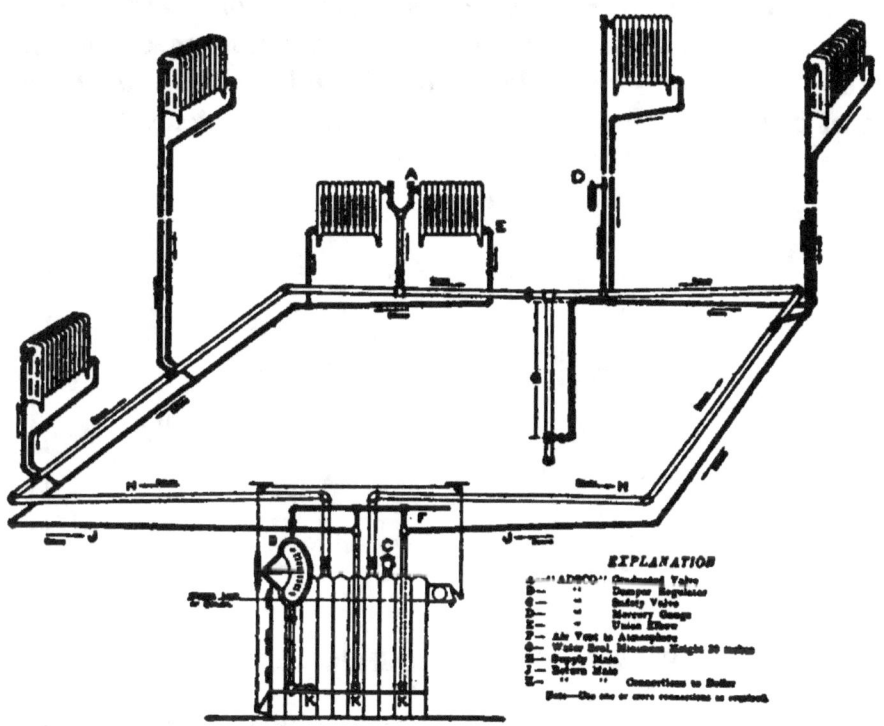

Fig. 237. Atmospheric System.

escapes through the atmospheric back pressure valve *G*. The heater may thus be cut out of service while the oil separator remains in use.

"The ventilation scheme provides for such rooms connected thereto, a supply of purified, humidified and heated fresh air. The air is partially heated in passing over the tempering heater *D*, and is drawn by the fan through the air washer *26* and re-heated to the proper temperature, passing over the re-heater *E* into the main air supply duct. The supply of steam to the tempering heater and re-heater coils is automatically governed by temperature control system valve *H*."

Atmospheric System of Steam Heating.

The "Atmospheric System" is a low pressure system developed by the American District Steam Company. It is a two-pipe gravity return system, operating with pressures of from five to eight ounces, and with very rapid circulation. Each radiator is a separate unit, and can be manipulated as desired without affecting the others. The regulating valves are made in $3/4$ inch size, and the radiator should be bushed to $3/4$ inch for both connections, with the inlet at the top of one end and the outlet at the bottom of the other end. The various principles involved, and the general arrangement of piping is shown in Fig. 237. The main steam line in the

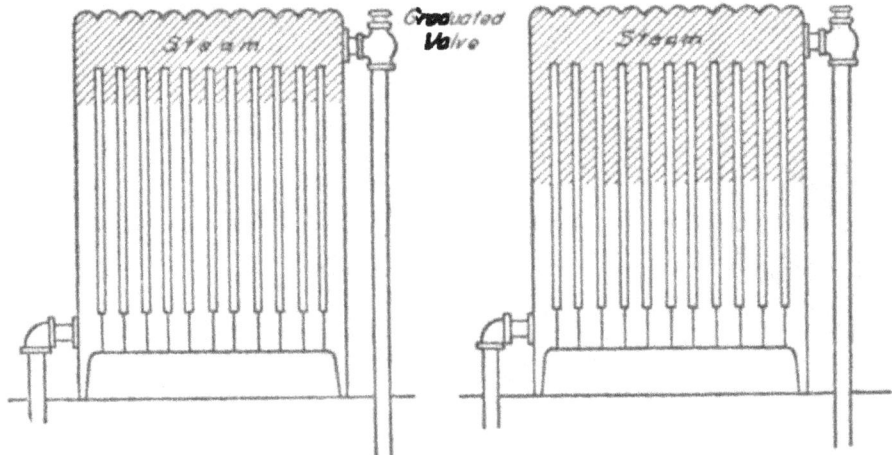

Fig. 238. Operation of Graduated Valve.

basement is laid out in a complete circuit to make certain of perfect circulation and equalization of pressure at all points in the system. The return pipes are under no pressure, and are used to provide gravity return of the water of condensation and as an outlet for the air in the system. Extra heating surface is used in each radiator and the return piping is vented to the atmosphere to allow air to freely enter or leave the system. This vent pipe may be $1 1/4$ inch on small installations, but a number of pipes may be required on large systems. Only one valve is used on the radiators, the inlet valve. This inlet valve is so arranged that the radiator may be one-quarter, one-half, or any desired part filled with steam, as shown in Fig. 238. The steam admitted displaces the air, and being lighter, remains at the top of the radiator. Sizes of pipe to install for various amounts of radiation, as recommended by the American District Steam Company are given in Table 80.

TABLE 80

PIPE SIZES FOR VARIOUS AMOUNTS OF RADIATION

Square Feet of Radiation	Pipe Supply	Return Pipe	Square Feet of Radiation	Pipe Supply	Return Pipe
30 feet	3/4 inch	3/4 inch	1400 feet	3 1/2 inches	1 1/2 inches
50 "	1 "	3/4 "	2200 "	4 "	2 "
100 "	1 1/4 inches	3/4 "	3600 "	5 "	2 "
200 "	1 1/2 "	1 "	6000 "	6 "	2 1/2 "
300 "	2 "	1 1/4 inches	8500 "	7 "	2 1/2 "
600 "	2 1/2 "	1 1/4 "	11000 "	8 "	2 1/2 "
900 "	3 "	1 1/2 "			

Central Station Heating. — There are many points in connection with district steam heating which are of value in relation to piping in general. Aside from this, however, the increase in the use of this method of heating, and its importance as a means of effecting economy are sufficient reasons for the inclusion of the following articles which describe the systems of the American District Steam Company as representing modern practice in this kind of piping. The information and drawings were very kindly supplied by the above company through Mr. H. E. Long, Chief Draftsman.

Central station heating consists of a central generating plant, from which steam is distributed through underground mains, carefully insulated and protected, to the radiation in homes and public or private buildings. Birdsall Holly invented the first system of this kind, and through him it was introduced in the city of Lockport, N. Y., in 1877. The source of supply of steam may be heating boilers, or the exhaust steam from electric or power plants may be utilized. Where exhaust steam is used it is necessary to have a live steam connection with a reducing valve to supply additional steam, should the exhaust be insufficient. The reducing valve should be provided with a by-pass for emergency use. A back pressure valve in the atmospheric exhaust pipe is necessary in order to maintain the desired back pressure. An oil and water separator should be connected in the main exhaust pipe which leads to the underground system. An initial pressure of from two to five ounces has been found sufficient to give proper circulation in the most extensive systems. A typical arrangement of piping as described above is shown in Fig. 239. As indicated, any engine can exhaust into a condenser, to the atmosphere, or to the heating system.

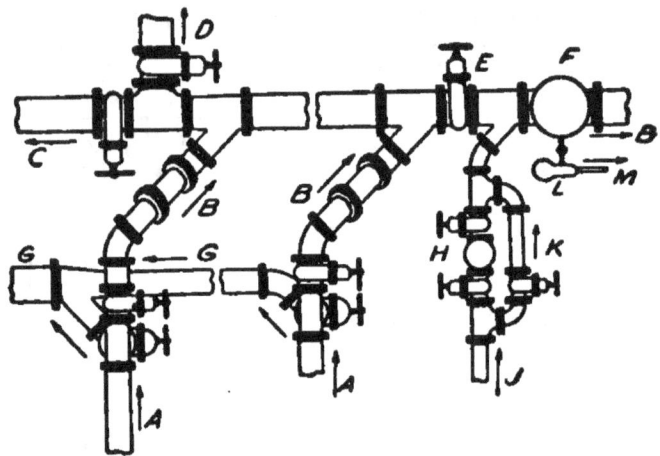

Fig. 239. Station Piping Connection for Exhaust Heating.

Underground Steam Mains. — The underground system of piping is a particularly important part of district heating. Some of the essential features of underground heating systems as installed by the American District Steam Company are: efficient insulation, perfect provision for expansion and contraction, provision for taking service connections from fixed points only, special attention to under-drainage, perfect grading and trapping of the mains, use of highest grade materials, and competent supervision of the work of installation. The methods of insulation found most efficient and durable by the above company are the wood

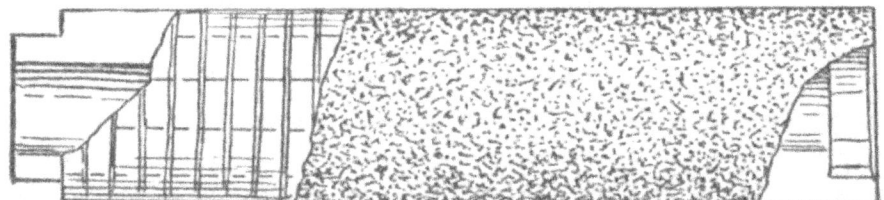

Fig. 240. "Standard" Steam Pipe Casing.

stave casing shown in Figs. 240 and 241, and the patented multi-cell construction shown in Figs. 242 and 243.

The wood casing is built up of staves of selected white pine, free from sap and thoroughly air and kiln dried. The staves have a

PIPING FOR HEATING SYSTEM

tongue and groove their length, which is locked by spirally wound banding wire. A four-inch mortise and tenon is cut on the ends, the mortise being one-half inch greater than the tenon to allow the joints to be firmly driven together. The casing is then coated with asphaltum pitch and rolled in sawdust. A tin and asbestos lining completes the casing. The lengths of sections vary up to eight feet. The standard thickness of the casing is four inches. The tin lining reflects the heat waves back to the pipe, and protects the casing. The standard practice of the American District Steam Company is to use a four-inch shell, tin and asbestos lined casing on low pressure steam lines and on hot water lines, and two-inch thickness, unlined, for return lines. The casing is made from two to three inches larger, inside diameter, than the iron pipe which it covers, thus providing an annular air space which is made into "dead air space" by the use of cast iron collars which also assist in anchoring the line. Cast iron guides and rollers placed about six feet apart are used to centre the pipe.

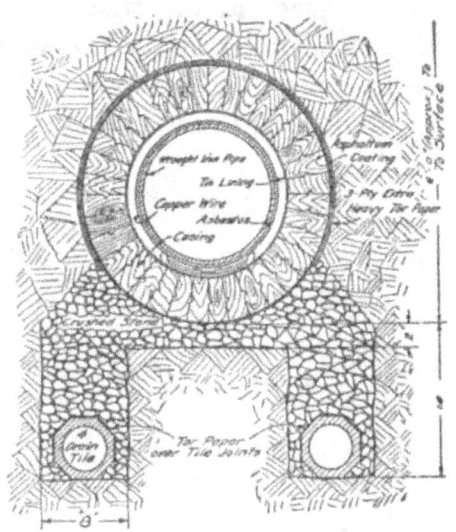

Fig. 241. "Standard" Steam Main Construction in Casing.

Fig. 242. Standard Steam Main Construction — Multi-cell.

Fig. 241 shows a cross section of the standard steam main construction in wood stave casing for mains six inches and larger. For mains five inches and smaller, one of the drains may be omitted.

The multi-cell construction, shown in Figs. 242 and 243, is built in place in the trench. A concrete base upon which rest supports for the pipe, is built on a layer of crushed stone. Hollow tile blocks on end rest upon this base, and form the side walls, the joints with the base and between tiles being made with cement. The tiles are then filled with shavings, and the tops closed with cement. The space above the piping insulation is also filled with shavings. Tile blocks with closed ends laid across the top close the conduit. All joints are carefully cemented. The closed tiles form a multi-cell insulation of dead air. The crushed stone at the sides provides for effective drainage to the drain tile. It will be noted that the piping is entirely separate from the conduit, which is thereby relieved from the effects of expansion of the piping. The cross section shown in Fig. 242 is for mains from six inches to sixteen inches inclusive. For smaller size mains only one-drain tile is used. For mains eighteen inches and larger, the arch form of construction is used, in order to secure the strength necessary on account of the increased width of the conduit, Fig. 243.

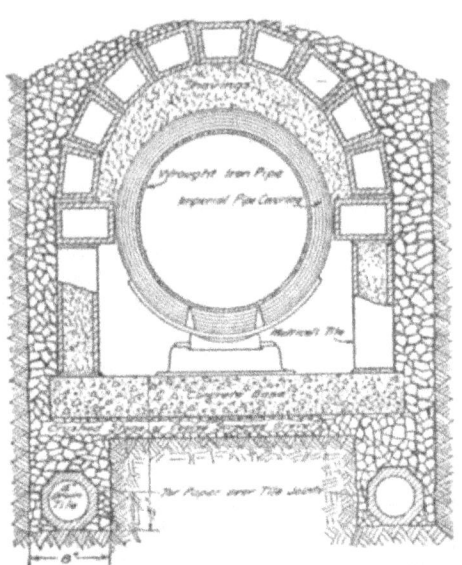

Fig. 243. Multi-cell Construction for Large Mains.

Underdrainage. — In addition to the insulation provided it is necessary to prevent any water from coming into contact with the steam pipe. The effect of water would be condensation of steam in the main, as well as ultimately affecting the durability of the insulation. This means that adequate underdrainage must be provided, regardless of the kind of insulation used.

The methods of underdrainage, as installed by the American district Steam Company, are shown in Figs. 241, 242 and 243. When the trench is dug, a properly graded and drained field tile or uncemented sewer pipe is installed. This pipe is connected at as frequent intervals as necessary with the sewer, using check valves. The drain tile and bottom of the trench are then covered with a heavy layer of broken stone, gravel or clean cinders. This forms a porous drain bed upon which the casing rests. The under-

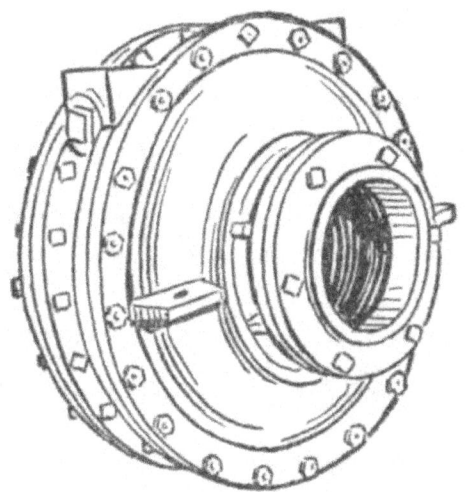

Fig. 244. Variator.

drainage shown in the figures is typical for ordinary clay soil. It is frequently installed in a different manner, depending upon the amount of moisture which may be held in suspension, due to the varying soils encountered in the trench. In every case, competent supervision by experienced engineers should be obtained.

Installation in Wood Casings. — After the piping is made up in place it is spirally wrapped with $1/32$-inch asbestos paper. This

Fig. 245. Double Expansion Joint, Showing Method of Anchoring.

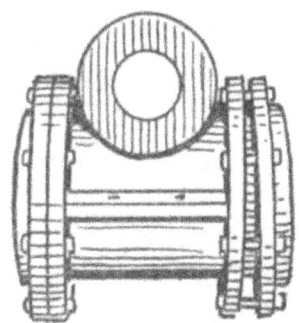

Fig. 246. Anchorage Fitting.

paper is held in place by binding with pliable copper wire. The casings are then forced together and the joints cemented with hot pitch. A protection of three-ply tar paper is placed over the

222   A HANDBOOK ON PIPING

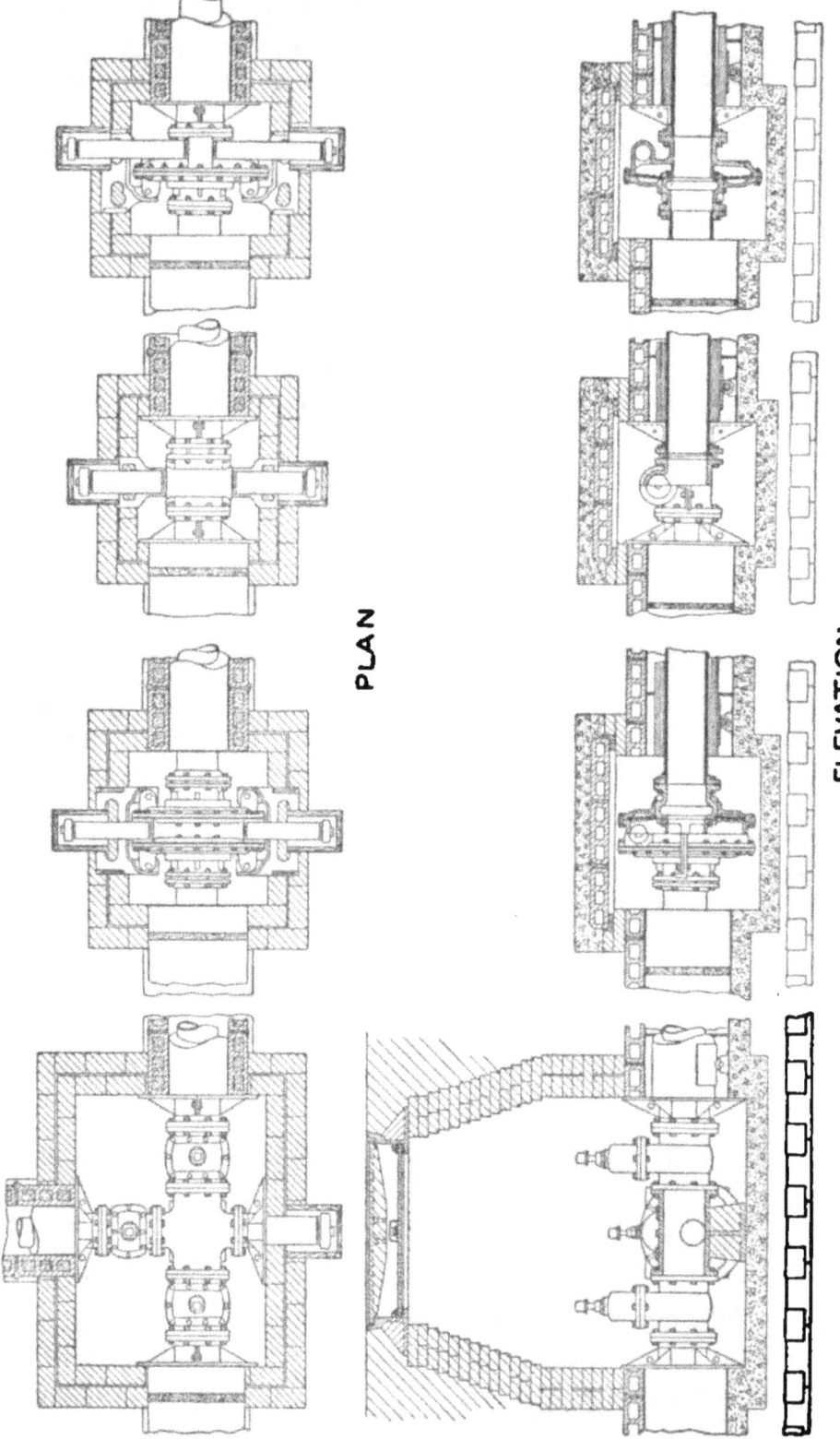

Fig. 247. Variator Construction, with Multi-cell Insulation. Plan and Longitudinal Section.

PIPING FOR HEATING SYSTEM

line and reaching to a point below the centre of the casing. The trench is then ready for filling.

Expansion and Contraction. — The two methods of caring for expansion and contraction, shown in Figs. 244 and 245, are devices made by the American District Steam Company. The "variator," Fig. 244, has two corrugated copper diaphragms. It is made with a fixed casing and two movable slips. The outer

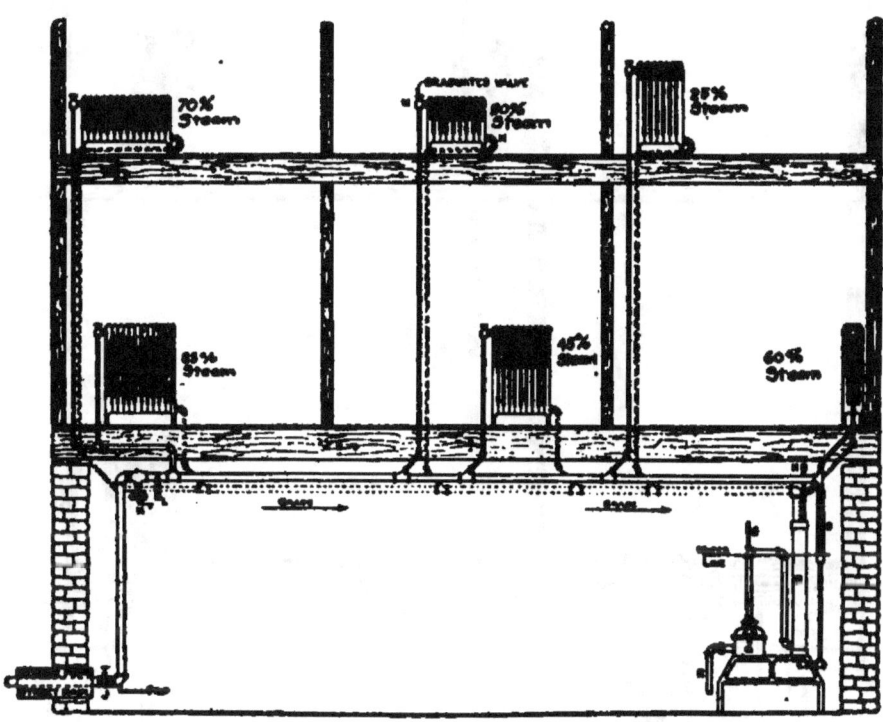

Fig. 248. Interior Piping and Meter Setting. Atmospheric System.

edge of the diaphragm is held in the casing, which casing is securely anchored; the inner edge of the diaphragm is fastened to the end of the slip. The casing of the variator and of the anchorage fitting, Fig. 246, are provided with service openings, so that branches are taken from fixed points. These variators are placed about 100 feet part, and have an anchorage fitting half way between them. Such an expansion device does not require packing or attention after being installed, and so avoids the expense due to the large number of manholes necessary to care for the slip joint expansion joints. When manholes can be used, the slip joint shown in Fig. 245 may be used. As shown, it is provided with service openings. The methods of installation, arrangement of manholes, anchorages, and other details are shown in Fig. 247 for the use of variators and multi-cell insulation. With expansion joints more manholes would be necessary.

Interior Piping for Central Station Heat. — If the building to be heated is piped for steam or hot water, necessary connections

Fig. 249. Interior Piping. One-pipe System.

can be made for using the existing piping. Any system of steam or hot water heating may be used in a new installation, but the atmospheric system previously described is advised as being most economical, Fig. 248. The interior piping for a one-pipe system is shown in Fig. 249. When hot water piping is already installed it may be continued by using a heater in which the water is heated by steam from the street.

PIPING FOR HEATING SYSTEM

The steam after being used is measured by a condensation meter, Fig. 250, which records the pounds of condensed steam. From the meter the condensation passes to the sewer. Unless

Fig. 250. Condensation Meter.

the district heated is very compact and close to the central station, it is generally better to allow the condensation to pass to the sewer than to attempt to return it to the plant. The cost of return lines and their up-keep generally makes such an investment unprofitable.

CHAPTER XIII

WATER AND HYDRAULIC PIPING

Water Piping. — The purpose of this chapter is not to treat extensively of the subject of water piping, but to give such information as it is believed will be of practical value to those who have piping to do around a building or plant.

The sizes and kinds of piping, valves, and fittings which are used for water have been treated in the earlier chapters. The following articles will deal with some of the special kinds of water piping.

Gravity Pipe Lines. — If a pipe is used to fill one reservoir from another at a higher level, the pressure in the pipe will de-

Fig. 251. Hydraulic Grade.

Fig. 252. Siphon.

crease uniformly from the higher to the lower level, this difference being due to friction. The pressures can be represented by the line $x-y$, Fig. 251, where the pressures at various points are proportional to the height of line $x-y$ above the pipe. If the pipe should rise above $x-y$ at any point, the pressure will be negative, and a partial vacuum will be formed, as at point A of the dotted pipe line, resulting in decreased flow. This may be relieved by an air cock, or the outlet of the pipe may be restricted. The line $x-y$ is called the hydraulic grade.

A pipe used to convey water from one container to another, arranged as in Fig. 252, is called a siphon. In order to start water flowing the air must be removed from the pipe, when the atmospheric pressure at x will cause the water to rise in the pipe to point z, from which it flows into container 2. The maximum theoretical vertical distance between x and z is 34 feet. The altitude of surface x and friction in the pipe will reduce this amount. Air from the water may collect at the point z and must be removed to keep the siphon in operation.

Flow of Water in Pipes. — The flow of water in pipes is too large a subject to be treated with any degree of completeness in this book, and the reader is referred to works on hydraulics. A few approximations and some common pipe data will be given, however.

The quantity of water delivered by a pipe will depend upon the head or pressure and the frictional resistances. At a given point the cubic feet of water passing will be equal to the area of the pipe times the velocity of the water.

$$Q = Av \quad \ldots \ldots \ldots \ldots \ldots \ldots \ldots \ldots \ldots (26)$$

when

Q = cubic feet per second.
A = area of cross-section of pipe, square feet.
v = velocity of flow, feet per second.

If the head or pressure is given the velocity may be figured and then the quantity obtained by using the above formula. Table 81 gives pressures equivalent to various heads of water.

TABLE 81

EQUIVALENT PRESSURES AND HEADS OF WATER

Feet Head	Press. per Sq. In.	Feet Head	Press. per Sq. In.	Feet Head	Press. per Sq. In.	Feet Head	Press. per Sq. In.
1	.43	50	21.65	170	73.64	290	125.62
2	.86	60	25.99	180	77.97	300	129.95
3	1.30	70	30.32	190	82.30	310	134.28
4	1.73	80	34.65	200	86.63	320	138.62
5	2.16	90	38.98	210	90.96	330	142.95
10	4.33	100	43.31	220	95.30	340	147.28
15	6.49	110	47.64	230	99.63	350	151.61
20	8.66	120	51.98	240	103.96	360	155.94
25	10.82	130	56.31	250	108.29	370	160.27
30	12.99	140	60.64	260	112.62	380	164.61
35	15.16	150	64.97	270	116.96	390	168.94
40	17.32	160	69.31	280	121.29	400	173.27

The theoretical velocity can be found from the formula for falling bodies, as given below:

$$v = \sqrt{2gh} \dotfill (27)$$

in which v = velocity of flow, feet per second.
h = head of water, feet.
g = 32.16.

Values given by the above formulas are theoretical, and if the length of the pipe is at all great will be very much reduced.

For clean straight pipe the quantity of water discharged and friction loss at different velocities of flow may be obtained from Fig. 253, which was plotted from Ellis and Howland's tables by Mr. Walter R. Clark, Ph.B., Mechanical Engineer with Bridgport Brass Company, using formulas (28) and (29).

v = velocity in feet per second.
G = gallons per minute.
F = pounds friction loss per 100 feet.
D = diameter of pipe in inches.

$$G = .245vD^2 \dotfill (28)$$
$$F = \frac{.03G^2}{D^5} \dotfill (29)$$

Formula (28) is taken for velocities greater than three feet per second. The method of using this chart may be understood from an example. A flow of 300 gallons per minute is required with a pressure loss of 25 pounds. The distance is 100 feet. Find the intersection of a vertical line from 300 gallons with a horizontal line through 25 pounds friction loss, which gives a $2^1/_2$ inch pipe and 19 feet per second velocity. The heavy lines show actual diameters, light lines show nominal diameters.

All fittings, meters, changes in direction, changes in the condition of the pipe and other factors produce friction and tend to reduce the flow so that they should be taken into account when estimating sizes of pipes. The length of pipe equivalent to an elbow for various sizes of pipe and velocities of flow may be found in Fig. 254, which shows results obtained from experiments by Professor F. E. Giesecke (Domestic Engineering, Nov. 2, 1912).

Pump Suction Piping. — The flow of water into the suction pipe is dependent upon atmospheric pressure, from which it follows that the piping should be direct and with as few valves and angles as possible so as to avoid friction. It is of course essential that the piping should be tight. Whenever possible, new

WATER AND HYDRAULIC PIPING

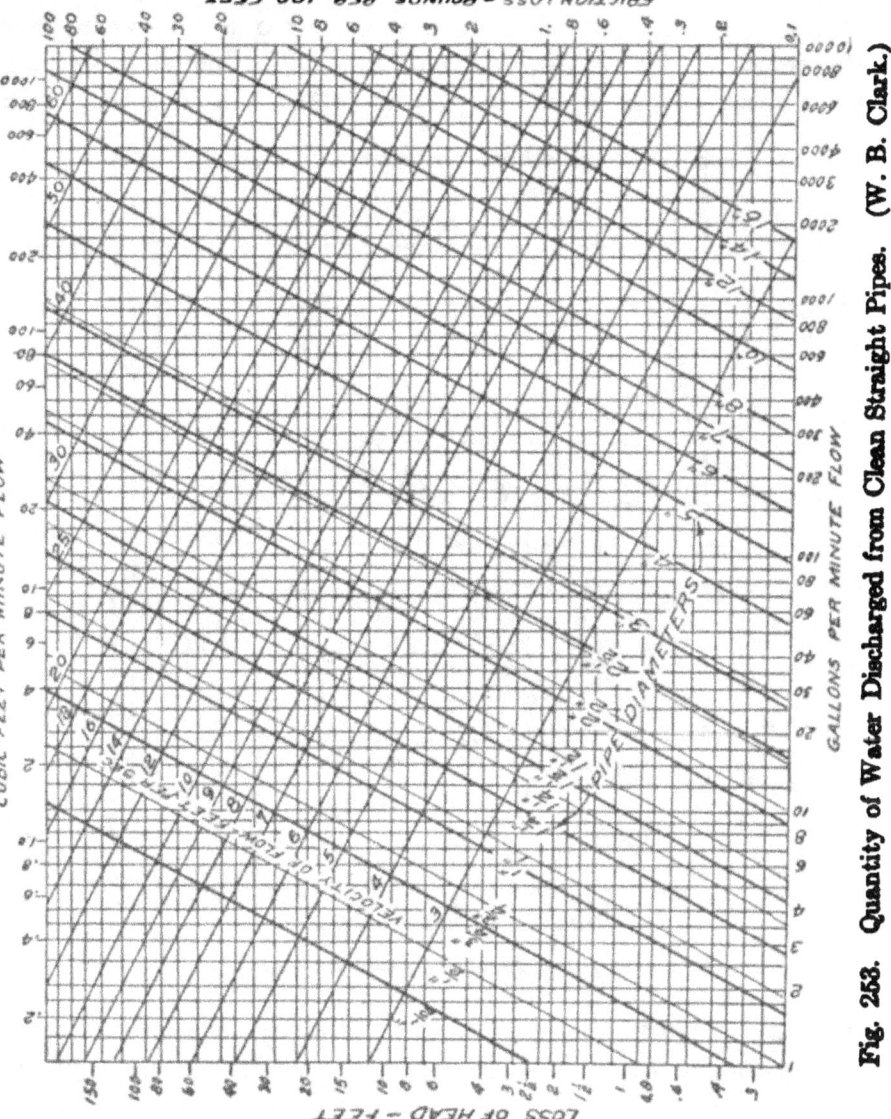

Fig. 253. Quantity of Water Discharged from Clean Straight Pipes. (W. B. Clark.)

230 A HANDBOOK ON PIPING

piping should be tested with water under a pressure of between 25 and 50 pounds per square inch.

The velocity of flow in suction piping under ordinary conditions may be from 150 to 200 feet per minute. For long pipes or high lifts a larger pipe should be provided to reduce the velocity of the water. This velocity depends upon the difference in pressures between that on the surface of the water and in the pump,

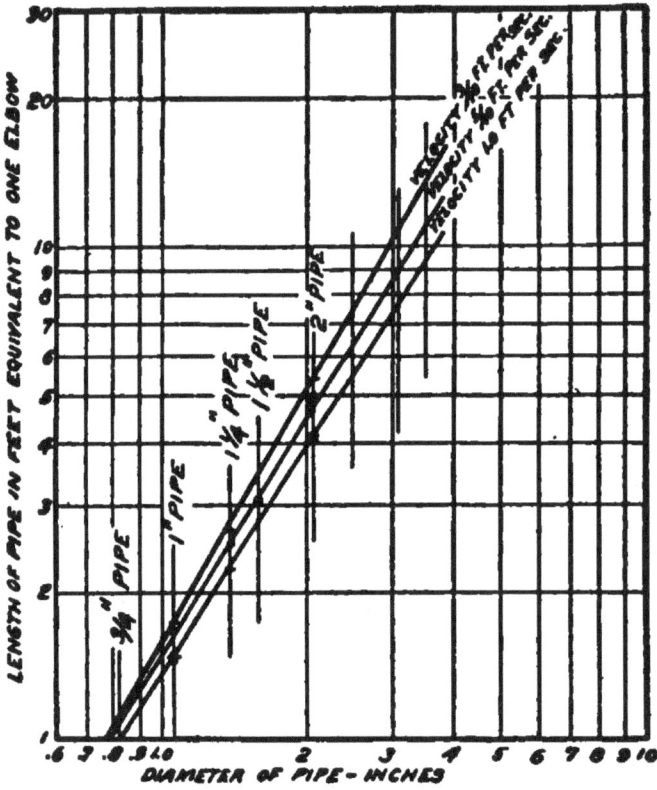

Fig. 254. Length of Pipe Equivalent to an Elbow, Tee, etc.

and cannot exceed that due to the vacuum less the head of water in the suction pipe. If this velocity is too low or the pipe too small, the pump cylinder will not fill on each stroke.

The column of water in the suction pipe must be stopped and started at the end of each stroke. This action can be modified by providing a vacuum chamber into which the water may continue to flow. In every case it should be so placed that the water may flow into it without changing its direction abruptly, Fig. 255. The proper position for the vacuum chamber is at the highest point in the suction pipe and as near the pump as possible, in order to obtain the full benefit of the regulating action.

With long pipe or high lifts a foot-valve should be provided at the lower end to keep the pipe full of water. In the arrangement shown in Fig. 256 a long pipe is avoided by the use of a well, supplied by a pipe through which water flows by gravity. The pump takes its water from this well. This method is frequently used for supplying condensing water.

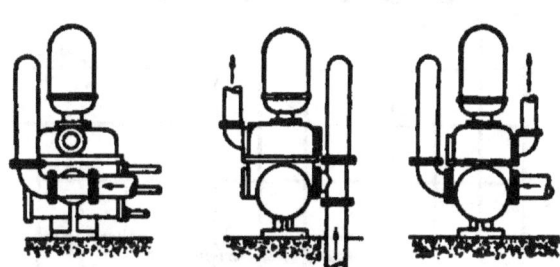

Fig. 255. Arrangement of Suction Piping.

The maximum theoretical height through which cold water can be raised by suction is 34 feet, at sea level. At higher levels this distance is less. Air leaks and friction reduce this so that the practical lift is about 26 feet. When water is heated it gives off vapor or steam at 212° F. under atmospheric pressure. At lower pressures this action takes place at lower temperatures. For this reason hot water cannot be raised as high as cold water by suction. The theoretical heights that hot water may be raised at different temperatures are shown in Fig. 257. It is al-

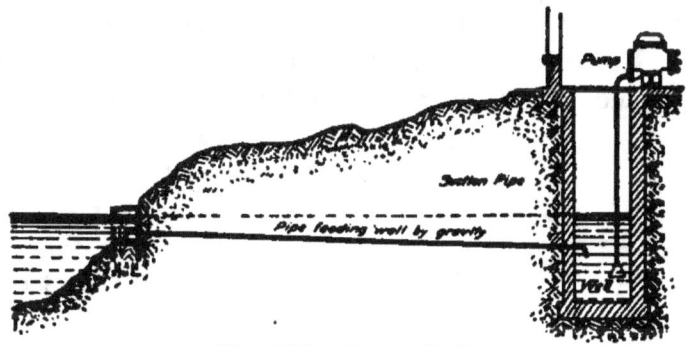

Fig. 256. Pump Well.

ways better to arrange to have hot water flow into the pump, especially if it is above 120° F.

Pump Discharge Piping. — Since the water delivered has the force of the pump pressure it may be given any velocity, and friction is not so serious as in the suction pipe. For this reason the discharge pipe is generally made smaller. A velocity of 250 to 300 feet per minute is a fair value for the discharge pipe, although velocities up to 400 feet per minute are allowable.

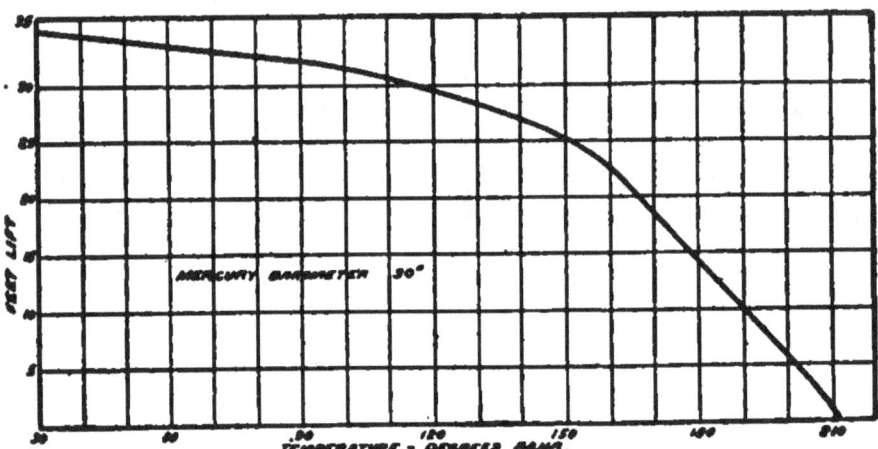

Fig. 257. Theoretical Heights that Hot Water may be Raised by Suction.

Whenever valves are used, either in the suction or discharge piping, the gate form should be adopted as it offers very little resistance to flow, while globe valves offer very large resistance.

Boiler Feed Piping. — Boilers of over 50 horsepower should have at least two methods of feed water supply in order to insure a supply at all times. When city mains are used for boiler feed a tank should be provided with a large capacity where water can be stored, as it is unsafe to depend upon outside sources of supply. The city pipes should feed into this tank and the boilers should be supplied by a pump or injector.

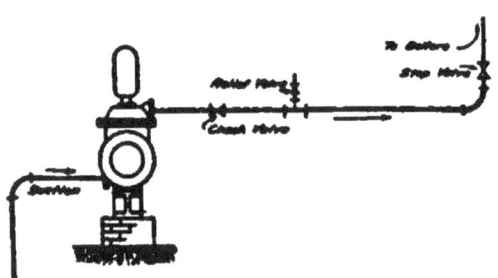

Fig. 258. Boiler Feed Pipe.

For hot feed water brass pipe is to be preferred, although extra heavy steel pipe may be used. The feed pipe to a boiler should be provided with a stop valve and check valve, the stop valve being nearer the boiler. A relief valve located between these two valves is desirable when a pump is used, as it will prevent an undue rise in pressure should the pump be started with the stop valve closed. This relief valve may be much smaller than the discharge pipe. Fig. 258 indicates the arrangement of valves.

Variations in pressure seriously affect the supply of water to the boilers. For this reason it is very desirable to have the boiler feed pipes independent of all other piping. Where a common pipe line is used to supply water for other purposes the opening of valves to draw off water changes the pressure in the pipe and the rate of feed to the boilers. The use of pump governors for maintaining a uniform pressure in the discharge line is described in Chapter VII.

Interior Water Piping. — When the water supply for a building is obtained from city or water company mains, a "corporation cock" is tapped into the street main. Connection is made between this cock and the service pipe leading into the building by lead pipe in order to provide flexibility, which is necessary to take care of any changes in alignment. The size of the cock will of course depend upon the amount of water to be supplied and may be from $3/8$ to $1 1/4$ inches for dwellings, larger sizes being used for public buildings and factories. The sizes of pipes used for delivering water to the different outlets in a building vary, but they may be the same as the fitting supplied if not over 25 feet long. The figures given in Table 82 show the usual range of sizes.

TABLE 82

Sizes of Water Supply Fittings

Fitting	Pipe Diameter
Corporation cocks	$3/8''$ to $2''$
Compression bibbs or faucets for basins, sinks, etc.	$1/4''$ to $2''$
Ball cocks	$1/2''$ to $2''$
Stop cocks	$1/2''$ to $2''$

The interior piping should be of the material best adapted to its use. Plain iron pipe should not be used for hot water as it corrodes very rapidly. Brass pipe is best, but galvanized iron is also suitable. For cold water either galvanized or lead pipe may be used.

Hydraulic Pipe and Fittings. — The dimensions of standard steel pipe are given in Chapter II. For hydraulic work extra strong pipe may be used for pressures up to 1000 pounds and double extra strong for pressures up to 6000 pounds per square inch. Extra strong and double extra strong screwed fittings may be used for making joints.

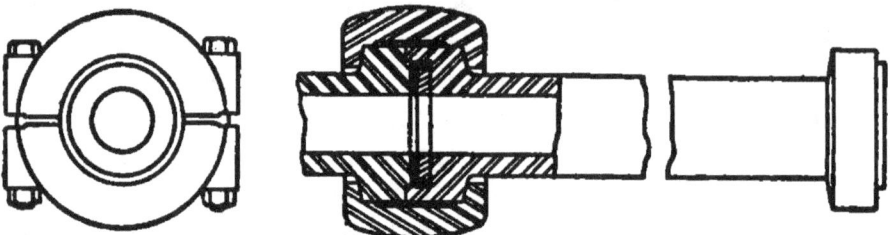

Fig. 259. Hydraulic Pipe and Coupling.

The pipe and couplings shown in Fig. 259 are made by the Watson-Stillman Company for pressures of 1000 and 3000 pounds. The internal diameter may be the same as either extra strong or double extra strong pipe. The flanges are made integral with

Fig. 260. Hydraulic Flange Union.

the pipe and are held together by a very heavy steel split ring, the two parts of which are drawn together by two bolts. A cup packing is used to prevent leakage. The pipe is made in lengths to suit the plans of the installation. Fittings are also made with flanges arranged to use the same clamp couplings. A form of flange union for screwed pipe as adopted by the same company in connection with pumps, presses and accumulators is shown in Fig. 260. It is recommended for pressures of 1000 to 3000 pounds in sizes from three to six inches.

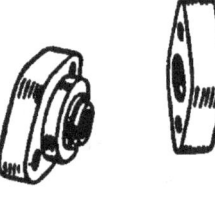

Fig. 261. Hydraulic Flange Fittings.

The two flanges are made of forged steel and have inside thread connections for the pipe. One part is recessed to receive a pro-

WATER AND HYDRAULIC PIPING

jection from the other, a leather washer being inserted between them. Fittings and companion flanges are made with similar joints as shown in Fig. 261.

In hydraulic systems where there is a possibility of shocks which may raise the pressure above the safe amount, or where

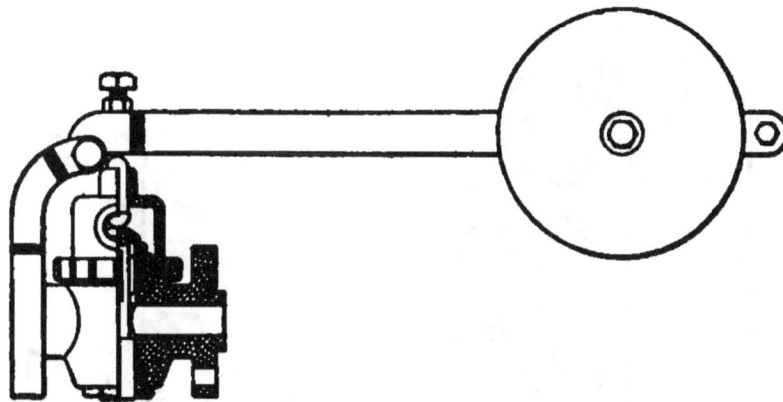

Fig. 262. Hydraulic Safety Valve.

the pressure from the pumps may become excessive due to closure of the discharge pipe, some form of safety valve should be used. These are made in both the spring-weighted form and the lever form. A Schutte hydraulic safety valve is illustrated in Fig. 262, which is made for pressures up to 6000 pounds.

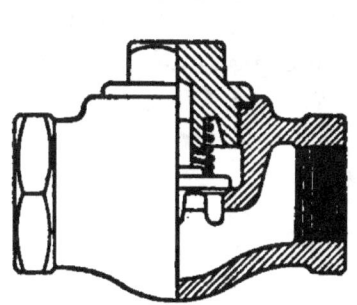

Fig. 263. Hydraulic Check Valve.

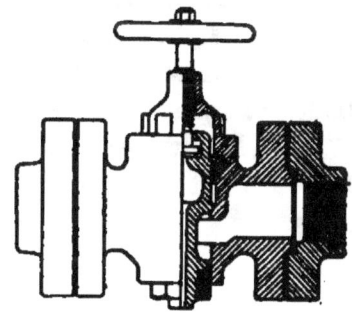

Fig. 264. Balanced Hydraulic Valve.

Hydraulic Valves. — The general forms of valves for hydraulic purposes are the same as those described in Chapter VI, but the construction is heavier. Several valves as made by Schutte & Koerting Company are shown in Figs. 263, 264 and 265. A hydraulic check valve as used for pressures up to 1500 pounds per square inch is shown in Fig. 263. The small spring is to assure seating. A valve for use at the same pressure is shown in Fig. 264. This valve is balanced above and below the seat, so that the flow may be from either end, and requires but a small effort

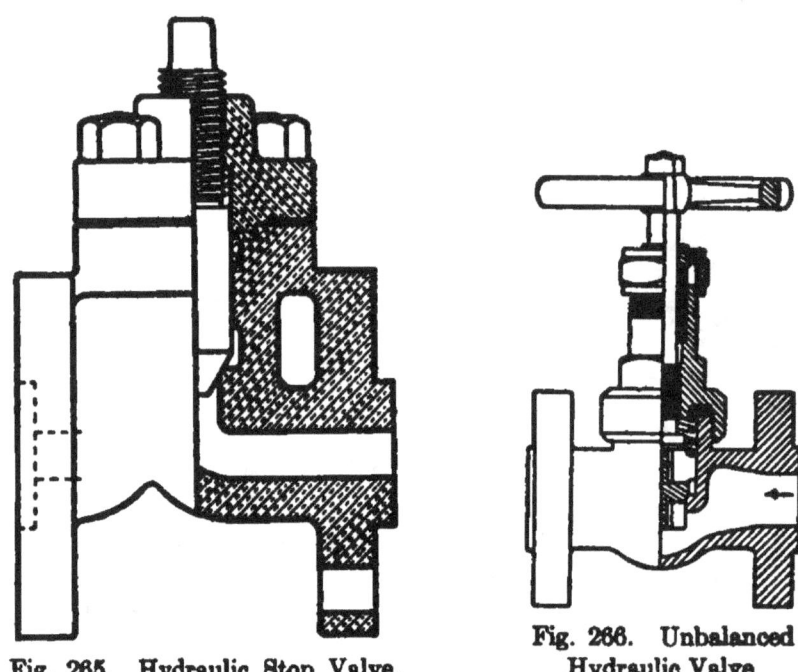

Fig. 265. Hydraulic Stop Valve.

Fig. 266. Unbalanced Hydraulic Valve

for operation. A hydraulic stop valve for pressures up to 9000 pounds per square inch is shown in Fig. 265. An unbalanced hydraulic stop and check valve for working pressures up to 1500 pounds per square inch is shown in Fig. 266. A fine pitch thread and large handwheel are necessary for ease of operation. Such valves are often used on high pressure oil lines for turbine bearings.

CHAPTER XIV

COMPRESSED AIR, GAS AND OIL PIPING

Compressed Air Piping. — Compressed air piping holds many features in common with steam piping. It should be arranged as direct as possible, and with provision for drainage and expansion. All pockets, loops or places where moisture might collect should be carefully drained. There is also the danger of freezing in cold weather unless drains are provided. For these reasons separators should be installed at the low points on the pipe line. Such separators can be similar to the usual steam separator, or can take the form of a receiver. In many cases it is well to have a receiver near the point where the air is used and especially when

Fig. 267. Expansion Joint Used for Compressed Air Line — Nicholson, Pa. Tunnel, D. L. & W. Cut-off.

there is a widely changing demand for air. Often a receiver is desirable at both ends of the pipe line, more especially with long lines.

Friction and air leakage are constant sources of loss with air piping and should be carefully avoided by making the line as tight as possible, and using long turn fittings. Gate valves and shut-off cocks should be extra heavy. Careful attention to supports will do much toward eliminating vibration and so help to maintain tight joints.

The author is indebted to the Ingersoll-Rand Company, for Fig. 267 which illustrates a short section of a pipe used on the contract for the Nicholson Pennsylvania Tunnel for the D., L. and W. cut-off, and shows a simple but effective form of expansion joint. This piping has been used on several jobs and is still in perfect condition, due to the care exercised in laying it and a special graphite mixture used on all joints. This pipe is laid so

Fig. 268. Values for Coefficient C.

as to drain toward the air receiver from which any moisture can be blown off.

Compressed Air Transmission. — In calculating pipe lines for compressed air, Unwin's formula for flow of fluids as stated below may be used.

Q = Volume in cubic feet per minute at pressure P_2.

P_1 = pressure at entrance in pounds per square inch.

P_2 = pressure at end of pipe in pounds per square inch.

d = diameter of pipe in inches.

L = length of pipe in feet.

w_1 = weight of air in pounds per cubic foot at pressure P_1.

C = an experimental coefficient.

$$Q = C\left[\frac{(P_1 - P_2)d^5}{w_1 L}\right]^{1/2} \dots\dots\dots\dots\dots\dots(30)$$

COMPRESSED AIR, GAS AND OIL PIPING

TABLE 83. — WEIGHT OF AIR AT VARIOUS PRESSURES AND TEMPERATURES

Based on an Atmospheric Pressure of 14.7 Pounds Absolute at Sea Level

Temp. of Air Deg. Fahr.	Gauge Pressure, Pounds																				
	0	5	10	20	30	40	50	60	70	80	90	100	110	120	130	140	150	175	200	225	250	300
	Weight in Pounds per Cubic Foot																				
−20	.0900	.1205	.1515	.2125	.2744	.3360	.3970	.4580	.5190	.5800	.6410	.702	.7635	.825	.886	.948	1.010	1.165	1.318	1.465	1.625	1.930
−10	.0882	.1184	.1485	.2090	.2685	.3283	.3880	.4478	.5076	.5674	.6272	.687	.747	.807	.868	.928	.989	1.139	1.288	1.438	1.588	1.890
0	.0864	.1160	.1455	.2040	.2630	.3215	.3800	.4385	.4970	.5555	.6140	.672	.731	.790	.849	.908	.968	1.114	1.260	1.406	1.553	1.850
10	.0846	.1136	.1425	.1995	.2568	.3145	.3720	.4292	.4863	.5433	.6006	.658	.716	.774	.832	.889	.947	1.090	1.233	1.376	1.520	1.810
20	.0828	.1112	.1395	.1955	.2516	.3071	.3645	.4205	.4770	.5330	.5890	.645	.701	.757	.813	.869	.927	1.067	1.208	1.348	1.489	1.770
30	.0811	.1088	.1366	.1916	.2465	.3015	.3570	.4121	.4672	.5221	.5771	.632	.687	.742	.797	.852	.908	1.046	1.184	1.322	1.460	1.735
40	.0795	.1067	.1338	.1876	.2415	.2954	.3503	.4038	.4576	.5114	.5652	.619	.673	.727	.781	.835	.890	1.025	1.161	1.296	1.431	1.701
50	.0780	.1045	.1310	.1839	.2367	.2905	.3432	.3960	.4487	.5014	.5541	.607	.660	.713	.766	.819	.873	1.006	1.139	1.271	1.403	1.668
60	.0764	.1025	.1283	.1803	.2323	.2840	.3362	.3882	.4402	.4927	.5447	.596	.649	.700	.752	.804	.856	.988	1.116	1.245	1.376	1.636
70	.0750	.1005	.1260	.1770	.2280	.2791	.3302	.3808	.4316	.4824	.5332	.584	.635	.686	.737	.788	.839	.967	1.095	1.223	1.350	1.604
80	.0736	.0988	.1239	.1738	.2237	.2739	.3242	.3738	.4234	.4729	.5224	.572	.622	.673	.723	.774	.824	.949	1.074	1.199	1.325	1.573
90	.0723	.0970	.1218	.1707	.2195	.2688	.3182	.3670	.4154	.4639	.5122	.561	.611	.660	.709	.759	.809	.932	1.054	1.177	1.300	1.544
100	.0710	.0954	.1197	.1676	.2155	.2638	.3122	.3602	.4079	.4555	.5033	.551	.599	.648	.696	.745	.794	.914	1.035	1.155	1.276	1.517
110	.0698	.0937	.1176	.1645	.2115	.2593	.3070	.3542	.4011	.4481	.4950	.542	.589	.637	.685	.732	.780	.899	1.017	1.135	1.254	1.491
120	.0686	.0921	.1155	.1618	.2080	.2549	.3018	.3481	.3944	.4403	.4866	.533	.579	.626	.673	.720	.767	.884	1.001	1.118	1.234	1.465
130	.0674	.0905	.1135	.1590	.2045	.2505	.2966	.3446	.3924	.4296	.4770	.524	.570	.616	.662	.708	.754	.869	.984	1.099	1.214	1.440
140	.0663	.0889	.1115	.1565	.2015	.2465	.2915	.3364	.3813	.4262	.4711	.516	.561	.606	.651	.696	.742	.855	.968	1.081	1.194	1.416
150	.0652	.0874	.1096	.1541	.1985	.2425	.2865	.3308	.3751	.4193	.4636	.508	.552	.596	.640	.685	.730	.841	.953	1.064	1.175	1.392
175	.0626	.0840	.1054	.1482	.1910	.2335	.2755	.3181	.3607	.4033	.4450	.488	.531	.573	.616	.658	.701	.808	.914	1.021	1.128	1.337
200	.0603	.0809	.1014	.1427	.1840	.2248	.2655	.3054	.3473	.3882	.4291	.470	.511	.552	.592	.633	.674	.776	.879	.982	1.084	1.287
225	.0581	.0779	.0976	.1373	.1770	.2163	.2555	.2949	.3344	.3738	.4129	.452	.491	.531	.570	.609	.649	.747	.846	.944	1.043	1.240
250	.0560	.0751	.0941	.1323	.1705	.2085	.2466	.2845	.3223	.3602	.3981	.436	.474	.513	.551	.589	.627	.722	.817	.912	1.007	1.197
275	.0541	.0726	.0910	.1278	.1645	.2011	.2378	.2745	.3111	.3478	.3844	.421	.458	.494	.531	.568	.605	.697	.789	.881	.972	1.155
300	.0523	.0707	.0881	.1237	.1592	.1945	.2300	.2654	.3008	.3362	.3716	.407	.442	.478	.513	.549	.585	.673	.762	.852	.940	1.118
350	.0491	.0658	.0825	.1160	.1495	.1828	.2160	.2492	.2824	.3156	.3488	.382	.415	.449	.482	.516	.549	.632	.715	.799	.883	1.048
400	.0463	.0621	.0779	.1090	.1405	.1720	.2030	.2348	.2601	.2974	.3287	.360	.391	.423	.454	.486	.517	.596	.674	.753	.831	.987
450	.0437	.0586	.0735	.1033	.1330	.1628	.1925	.2220	.2515	.2810	.3105	.340	.369	.399	.429	.458	.488	.562	.637	.711	.786	.934
500	.0414	.0555	.0696	.0978	.1260	.1540	.1820	.2100	.2380	.2660	.2940	.322	.351	.379	.407	.435	.463	.534	.604	.675	.746	.885
550	.0394	.0528	.0661	.0930	.1198	.1464	.1730	.1996	.2262	.2528	.2794	.306	.333	.359	.386	.413	.440	.507	.573	.641	.709	.841
600	.0376	.0504	.0631	.0885	.1140	.1395	.1650	.1904	.2158	.2412	.2668	.292	.317	.343	.368	.393	.419	.483	.547	.611	.675	.801

TABLE 84.—COMPRESSED AIR TRANSMISSION [*Copyright, 1906, by Ingersoll-Rand Company*]

Size of Pipe, Inches
Loss of Pressure, in Pounds, by Friction in 1000 Feet Lengths of Pipe
[At 60 Pounds Gauge]

Delivery in Cu. Ft. of Compressed Air per Min.	Equiv. Delivery in Cu. Ft. of Free Air per Min.	1	1¼	1½	2	2½	3	3½	4	4½	5	6	7	8	9	10	12	14	16
9.84	50	18.24	5.06	1.95	.42	.13	.05												
14.73	75		11.34	4.33	.95	.29	.11	.05											
19.64	100		20.16	7.79	1.69	.52	.19	.08	.04										
24.60	125			12.23	2.65	.81	.30	.13	.07	.03									
29.45	150			17.53	3.80	1.16	.44	.19	.09	.05	.03								
34.44	175				5.17	1.58	.59	.26	.13	.07	.04								
39.35	200				6.77	2.09	.78	.36	.17	.09	.06								
49.20	250				10.61	3.24	1.22	.55	.27	.15	.08	.01							
58.90	300				15.20	4.65	1.78	.78	.38	.21	.12	.02	.01						
68.6	350					6.31	2.37	1.07	.53	.29	.17	.03	.01						
78.6	400					8.28	3.11	1.40	.69	.39	.22	.05	.02	.01					
88.4	450					10.47	3.94	1.77	.88	.48	.28	.06	.03	.01	.01				
98.4	500						4.88	2.20	1.08	.60	.34	.08	.04	.01	.01				
118.1	600						7.03	3.17	1.56	.87	.49	.11	.05	.02	.02	.01			
137.5	700						9.52	4.29	2.12	1.17	.67	.14	.06	.03	.03	.02			
156.6	800							5.57	2.75	1.52	.87	.19	.09	.04	.04	.03	.01		
176.5	900							7.08	3.49	1.94	1.17	.27	.12	.06	.05	.03	.01		
196.4	1000							8.77	4.33	2.40	1.37	.34	.15	.08	.06	.04	.02	.01	.01
294.5	1500								9.73	5.39	3.08	.43	.19	.09	.06	.05	.02	.01	.01
393.7	2000										5.51	.54	.24	.12	.12	.09	.03	.01	.02
492	2500									9.65	8.61	1.20	.55	.27	.15	.16	.06	.03	.03
589	3000											2.16	.98	.41	.27	.25	.09	.04	.04
686	3500											3.36	1.53	.77	.42	.36	.14	.06	.05
786	4000											4.82	2.19	1.11	.61	.48	.19	.09	.07
884	4500											6.54	2.97	1.50	.83	.63	.25	.11	.90
													3.91	1.98	1.08	.79	.32	.15	
984	5000												4.94	2.51	1.36	.99	.39	.18	
													6.19	3.10	1.69				

For longer or shorter pipes the friction loss is proportional to the length, *i.e.*, for 500 feet ½ of the above; for 4000 feet four times the above, etc.

Loss of Pressure, in Pounds, by Friction in 1000 Feet Lengths of Pipe
[At 80 Pounds Gauge]

7.74	50	14.31	3.96	1.53	.33	.10	.03	.01								
11.3	75		8.46	3.26	.71	.21	.06	.03								
15.2	100		15.31	5.92	1.28	.39	.14	.06	.01							
19.4	125			9.64	2.09	.64	.24	.11	.03							
23.2	150			13.79	2.99	.91	.34	.15	.05							
27.2	175				4.09	1.25	.47	.21	.07							
31.0	200				5.34	1.63	.61	.27	.10							
38.7	250				8.32	2.54	.96	.43	.13							
46.5	300				12.01	3.67	1.38	.62	.21							
54.2	350					4.99	1.88	.84	.30	.01	.01					
62.0	400					6.53	2.45	1.11	.41	.01	.01					
69.7	450					8.25	3.13	1.40	.54	.01	.01					
77.4	500					10.81	3.83	1.73	.69	.02	.01					
92.9	600						5.61	2.46	.85	.03	.02	.01				
108.2	700						7.46	3.37	1.22	.05	.03	.01				
124.0	800						9.36	4.42	1.66	.06	.04	.02	.01			
139.5	900							5.61	2.18	.08	.05	.03	.01			
152	1000							6.64	2.77	.10	.06	.03	.02	.01		
232	1500							15.41	4.24	.15	.09	.04	.02	.01		
310	2000								7.62	.20	.12	.06	.03	.02		
387	2500								11.79	.27	.15	.08	.04	.02	.01	
465	3000									.34	.18	.09	.05	.03	.01	.01
542	3500									.40	.22	.12	.06	.03	.01	.01
620	4000									.95	.43	.21	.12	.06	.02	.02
697	4500									1.69	.77	.33	.19	.09	.03	.02
774	5000									2.64	1.19	.48	.28	.12	.05	.03

TABLE 84 (CONTINUED) — COMPRESSED AIR TRANSMISSION

Loss of Pressure, in Pounds, by Friction in 1000 Feet Lengths of Pipe
[At 100 Pounds Gauge]

Delivery in Cu. Ft. of Compressed Air per Min.	Equiv. Delivery in Cu. Ft. of Free Air per Min.	Size of Pipe, Inches														
		1	1¼	1½	2	2½	3	3½	4	4½	5	6	7	8	9	10	12	14	16
6.41	50	11.89	3.29	1.28	.27	.08	.03	.01											
9.61	75		7.42	2.87	.62	.19	.07	.03	.01										
12.81	100		13.20	5.11	1.15	.34	.12	.05	.02	.01									
15.81	125			7.75	1.68	.52	.19	.08	.04	.02	.01								
19.22	150			11.42	2.48	.76	.29	.13	.06	.03	.02								
22.39	175				3.36	1.03	.39	.17	.09	.04	.03								
25.62	200				4.43	1.36	.51	.23	.12	.06	.04	.01	.01						
31.62	250				6.72	2.06	.77	.35	.17	.09	.05	.02	.01						
38.44	300				9.95	3.04	1.14	.51	.25	.14	.08	.03	.01	.01					
44.78	350				13.41	4.11	1.54	.69	.34	.19	.11	.04	.02	.01					
51.24	400					5.40	2.06	.92	.45	.25	.15	.05	.03	.01	.01	.01			
57.65	450					6.85	2.57	1.16	.57	.32	.18	.07	.03	.02	.01	.01			
63.24	500					8.21	3.08	1.39	.68	.38	.22	.08	.04	.02	.02	.02	.01		
76.88	600					12.21	4.58	2.14	1.03	.57	.33	.12	.05	.03	.02	.02	.01		
89.56	700						6.19	2.79	1.38	.77	.44	.17	.07	.04	.03	.03	.01		
102.5	800						8.13	3.67	1.81	1.00	.57	.22	.10	.05	.04	.04	.02	.01	
115.3	900						10.23	4.64	2.29	1.27	.76	.28	.13	.06	.04	.04	.02	.01	
126.5	1000						12.39	5.00	2.76	1.23	.88	.34	.16	.08	.04	.04	.03	.01	.01
192.2	1500							12.81	6.68	3.51	2.03	.78	.36	.18	.09	.05	.05	.02	.01	.01
256.2	2000								11.35	6.61	3.62	1.41	.67	.33	.18	.10	.10	.04	.02	.01
316.2	2500									9.50	5.51	2.14	.97	.49	.27	.16	.16	.06	.03	.01
384.4	3000									14.04	8.11	3.16	1.44	.76	.39	.23	.23	.09	.04	.02
447.8	3500										10.95	4.26	1.93	.98	.53	.31	.31	.12	.05	.03
512.4	4000										14.48	5.59	2.55	1.30	.72	.41	.41	.16	.07	.04
576.5	4500											7.04	3.22	1.84	.89	.52	.52	.21	.09	.05
632.4	5000											8.51	3.88	1.93	1.07	.63	.63	.25	.11	.08

For longer or shorter pipes the friction loss is proportional to the length, i.e., for 500 feet ½ of the above; for 4000 feet four times the above, etc.

Loss of Pressure, in Pounds, by Friction in 1000 Feet Lengths of Pipe

[At 125 Pounds Gauge]

5.26	50	9.88	2.70	1.05	.23	.07	.03	.01											
7.89	75	22.20	6.07	2.37	.51	.16	.06	.03	.01										
10.51	100	39.50	10.82	4.22	.91	.28	.10	.05	.02	.01									
13.15	125	......	16.88	6.58	1.42	.43	.16	.07	.04	.02	.01								
15.79	150	......	24.33	9.47	2.04	.63	.23	.11	.05	.03	.02	.01							
18.41	175	......	33.05	12.90	2.78	.85	.32	.14	.07	.04	.02	.01							
21.05	200	......	......	16.84	3.63	1.11	.42	.19	.09	.05	.03	.01							
26.30	250	......	......	26.30	5.68	1.73	.65	.29	.15	.08	.05	.02	.01						
31.58	300	......	......	37.90	8.18	2.51	.94	.42	.21	.12	.07	.03	.01						
36.81	350	......	......	......	11.08	3.39	1.27	.58	.28	.16	.09	.04	.02	.01					
42.10	400	......	......	......	14.51	4.44	1.67	.75	.37	.21	.12	.05	.02	.01					
47.30	450	......	......	......	18.38	5.61	2.11	.95	.47	.26	.15	.06	.03	.01	.01				
52.60	500	......	......	......	22.68	6.95	2.61	1.18	.58	.32	.18	.07	.03	.02	.01				
63.20	600	......	......	......	......	10.00	3.76	1.69	.84	.46	.27	.10	.05	.02	.01	.01			
73.70	700	......	......	......	......	13.60	5.11	2.31	1.14	.63	.36	.14	.06	.03	.02	.01			
84.20	800	......	......	......	......	17.80	6.68	3.01	1.49	.83	.47	.18	.08	.04	.02	.01	.01		
94.70	900	......	......	......	......	......	8.45	3.81	1.88	1.04	.60	.23	.11	.05	.03	.02	.01		
105.1	1000	......	......	......	......	......	10.42	4.71	2.32	1.29	.74	.29	.13	.07	.04	.02	.01		
157.9	1500	......	......	......	......	......	23.48	10.59	5.23	2.90	1.65	.64	.29	.15	.08	.05	.02	.01	
210.5	2000	......	......	......	......	......	......	18.81	9.20	5.15	2.94	1.15	.52	.26	.15	.08	.03	.02	.01
263.0	2500	......	......	......	......	......	......	29.40	14.52	8.05	4.60	1.80	.82	.41	.23	.13	.05	.02	.01
315.8	3000	......	......	......	......	......	......	......	20.90	11.59	6.63	2.59	1.18	.60	.33	.19	.07	.03	.02
368.1	3500	......	......	......	......	......	......	......	28.51	15.78	9.01	3.53	1.61	.81	.45	.26	.10	.04	.02
422.0	4000	......	......	......	......	......	......	......	......	20.61	11.80	4.61	2.19	1.06	.58	.34	.13	.06	.03
473.0	4500	......	......	......	......	......	......	......	......	26.10	14.90	5.83	2.65	1.34	.73	.43	.17	.08	.04
526.0	5000	......	......	......	......	......	......	......	......	32.20	18.45	7.20	3.27	1.65	.90	.53	.21	0	.05

Values of w_1 are given in Table 83 which is from Ingersoll-Rand Company's catalog. The coefficient C may be taken from the curve, Fig. 268, where a number of values have been plotted.

For computations having to do with compressed air transmission, the information given in Tables 84 and 85 which are from the catalog of Ingersoll-Rand Company, may be used.

TABLE 85

MULTIPLIERS FOR DETERMINING THE VOLUME OF FREE AIR

At Various Altitudes which, when Compressed to Various Pressures, is Equivalent in Effect to a Given Volume of Free Air at Sea Level

Altitude in Feet	Barometric Pressure		Gauge Pressure				
	Inches of Mercury	Pounds per Square Inch	60 Lbs.	80 Lbs.	100 Lbs.	125 Lbs.	150 Lbs.
			Multipliers				
0	30.00	14.75	1.000	1.000	1.000	1.000	1.000
1000	28.88	14.20	1.032	1.033	1.034	1.035	1.036
2000	27.80	13.67	1.064	1.066	1.068	1.071	1.072
3000	26.76	13.16	1.097	1.102	1.105	1.107	1.109
4000	25.76	12.67	1.132	1.139	1.142	1.147	1.149
5000	24.79	12.20	1.168	1.178	1.182	1.187	1.190
6000	23.86	11.73	1.206	1.218	1.224	1.231	1.234
7000	22.97	11.30	1.245	1.258	1.267	1.274	1.278
8000	22.11	10.87	1.287	1.300	1.310	1.319	1.326
9000	21.29	10.46	1.329	1.346	1.356	1.366	1.374
10000	20.49	10.07	1.373	1.394	1.404	1.416	1.424

The Air Lift Pumping System. — The use of compressed air as a means of raising water is illustrated in Fig. 269. This form of air lift pump was patented by Dr. E. S. Pohle in 1886. Several arrangements for the lower end of the air pipe are shown. The system is composed entirely of piping and the operation is as follows: air is piped to the lower end of the water pipe where it mixes with the water. As this mixture is lighter than the water it is forced up the pipe and out at the discharge. The lift of course is the distance from the water level to the discharge opening. The distance from the water level to the bottom of the pipe where the air is introduced is called the submergence. The amount of submergence to give most efficient results varies greatly and is often determined by trial for a given installation.

Concerning the proportions of air lift wells, *Practical Engineer*, January 1st, 1916, gives Table 86, and says: "there are two classes of submergence, starting submergence, which is temporary, and running submergence, which is the important factor in any pumping proposition. It is usually expressed as a percentage of the total length of the water column from the point where the air is introduced to the point of discharge.

Necessary percentage of submergence varies in accordance with the lift, low lifts requiring proportionately more submer-

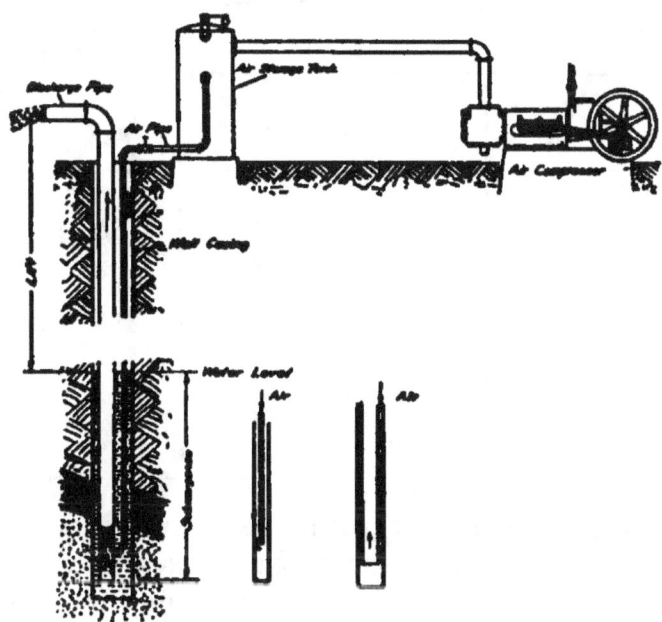

Fig. 269. Air Lift Pumping System.

gence than high lifts. The range of these percentages is found within the following limitations: for a lift of 20 feet, 66 per cent.; for a lift of 500 feet, 41 per cent.

Knowing the total lift and running submergence, the approximate amount of free air required can be calculated from the following formula:

$$V = \frac{L}{Log\left[\frac{s+34}{34}\right] \times C} \quad \ldots \ldots \ldots \ldots \ldots \ldots (31)$$

where

V = volume of free air to raise one gallon of water.
L = total lift in feet.
s = running submergence in feet.
C = constant found in the following table.

Lift in Feet (L)	Constant
10 to 60 inclusive	245
61 to 200 "	233
201 to 500 "	216
501 to 650 "	185
651 to 750 "	156

TABLE 86

Well Pipe Sizes

Well Casing Inches	Side Inlet			Center Air Pipe	
	Water Pipe	Air Pipe	Capacity Gallons per Minute	Air Pipe	Capacity Gallons per Minute
3½	...	...	...	1¼	115
4	1½	¾	25	1½	150
4½	2	1	50	...	...
5	2½	1	75	2	240
6	3	1¼	105	2	360
7	3½	1½	145	...	...
8	4	1½	190	...	...
9	5	2	300	...	...
10	6	2	425	...	...

Gas Fitting. — Piping for gas inside of buildings is generally spoken of as gas fitting. It is not the purpose of this chapter to cover thoroughly the field of gas fitting, but to give only such information as might be of use to those who occasionally have some gas fitting to do. Gas piping should always be carefully done and thoroughly tested.

The various pipes used in conveying gas from the source of supply to points where it is burned are distinguished by different names, depending upon their particular purpose. The cast-iron pipes used to convey the gas through the streets are called mains. From the mains, service pipes of cast-iron or wrought iron lead to the building. These should be taken from the top of the mains. Inside of the building the distributing pipes carry the gas to the lights, heaters, etc. A riser is a vertical pipe through which the gas flows upward. A drop is one in which the gas flows downward.

Materials. — Cast iron and standard steel or wrought iron pipe are used for gas piping. Fittings should be of malleable iron and galvanized. Cast-iron fittings are heavier than malleable iron and are more easily cracked or otherwise damaged. Gas fittings in addition to those shown in the chapter on Pipe Fittings are illustrated in Fig. 270. For turning on and off gas in service pipes gas cocks are used, as shown in Fig. 271. Fig. 272 is a meter cock, and Fig. 273 is a gas stove cock.

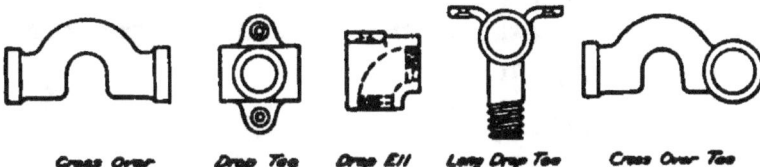

Fig. 270. Gas Fittings.

Location of Piping. — The gas supply pipe from the street should incline upward from the main in order that any condensation will drain back into the main. The amount of slope is not material but should be sufficient to prevent the possibility of water pockets forming, due to the settling of the pipe. In every case the pipe should be firmly supported, and should be tested for leaks before filling in the trench.

The piping in the building should be run to the fixtures with as few fittings as possible, and should be pitched to provide drainage.

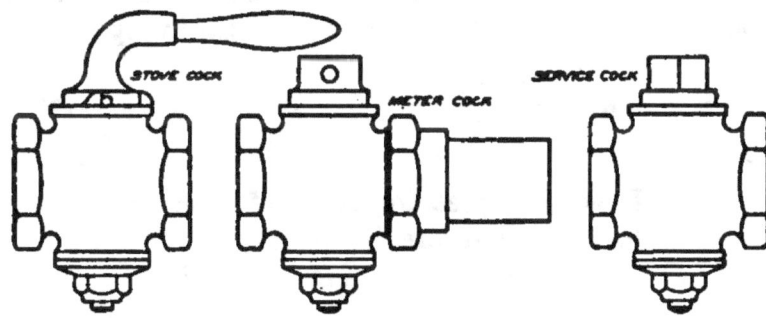

Figs. 271, 272, and 273. Stove Cock, Meter Cock, Service Cock.

For this reason it is better to supply burners and fixtures by risers rather than by drops.

Sizes of Pipes. — The size of pipes should be based upon the maximum quantity (cubic feet) which is likely to be used. For lights the meter rating of five cubic feet per hour may be used in estimating the sizes of pipes. For cook stoves the size of pipe will vary from $3/4$ inch up to $1 1/2$ inches or more, depending upon the size of the stove. Service pipes should never be less than $3/4$

inch, regardless of length, and in cold climates where there is a possibility of frost forming it is better to use at least one inch pipe. Insulating material may be used to protect the piping from extreme cold. Alcohol may be poured into the pipe and allowed to melt the frost which forms due to moisture in the gas. The size of pipe for a given quantity of gas may be figured by Molesworth's formula for maximum supply in cubic feet per hour.

$$V = 1000 \left[\frac{d^5 h}{GL}\right]^{1/2} \quad \quad \quad \quad \quad (32)$$

in which

V = maximum cubic feet per hour.
d = diameter of pipe, inches.
h = pressure, inches of water.
G = specific gravity of gas (air = 1).
L = length of pipe in yards.

The value of G may be taken at from .40 to .65 based on a value of 1 for air.

A series of articles "Instructions for Gas Company Fitters," by Mr. George Wehrle, published in *The Gas Age*, New York, permission of which has been given the author under the copyright of the former, give complete particulars of the above subject. Mr. Wehrle uses formula (33) for the flow of gas in pipes.

$$V = 1350 \left[\frac{d^5(P_1 - P_2)}{GL}\right]^{1/2} \quad \quad \quad \quad (33)$$

in which

P_1 = initial pressure, inches of water.
P_2 = final pressure, inches of water.
other letters as in formula (32).

Quoting further from Mr. Wehrle's articles on the subject of "Conductivity of Pipes," he says:

"The conductivity of a pipe is its carrying capacity in volumes of gas, which is variable under certain conditions of pressure, length and gravity of gas.

"All fitters know that the elimination of 'dead ends' in gas pipes is favorable to the carrying capacity of the pipe, but to just what extent, and the cause, should be understood.

"In the accompanying table, Fig. 274, explanation is given of results to be obtained under different conditions representing different methods of installing pipes. The first illustration represents a pipe supplied from one end, discharging from the other, as a service pipe. This condition is represented as unity in all quantities. The second illustration represents a pipe fed from both ends, discharging from a point midway between the ends. Here a comparison with the first illustration shows that such an installation, considering the specific gravity of the gas to be the same, will pass 2.8 times the amount of gas in a given time for the same length, diameter and pressure drop; will pass the same amount of gas for the same length and diameter with a pressure

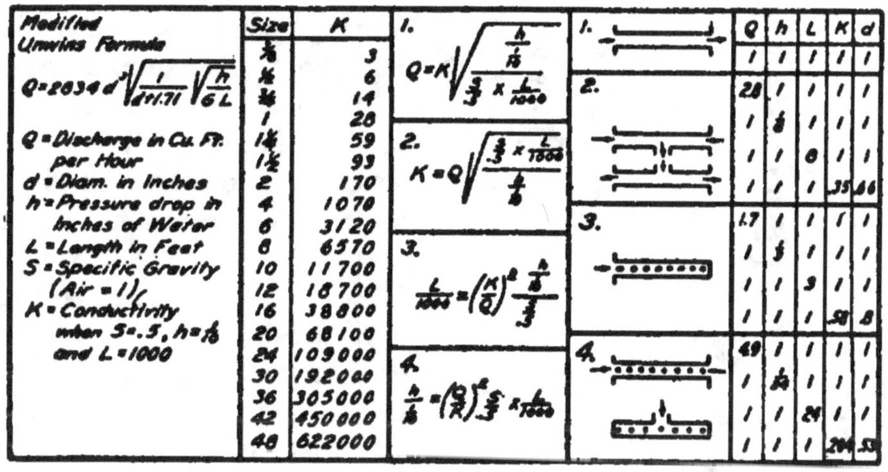

Fig. 274. Pipe Conductivity Chart.

drop of $1/8$ as much; will pass the same amount of gas with the same pressure drop and diameter with a length eight times as great; or will pass the same amount of gas for the same length and pressure drop with a diameter .66 as great.

"The third and fourth illustrations bear the same relative value to each other as the first and second, but have different values, as shown, when compared with the first. Comparisons of any one with another are easily made. Exactly the same values are obtained if the pipe is fed from the center, discharging from both ends.

"Problems very often occur in house piping where a knowledge of the above information proves of great value. In meter header installations, illustration No. 4, or its counterpart (fed from the middle, discharging both ways) should in all cases be used,

250 A HANDBOOK ON PIPING

even at the expense of additional pipe over No. 3. In street main work, it is general practice to tie in all dead ends possible, most companies allowing a considerable expense to be used for that purpose."

Testing. — Gas Pipe systems should be tested before turning on the gas in order to be certain that all the joints are gas tight.

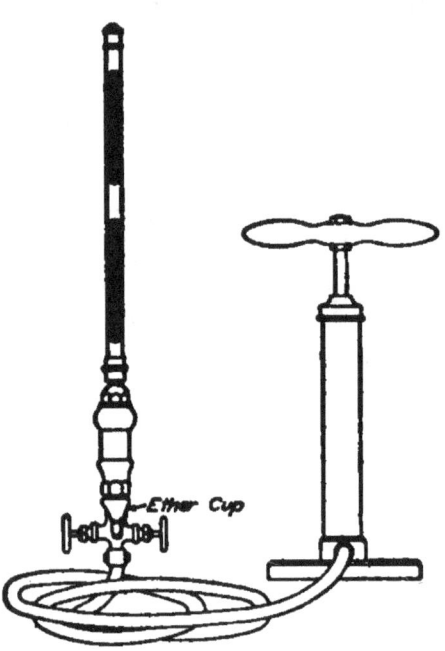

Fig. 275. Gas Proving Pump.

Gas fitters' proving pumps are made for this purpose. They may be used with a mercury column or a spring gage, the former being preferred. A proving pump is shown in Fig. 275. The pump is used to force air into the system, and a pressure should be maintained for one hour, with a pressure loss of not more than $1/4$ inch of mercury (about $1/8$ pound per square inch pressure). The rate of drop in pressure is an indication of the extent of leakage. In order to locate the leaks an ether cup is attached to the pump through which ether may be introduced into the piping, and by its odor indicate the points of leakage.

Gas Meters. — Gas is ordinarily measured in cubic feet. The usual form of meter for measuring gas is illustrated in Figs. 276 and 277. This form is called a dry gas meter, and generally consists of two chambers which are separated from each other by partitions and flexible diaphragms. The operation may be understood by reference to the diagram, Fig. 277. The gas from the street enters through pipe *1* to the space *A*, and then through opening *2* to spaces *B, B*, where it exerts pressure against the diaphragm *3, 3* and so forces the gas from spaces *C, C*, out through *4, 5* and *6* to the piping system. When the spaces *B, B* are filled the slide valve *7* is moved so as to open port *4* to space *A* and to connect port *2* with the outlet pipe. Gas then flows into spaces *C, C* and moves the diaphragm, expelling the gas from spaces

B, B. This operation is automatic and is communicated to the recording discs which record the amount of gas measured by the meter. A view of the recording dials is shown in Fig. 278. To read the meter begin with the dial at the left and read the smaller of the two numbers on each side of the hand on each of the three dials, and add two ciphers. The reading as illustrated is 66200. Such a reading, subtracted from the previous reading will give the amount of gas consumed in the interval. The small dial

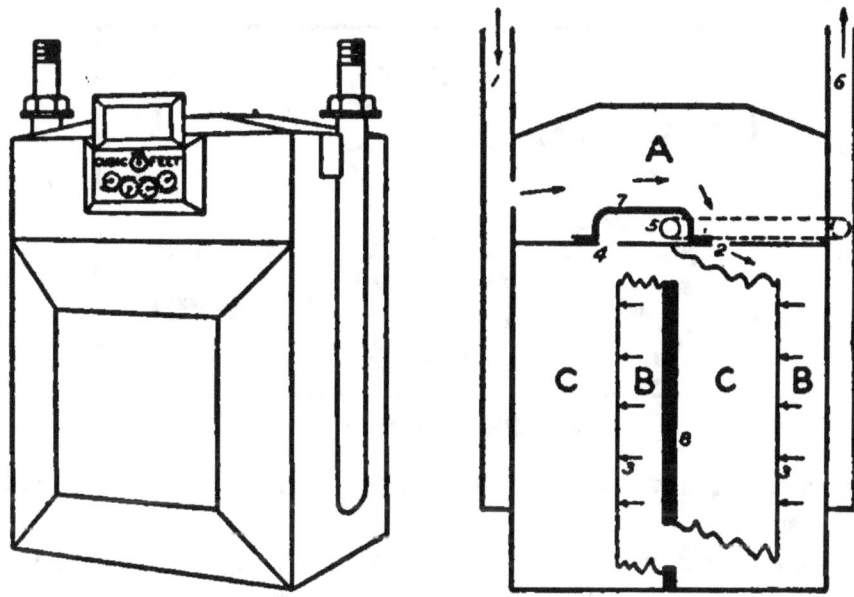

Fig. 276. Dry Gas Meter.   Fig. 277. Dry Gas Meter Diagram.

may be used to observe the rate of consumption as well as to indicate leaks in the system. Before connecting a gas meter it is advisable to be sure that the pipes are all clean and that no undue pressure can come upon the diaphragm. The connection to the meter should be one size larger than the pipes through which the gas is supplied. A meter should be placed level on a solid support and not in a damp place or where it will be subject to extreme temperatures. Sizes of gas meters are sometimes based upon a consumption of five cubic feet of gas per hour per burner, so that a 100-light meter would have a capacity of 500 cubic feet per hour.

The report of the Committee on Meter Connections of the American Gas Institute gives much valuable information on this matter, and reports standard methods of many different companies. No standard is recommended, as the requirements are not the same in different cities. The following matter is abstracted from the above report.

"The tendency is for gas companies to discontinue the use of lead outlet connections, especially above the 10-light size, and to discontinue the use of lead inlet connections for all sizes, and to use all-iron connections and suitable swing joints, and, in addition, a solid or a split tie-in between the inlet and the outlet piping

Fig. 278. Gas Meter Dial.

in order to relieve the meter screws and column seams of all avoidable strain."

Philadelphia practice as described in the report follows, illustrated by drawings from the United Gas Improvement Company. Fig. 279 shows the standard meter connections for all meters except those with flange connections.

"The method of connecting 3- to 200-light meters, as shown by the following sketches (Fig. 279) calls for the use of all-iron inlet and outlet connections having two double swing joints on the inlet side, and one double swing joint on the outlet side.

"The two-piece, cast iron tie-in between the inlet and outlet meter unions is first adjusted, when setting three- and five-light meters, to the meter to be set; the two parts are bolted together and then attached to the inlet piping after which the outlet piping connections are made up.

"When a meter is changed on this type of connection, the two-piece tie-in is removed and refitted to the screws of the meter to be set, after which it is replaced in position on the piping connections.

COMPRESSED AIR, GAS AND OIL PIPING

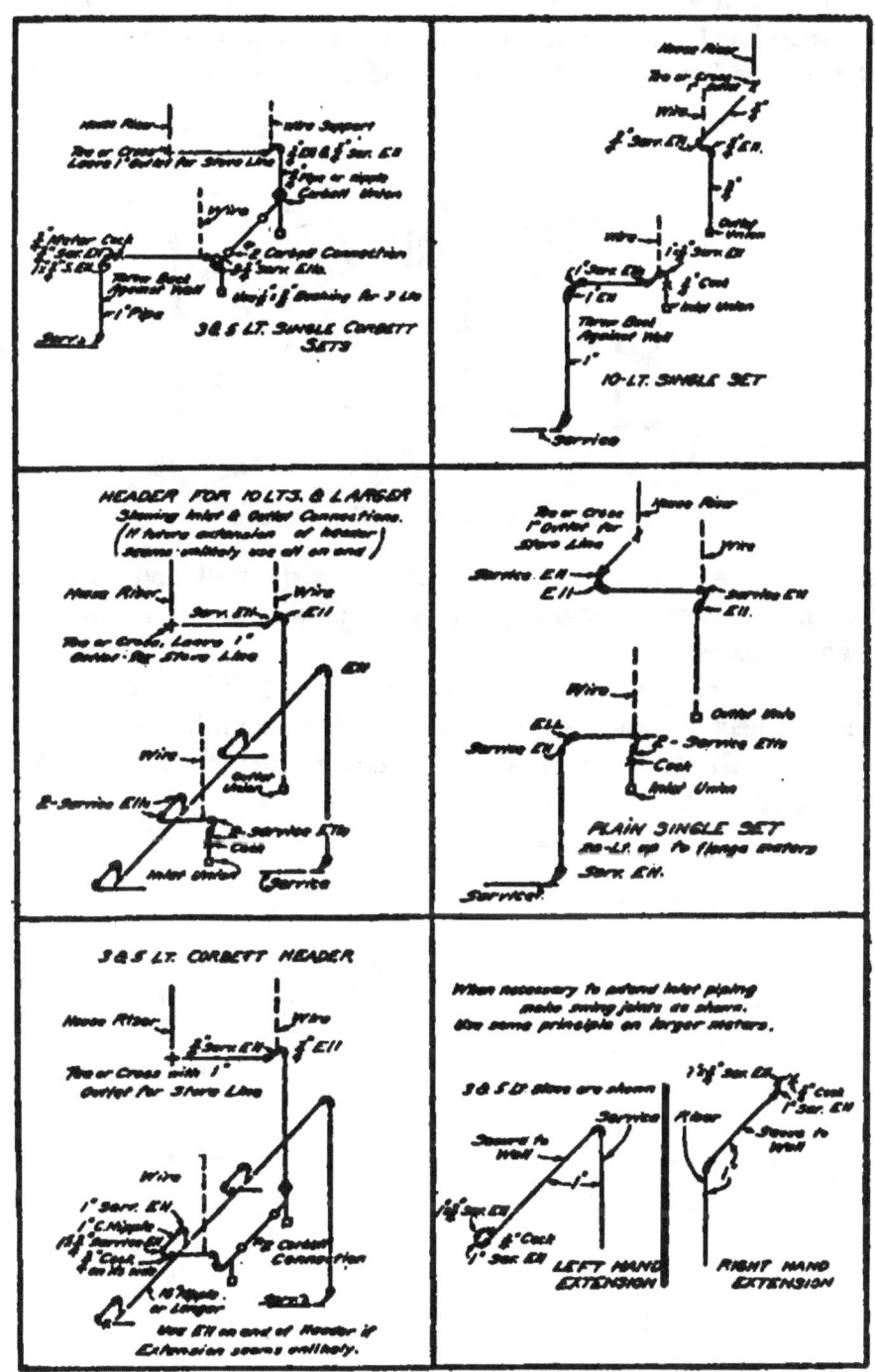

Fig. 279. Standard Meter Connections.

"The connection shown for meters larger than the five-light size permits of all necessary adjustment of the piping to the variable widths between meter screws, and enables the fitter to face

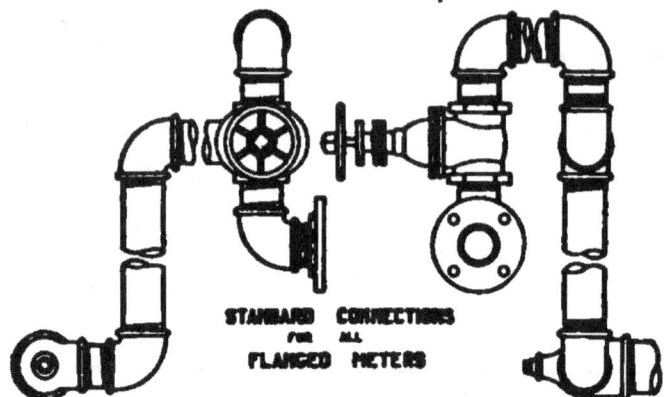

Fig. 280. Flanged Meter Connections.

up the meter screws and meter unions fairly well and to level the meter without straining the connections, meter screws, or column seams.

"All meters set in Philadelphia are supported by means of hanger shelves, two-piece adjustable shelves mounted on a back board placed on the wall below the meters, or are set on meter tables, or on the floor."

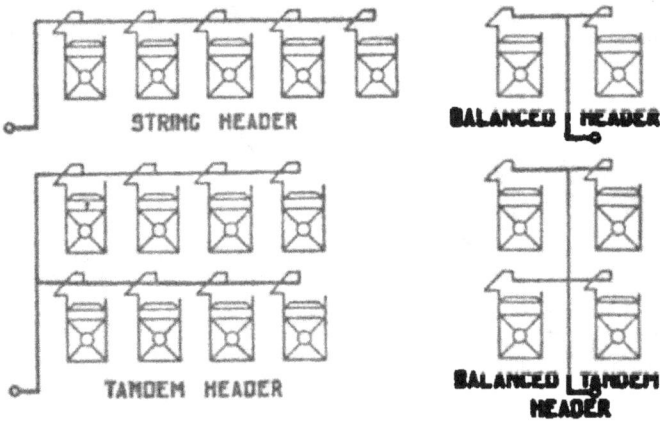

Fig. 281. Types of Header Construction.

The drawings reproduced in Figs. 280 and 281 show the practice of the United Gas Improvement Company at Philadelphia for flanged meter connections and types of header construction.

COMPRESSED AIR, GAS AND OIL PIPING

Gas Piping Specifications. — Many cities and gas companies have rules and regulations governing the installation of gas piping. Good practice is represented by the following quotations from the specifications for fuel and illuminating gas of the United Gas Improvement Company, Philadelphia Gas Works, and the accompanying drawings, Figs. 279, 280, 281 and 282. This matter was supplied through the kindness of Mr. Walton Forstall, Assistant Engineer of distribution.

6. Pressure Test. — The pipe should stand a pressure of 3 pounds per square inch, or 6 inches of mercury column, without showing any drop in the mercury column of the gauge, for a period of at least ten minutes. Leaky fittings or pipe should be removed; cold-caulked or cement-patched material will be rejected.

8. Obstructions and Jointing. — The piping should be free from obstructions. Every piece of pipe should be stood on end and thoroughly hammered, and also blown through, before being connected. White lead or other jointing material should be used sparingly, to avoid clogging the pipe. Jointing material should always be put on the male thread on end of pipe, and not in the fitting. The use of gas fitters' cement on joints is prohibited. After being connected, all piping should be blown through from the last outlet on each floor to the lower end of the riser, to make sure it is clear. No piping should be coated or painted until inspected and passed by the company. The use of unions in concealed work is not permitted; long screws or right-and-left couplings should be used.

9. Slope of Piping. — The piping should slope toward the meter, or toward an outlet from which condensation can be removed if necessary; or it may be laid level. Piping with a perceptible sag, which might hold condensation, will be rejected. It is especially important that underground piping be laid in such a way that condensation may be readily removed.

12-A. Protection of Piping. — When necessary to imbed a pipe in direct contact with neat cement or ordinary concrete, black iron pipe may be used. If cinders, salt, sea water, or other substance which has a corrosive action on the piping, is to be used in the fabrication of the cement or concrete, or if the concrete or cement in which the pipe is laid is to be exposed to brine, acid pickling-bath liquor, or other liquids of corrosive nature, or if the pipe is to be in contact with composition flooring, or similar structural material, the piping shall be made up of pipe and fittings galvanized on the outside, and shall be painted with two coats of a pure red-leaded paint, a bituminous paint, or an equivalent protective coating. It is preferable that it also be wrapped or coated with an approved material for protection against corrosion.

13. **Outlets.** — Ceiling outlets should project not more than 2 inches, nor less than $5/8$ inch, and should be firmly secured and perfectly plumb. Side-wall outlets should project not more than $7/8$ inch, nor less than $5/8$ inch, and should be at right angles to the wall and firmly secured.

14. **Gas Engine Connection.** — (a) The gas piping should be of sizes in accordance with the following schedule:

Size of Engine	Size of Connection
1 to 5 H. P.	1 inch
6 " 10 "	$1\frac{1}{4}$ "
11 " 20 "	$1\frac{1}{2}$ "
21 " 30 "	2 "
31 " 40 "	$2\frac{1}{2}$ "

These sizes apply only where the length of piping from meter to engine is 50 feet or less, and where the piping supplies the gas engine alone. If other fuel or illuminating appliances are to be supplied from the same piping, the sizes given above should not be used.

16. **Explanation of Piping Schedule.** — (a) This schedule is based on the standard formula for the flow of gas through pipes. The friction, and, therefore, the pressure necessary to overcome the friction, increases with the quantity of gas flowing through and as the aim of the table is to have the loss in pressure not to exceed one-tenth of an inch water pressure, per 30 feet of length of piping, the size of the pipe increases from an extremity towards the meter, as each section has an increasing number of outlets to supply. The quantity of gas the piping may be called on to deliver is stated in terms of $3/8$ inch outlets, instead of cubic feet, outlets being used as a unit instead of burners, because at the time of first inspection the number of burners may not be definitely determined. The consumption of gas through an outlet is assumed to be 10 cubic feet per hour, this being rather less than would be used by two ordinary burners.

TABLE 87. — 15. PIPING SCHEDULE

Required Sizes of Piping for Various Lengths and Numbers of Outlets

No. of 3/8 Inches Outlets	Size of Pipe in Inches									
	3/8	1/2	3/4	1	1 1/4	1 1/2	2	2 1/2	3	4
	Length of Pipe in Feet									
1	20	30	50	70	100	150	200	300	400	600
2	......	27	50	70	100	150	200	300	400	600
3	......	12	50	70	100	150	200	300	400	600
4	....		50	70	100	150	200	300	400	600
5	....		33	70	100	150	200	300	400	600
6	....		24	70	100	150	200	300	400	600
7	....		18	70	100	150	200	300	400	600
8	....		13	50	100	150	200	300	400	600
9	....			44	100	150	200	300	400	600
10	....			35	100	150	200	300	400	600
11	....			30	90	150	200	300	400	600
12	....			25	75	150	200	300	400	600
13	....			21	60	150	200	300	400	600
14	....			18	53	130	200	300	400	600
15	....			16	45	115	200	300	400	600
16	....			14	41	100	200	300	400	600
17	....			12	36	90	200	300	400	600
18	....				32	80	200	300	400	600
19	....				29	73	200	300	400	600
20	....				27	65	200	300	400	600
21	....				24	58	200	300	400	600
22	....				22	53	200	300	400	600
23	....				20	49	200	300	400	600
24	....				18	45	190	300	400	600
25	....				17	42	175	300	400	600
30	....				12	30	120	300	400	600
35	....					22	90	270	400	600
40	....					17	70	210	400	600
45	....					13	55	165	400	600
50	....						45	135	330	600
65	....						27	80	200	600
75	....						20	60	150	600
100	....							33	80	360
125	....							22	50	230
150	....							15	35	160
175	....								28	120
200	....								21	90
250	....								14	59
300	....									39
350	....									29
400	....									22
500	....									14

If *any outlet is larger than* ³/₈ *inch*, it must be counted as more than one, in accordance with the schedule below:

Size of outlet in inches	½	¾	1	1¼	1½	2	2½	3	4
Value in ³/₈ inch, outlets	2	4	7	11	16	28	44	64	112

17. Use of Piping Schedule. — In using the schedule observe the following rules:

(a) No piping between the meter and the first branch line should be smaller than ³/₄ inch.

(b) No piping should be smaller than ³/₈ inch.

(c) No independent line in the cellar or on the first floor, from the meter to a gas range should be smaller than 1 inch, but when the range is supplied from the house piping, a ³/₄ inch outlet will suffice. Above the first floor, an independent line from the meter to a gas range on an upper floor should not be smaller than ³/₄ inch. No pipe laid under ground should be smaller than 1¼ inches. No pipe extending outside of the main wall of a building should be smaller than ³/₄ inch.

(d) No ceiling outlet where the height of ceiling is 20 feet or more should be smaller than ³/₄ inch.

(e) Piping for any type of room heater, except gas logs, over 18 inches in length, and where line does not exceed 9 feet in length, should not be less than ½ inch. For similar installations with line exceeding 9 feet, the size should not be less than ³/₄ inch, but a short riser through the floor may be ½ inch. In other cases, and where the house piping is to supply fuel appliances, other than ranges, application should be made to the district shop to ascertain the proper size of piping. In any case the capped outlet should not be more than 2 inches nor less than ⅝ inch above the floor level.

(f) In determining the sizes of piping, always start at the extremities of the system and work toward the meter.

(g) The lengths of piping to be used in each case are the lengths measured from one branch or point of junction to another, disregarding elbows or turns. Such lengths will be hereafter spoken of as "sections" and are ordinarily of one size of pipe. There are only two reasons for which a change in size of piping will be allowed in a section. *First*: where the length of a section is greater than the length allowed for the outlets being supplied, as for example, if a section supplying two outlets is 33 feet long, 27

COMPRESSED AIR, GAS AND OIL PIPING

feet of this could be $^1/_2$ inch, and the remaining 6 feet of $^3/_4$ inch. *Second*: where the required length for the outlets being supplied will cause a violation of clause (*j*) unless the size is changed.

(*h*) If the exact number of outlets under consideration cannot be found in the schedule, take the next larger number. For example, if 27 outlets are required, the next larger number in the schedule, which is 30, should be taken.

(*i*) For any given number of outlets, do not use a smaller size pipe than the smallest size in the schedule for that number of outlets. Thus, to supply 17 outlets, no smaller size pipe than 1 inch may be used, no matter how short the section may be.

(*j*) In any piping plan in any continuous run from an extremity to the meter, there should not be used a longer length of any size pipe than shown for that size in the line opposite 1 outlet, as 50 feet for $^3/_4$ inch, 70 feet for 1 inch, etc. Exceptions to this rule are: *First*: when larger piping than called for by the schedule is run in following (*k*) of this paragraph. *Second*: when fitter voluntarily runs a larger pipe than is necessary, as for example, if three outlets are to be supplied by 60 feet of piping, instead of 50 feet of $^3/_4$ inch and 10 feet of $^1/_2$ inch being required, the entire 60 feet may be of $^3/_4$ inch piping. When two or more successive sections work out to the same size of piping, and their total length or sum exceeds the longest length shown for that size piping, the change in size to a larger pipe should be made as soon as the limiting length has been reached. For example, if 5 outlets are to be supplied through 30 feet of piping, and then these 5 and 1 more, making 6 in all through 24 feet of piping, it would be found by the schedule that 5 outlets through 30 feet require $^3/_4$ inch piping, and that 6 outlets through 24 feet require $^3/_4$ inch piping, but as the sum of the two sections, 30 plus 24 equals 54 feet, is 4 feet longer than the amount of $^3/_4$ inch piping that may be used in any continuous run, the 24 foot section must be changed from $^3/_4$ inch to 1 inch, 4 feet from the end nearest the meter.

(*k*) Never supply gas from a smaller size pipe to a larger one. If 25 outlets are to be supplied through 300 feet of piping and these 25 and 5 more, making 30 in all, through 100 feet of piping, it would be found by the schedule that 25 outlets through 300 feet require $2^1/_2$ inch pipe, and 30 outlets through 100 feet require 2-inch pipe, but as under this condition a 2-inch pipe would be supplying a $2^1/_2$ inch, the 100 foot section should be made $2^1/_2$

inches. This does not apply to the case of a small pipe inside of a building supplying one outside of a building, which has been made larger as per (c) of this paragraph, because it is exposed to out-door temperatures.

18. Plan of Piping. — In preparing a plan, Fig. 282, the following instructions should be strictly adhered to:

(a) Vertical piping should be drawn parallel to the short side of the sheet.

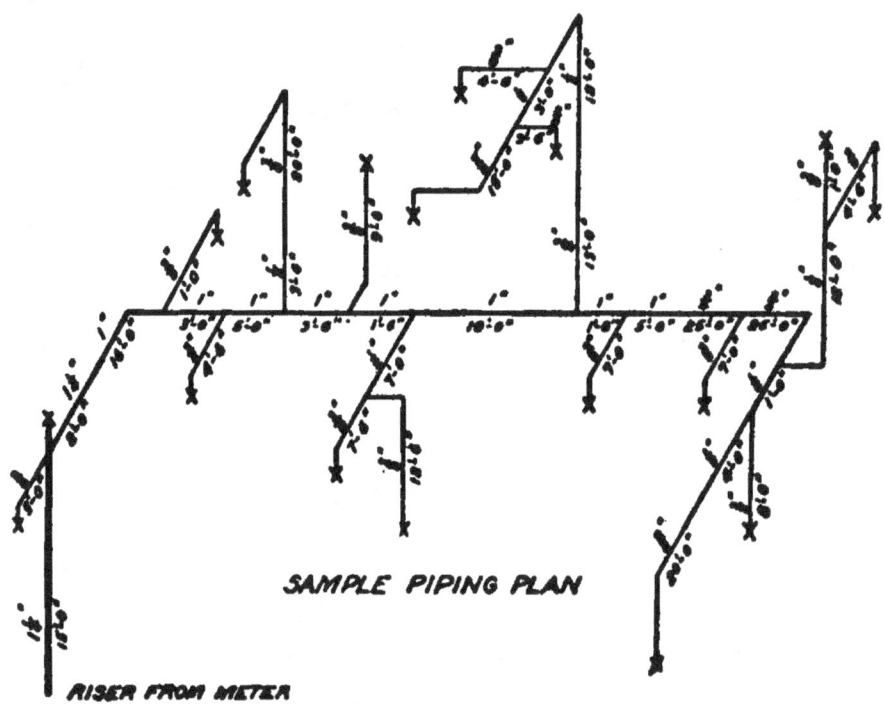

Fig. 282. Gas Piping Drawing.

(b) Piping through the length of the building should be shown parallel to the long side of the sheet.

(c) Piping across the width of the building should be shown diagonally on the sheet.

(d) State length and size of each section of piping. A section designates the length of piping existing between outlets, fittings and points of changes in piping sizes.

(e) On horizontal piping, mark the length under the line, and the size over the line.

(f) On vertical piping, including drops, mark the length to the right of the line, and the size to the left of the line.

COMPRESSED AIR, GAS AND OIL PIPING

(g) Mark each outlet ×, and in case of a plugged outlet, state its size.

28. Stems. — (a) The sizes of pipe or tubing in the stems of pendants, or of wall brackets, should be not smaller than those in the following schedule:

Length of Stem	Number of Burners	When made of Iron Pipe		When made of Brass Tubing		Gas-way through Cock
		Cased	Uncased	Plain	Chain	
30″ and under	1–2	1/8″	1/4″	3/8″	3/8″	1/8″
Over 30″ and under 42″	1–2	1/4″	1/4″	7/16″	3/8″	1/8″
42″ and over	1–2	1/4″	3/8″	7/16″	3/8″	1/8″
Any Length	3–6	1/4″	3/8″	1/2″	3/8″	3/16″
" "	7–12	3/8″	1/2″	1/2″	3/8″	3/16″
" "	13 and over	1/2″	3/4″	3/4″	1/2″	1/4″

"Length of stem" is understood to be a distance measured as follows:

In pendants, a straight line from the stiff joint to the lowest point of the pendant.

In brackets that carry more than one burner, a straight line from the stiff joint to the point where the arms diverge.

(b) A one-piece or harp pendant, if not over 33 inches long, may be made of 1/8 inch iron pipe, or 3/8 inch brass tubing. If longer than 33 inches, it should conform to the schedule.

(c) In the case of wall brackets, that carry more than one burner, the sizes given in the schedule are correct, except that no pipe of smaller size than 1/4 inch should be used.

29. Arms. — (a) Arms of pendants or of wall brackets, that is, those parts which carry gas for only one burner, should be made not smaller than the sizes in the following schedule:

Length of Arm	When made of Iron Pipe		Brass Tubing
	Cased	Uncased	
12 inch and under	1/8″	1/8″	3/8″
Over 12 inch and under 18 inch	1/8″	1/4″	7/16″
Over 18 inch	1/4″	3/8″	1/2″

"Length of arm" is understood to be a distance measured as follows:

In pendants, a straight line from the centre of the stem to the centre of the burner nozzle;

In stemless wall brackets, as stiff, single-swing or double-swing brackets, carrying but one burner, a straight line from the stiff joint to the centre of the burner nozzle, measured when the bracket has its maximum reach;

In stemmed wall brackets, a straight line from the point of divergence of the arm to the centre of the burner nozzle.

(b) In the case of cast wall brackets, the area of the gas-way in stems and arms should be not less than the area of the pipe, or tubing, of equivalent lengths, of the sizes already specified.

30. General. — (a) Special precautions should be taken in the construction to prevent the obstruction of the gas-way by foreign matter, such as solder, other jointing material, or metal chips. The ends of tubing should be reamed to remove burs. When duplex tubing is used, care should be exercised to prevent faulty alignment of gas-ways.

(b) Gas fitters' cement should not be used on any part of the fixture where it may be affected by the heat from the burners. Where solder is used, it should be of such a mixture that it will not be affected by the heat from the burners.

(c) Fixtures for out-door use, or in exposed situations, should be provided with suitable drips, or means for the convenient removal of condensation from any part of the fixture in which such condensation may accumulate.

(d) Globe rings should fit snugly over the threads of the burner nozzles, and should be so constructed that the screwing on of the burner will be certain to bind the globe ring firmly between the burner and the shoulder of the burner nozzle. Globe rings should have ample openings for the admittance of the proper quantity of air to the burners.

Oil Piping. — The following articles contain a few general ideas upon various kinds of oil piping. The problems to be met are about the same as those common to all kinds of piping. For lubricating oil almost any material may be used. Small oil pipes may be of copper, as it is easily bent to conform to the shape of the machine to which it may be attached. Brass and steel pipe and tubes are generally used with screwed fittings, brazed joints, and special connections. For fuel oil, steel piping and galvanized fittings are advisable. Oil pipes should always be sufficiently large to prevent choking, especially returns from a lubricating system.

Oil Piping for Lubrication. — Various methods of supplying oil to machinery are in use. In some cases it is advisable to use a simple oil or grease cup and allow such oil as is not used to go to waste. For steam engines the splash system may be employed. For oiling the valves and cylinders of steam driven machinery the oil may be supplied by a sight-feed lubricator. In other cases a force feed or sight-feed system may be employed and the oil collected, filtered and used over again automatically. Such

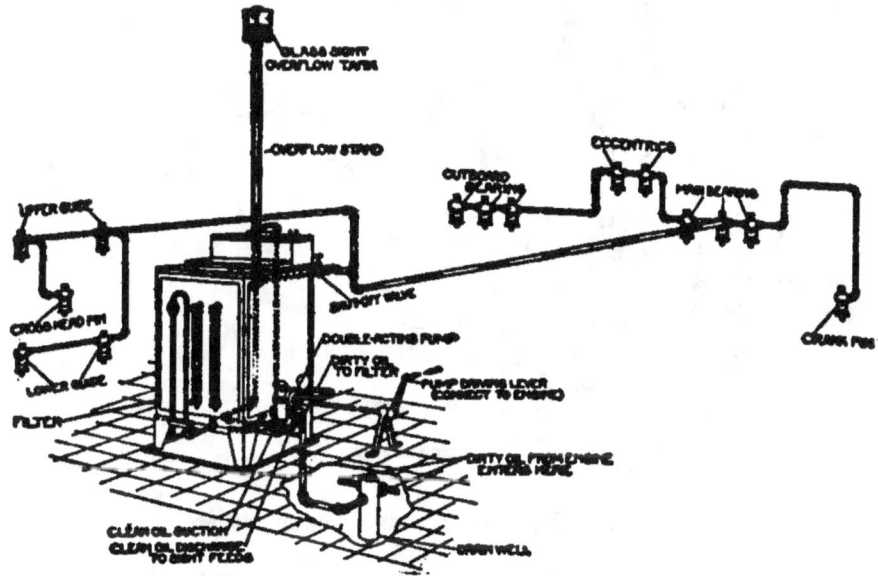

Fig. 283. Richardson Individual Oiling System.

systems involve pumps, piping, filters, etc., but cut down the amount of oil required. For an efficient lubricating system the following considerations should be given attention. The oil should be supplied in a continuous stream at the exact points where it is needed. The oil should be sufficiently cool so that it can carry away heat. There should be a carefully designed system for collecting the oil which drains from the lubricating system. There should be an efficient filter for removing dirt, particles of metal, and water from the oil.

Richardson Individual Oiling System. — This is one of the systems of the Richardson-Phenix Company, and is illustrated in Fig. 283, which shows the application to a simple engine. The oil, after being used, flows by gravity to a cast-iron drain well.

One end of a double-ended plunger pump raises this dirty oil from the well and discharges it into the filter where the oil is purified. It is then pumped through a system of piping on the engine, or other machine, and delivered to the sight feed oilers located at each of the points to be lubricated. A constant oil pressure is maintained by means of a glass overflow stand.

Phenix Individual Oiling System. — This method is similar to the Richardson system, but the apparatus is differently arranged, adapting it for small engines, air compressors, and ice machines.

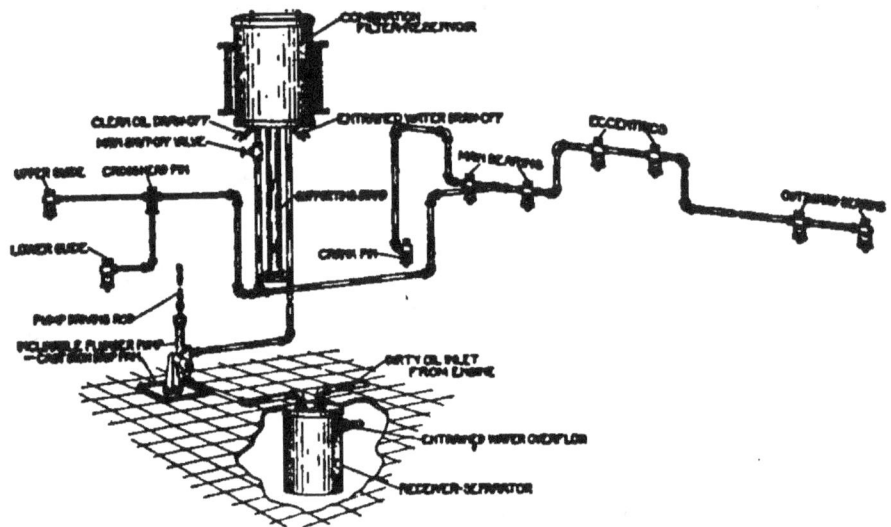

Fig. 284. Phenix Individual Oiling Stysem.

The principle of operation and the required parts are shown in Fig. 284. The dirty oil flows into a receiver-separator, where heavy foreign matter and entrained water are removed. It is then pumped up to the filter-reservoir where it is purified. From the final purification the oil flows by gravity to the various sight feed oilers.

Oil Pipe Fittings. — In addition to the regular pipe and fittings there are several special forms of fittings used with oil piping, a number of which are illustrated in Fig. 285. Feed valves are shown at A, B and C, where A is a plain feed valve, straight form, B is a sight feed valve, angle form, and C is a cross sight feed valve with a lever for stopping the feed. Regulation is obtained by the nut 1 and the valve is closed by throwing the lever 2 down into the position shown by dotted lines. These may be in the form of straight, angle, cross, or corner fittings, and are made

COMPRESSED AIR, GAS AND OIL PIPING 265

in sizes to fit ⅛, ¼, ⅜ and ½ inch pipe threads. A plain metal wiper cup is shown at *D*, Fig. 285; an adjustable wick wiper at *E*; a drip trough at *F*, and an oil cup wiper tip at *G*.

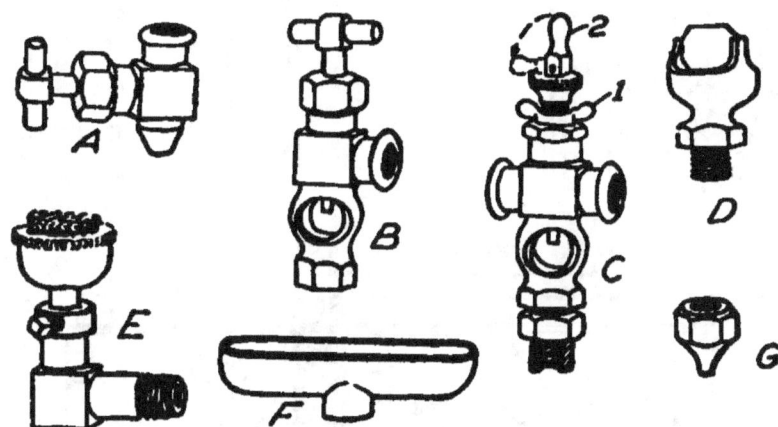

Fig. 285. Oil Pipe Fittings.

Some fittings are so made that no threading is required, such as the "Union-Cinch" fittings shown in Fig. 286. Each fitting is a union, thus making it very easy to assemble or take down the piping. Referring to the figure *A* is a soft brass cinch ring which is slipped over the end of the pipe *B*. The end of the pipe is then inserted in the fitting until it comes against the shoulder *C* and

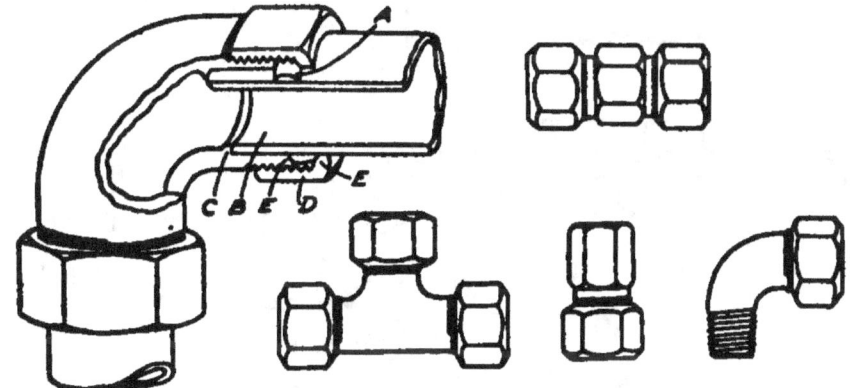

Fig. 286. Union Cinch Fittings.

the nut *D* screwed on, making a double joint at *E–E*, thus insuring tightness.

Oil Piping Drawing. — A drawing showing the pipe layout for an oiling system is reproduced in Fig. 287. The engine frame is indicated by very fine lines, heavy full lines being used for the

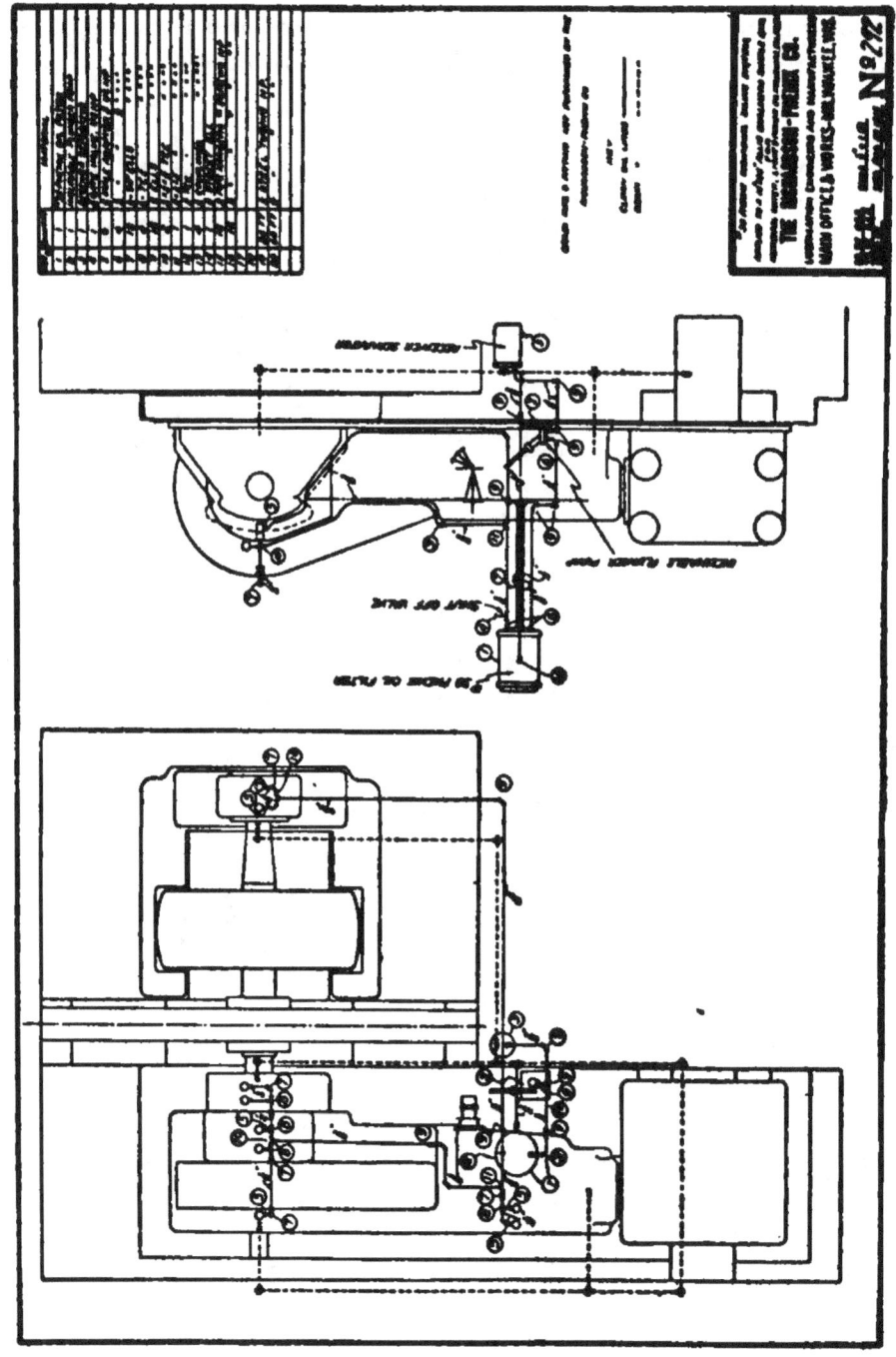

Fig. 287. Drawing Showing Piping for an Oiling System.

clean oil lines, and dotted lines for the drain oil pipes. Each fitting and part is numbered and the quantity required is given in the material list opposite the reference number. The conventional symbols as shown in the chapter on piping drawings are used as the scale is small.

Sight Feed Lubricator Connections. — The method of piping a double connection sight feed lubricator for steam cylinders is shown in Fig. 288. The operation of the lubricator depends upon having a greater pressure upon the oil than exists in the steam pipe. This is accomplished by a condenser pipe tapped into the steam pipe above the lubricator. The pressure inside the lubricator will then exceed the pressure in the steam pipe by an amount equal to the head of water (condensed steam) in the condenser pipe, and so force the oil from the lubricator into the steam pipe. It is necessary that the condenser pipe should be about 18 inches long, in order to be sure of a sufficient difference in pressure. If connection is made to a horizontal pipe the condenser pipe should extend above the steam pipe and then descend to the lubricator. The size of the connection A will

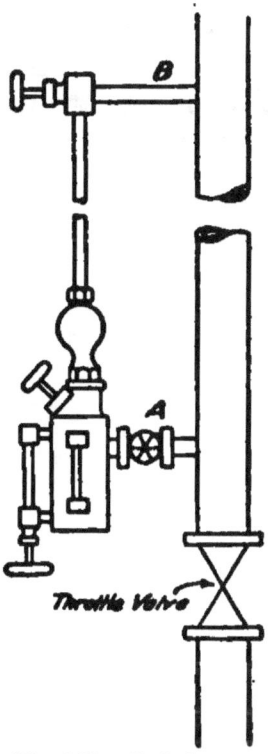

Fig. 288. Lubricator Connections.

depend upon the size of the lubricator, the pipe B may be $1/4$ inch steel pipe or brass tubing, iron pipe size.

Oil Fuel Piping. — Piping for oil fuel is not essentially different than for other purposes. Extra heavy standard pipe with screwed joints may be used. Tight joints may be obtained with flanges screwed on and packed with manilla paper, cardboard or prepared oilproof packing. Rubber or other material affected by the sulphur in the oil must be avoided. On this account copper piping should not be used. Fittings for oil piping may be extra heavy galvanized iron, brass or composition. Valves in the suction line to oil pumps should be of the gate pattern, as they offer less resistance to flow, but globe valves may be used in the delivery pipes. The velocity of flow in oil pipes ranges from a maximum of 20 feet per minute in suction pipes to 100 feet per minute in delivery pipes.

For the United States Navy service oil piping is specified as seamless-drawn steel, with steel flanges, pipes for heating coils, seamless-drawn steel, and suction oil piping, lap-welded steel or wrought iron. Service oil fittings are of cast steel or composition and suction oil fittings are cast steel or cast-iron screwed fittings. All joints and fittings in pressure piping must be oil tight under test, without the use of gaskets. On suction lines paper gaskets may be used.

CHAPTER XV

ERECTION — WORKMANSHIP — MISCELLANEOUS

Handling Pipe. — In the handling of pipe it is advisable to keep wrenches and tools off from the threads and to protect them from injury in other ways. Clean, sharp threads form the very best means of obtaining tight joints.

The countersinking of tapped holes is of advantage in making certain that the pipe will enter squarely, and that the threads will not cross. Fittings and valves are manufactured with such counterbores as shown at A and B in Fig. 289. The end of the pipe may also be chamfered.

Pipe may be cut off in a machine with a cutting-off tool, or by hand with a wheel pipe cutter, or with a hack saw. When cut

Fig. 289. Counter-bored Fittings.

by hand with a pipe cutter, the edge is almost always rough and turned in toward the centre, thus reducing the area of the pipe. In such cases a pipe reamer should be used to remove the turned in edge and restore the full diameter of the pipe. When a hack saw is used the pipe should be revolved away from instead of toward the worker in order to avoid stripping the teeth from the saw.

Putting Up Pipe. — The inaccuracies of "making up" render it undesirable to run piping without some means of allowing for differences in length. With screw fittings this is best done by means of elbows as shown in the simple case of Fig. 290. In case A the dimensions must be very exact to obtain good joints at x and y. In cases B and C the elbows allow the flanges at x and y to meet squarely as the pipe can turn on the elbows and so be brought into line. A small amount of lack of alignment can often be taken care of by a union. The bolt holes of flanges can allow a small amount of latitude if they are made $1/8$ inch larger than the bolts.

This of course would not apply to recessed flanges of any kind. Unions, or right and left couplings should always be placed in such positions that piping can be disconnected without taking down a long line of piping. Tees and plugs used instead of elbows make it easy to take off new branch lines should they be necessary. Valves are both a convenience and a nuisance. They should be used where necessary but not promiscuously, for they offer resistance to flow and must be kept in repair. Sometimes steam pipe is cut short in order to decrease the strain due to expansion. In such cases allow one half the change in length due to expansion and spring the pipe into place. This will be relieved when the pipe is heated and there will be only one half as much

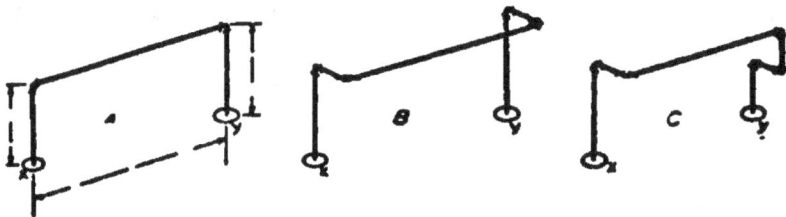

Fig. 290. Putting Up Pipe.

strain as there otherwise would be. When long pipe coils are used for heating there should be allowance for expansion — let the pipe slide on supports and leave room at the end between the coil and the building wall. Provide unions for convenience in disconnecting, especially near valves. Piping should be arranged so that repairs can be conveniently made; so that units can be readily cut out; and, so that various necessary combinations can be made in times of accident or repairs.

Pipe Dopes. — There are a number of prepared pipe dopes which may be used for making tight pipe joints by smearing on flange faces or on pipe threads. For most purposes flake graphite and oil is one of the best dopes, as joints made with it can be taken apart. A mixture which is suitable for either steam or water pipes is composed of 10 pounds of finely ground yellow ochre; 4 pounds of ground litharge; 4 pounds of whiting; and, $1/2$ pound of finely cut hemp; all of which is mixed with linseed oil until it is about the consistency of putty. White lead and red lead are also used. For permanent joints, red lead makes a tight joint which is satisfactory. A mixture for ammonia pipe joints is composed of litharge and glycerine mixed up in small quantities for each joint. As this substance sets very quickly, a joint made with it should not be changed.

Gaskets. — A great many materials are in use for maintaining tight joints between pipe flanges. In addition to selecting the proper materials for the fluid to be transmitted or the temperature to be withstood, the proper design of the flanges is very important. Rough or uneven surfaces are difficult to make tight with any substance. Bolts should be uniformly spaced and not too far apart for the thickness of flange. Water hammer and vibration are other frequent causes of leaky joints. With good true surfaces which are parallel, thin packing material may be used. Under such conditions a good quality of paper soaked in oil is suitable. In some plants, used drawing paper is kept and made into gaskets. Sheet rubber, with either cloth or wire insertion may be used for water or for saturated steam, the wire insertion being better for high pressures. Such packings may be had in sheets with thicknesses of from $1/32$ inch up to $1/4$ inch. Rough or uneven flanges require a thick packing. These expose a greater area to pressure than thin ones, and are, therefore, undesirable. For high pressure steam or water, or superheated steam, gaskets may be made of asbestos, corrugated metal, or corrugated metal and asbestos. Corrugated steel makes a gasket suitable for superheated steam. Fig. 291 shows a Goetze's corrugated copper gasket, with asbestos lining. Such gaskets are made in a large number of forms, suiting them to different purposes. For ammonia, sheet lead is often used. Asbestos packing may be used for acids, ammonia, or oils.

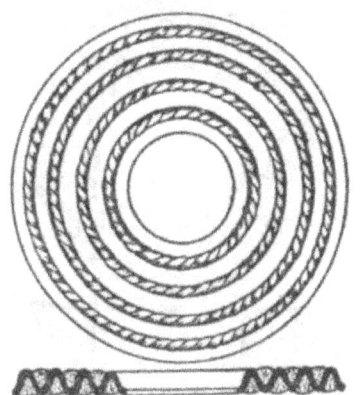

Fig. 291. Copper and Asbestos Gasket.

When rubber gaskets are used they can be prevented from sticking if the flanges are treated with plumbago, pulverized soapstone or chalk. The joint can then be broken and the gasket removed whole and used over again.

A full face gasket is one which extends over the whole face of the flange, Fig. 292; a ring gasket is one which fits inside of the bolts, Fig. 293. It is well to have the hole through the gasket

272 A HANDBOOK ON PIPING

slightly larger in diameter than the pipe as some kinds of gaskets spread when tightened or after use, and so decrease the size of the opening. Gaskets of rubber or asbestos may be cut by placing the sheet on the flange and striking around the edges with a hammer. The bolt holes can be cut in the same manner with a ball peen hammer. When a gasket has been put in place the flanges should be drawn together by tightening up the bolts uniformly — lightly at first, and then going over them again until they are all under the same tension. Graphite and oil placed on

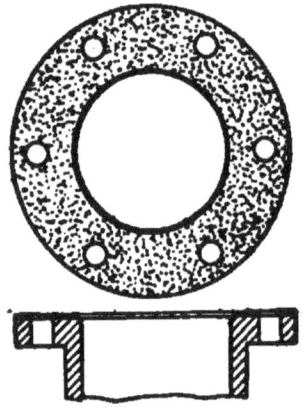
Fig. 292. Full Face Gasket.

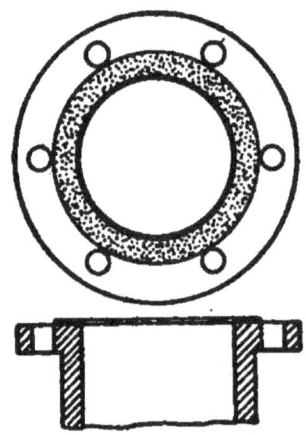
Fig. 293. Ring Gasket.

the bolt threads will make them easier to take down again when necessary.

Valves. — The importance of valves should be fully realized when piping is being assembled, as they are the means of control for the system. For this reason they should be carefully examined and cleaned out, lightweight valves should be avoided, and all valves should receive care in handling. It is important that steam lines be thoroughly blown out after erection in order to make sure that scale, iron filings, bits of metal and other objects are removed. The valve seats should then be examined before closing to see that nothing has been deposited on them. It sometimes happens that valves are ruined by cutting too long a thread on the pipe and then screwing the pipe too far into the valve allowing it to come against the seat. Valve seats may be sprung out of place by holding the valve on the end farthest away from the pipe to which it is being connected. When a valve leaks the seat should be reground at once, lest it be damaged beyond repair.

ERECTION — WORKMANSHIP — MISCELLANEOUS

Putting a wrench on the handwheel is a very poor way of attempting to make a valve tight.

Vibration and Support. — The question of vibration should be considered in connection with supports for piping. The use of

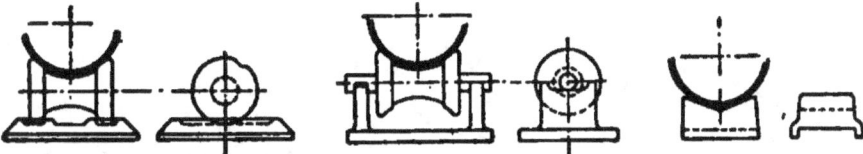

Fig. 294. Pipe Supports.

high velocities and small pipes makes it necessary to use care in the selection of supports or vibration with its consequent leaking joints will result. Other causes of vibration: too large steam pipes connected to an engine taking an intermittant supply from them, with the consequent surging of a large volume of steam; improper foundations for the machines to which the piping is connected. The distance between supports should generally be about 12 feet but this will vary with the kind of piping and number of valves and fittings. Supports should be provided near changes in direction, branch lines and particularly near valves. The weight of piping should not be carried through valve bodies if they are to be kept tight. Expansion and contraction due to changes in temperature require provision for movement of the piping. Hangers by which the pipe is suspended allow it to move freely but also admit of vibration being set up, especially where there

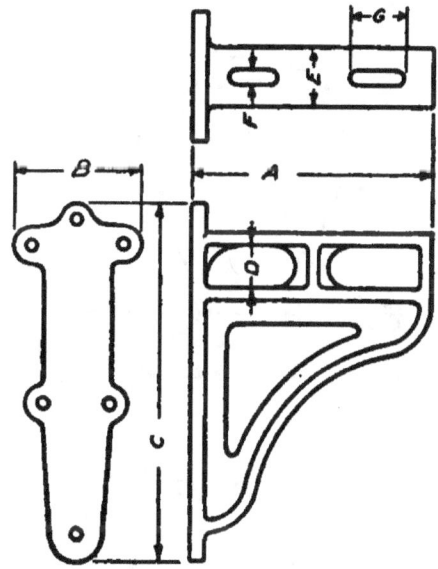

Fig. 295. Pipe Bracket.

are a number of changes in the direction of the line. Sometimes this vibration can be stopped by fastening one of the lengths of pipe. Various forms of brackets upon which the pipe can rest, free to move both lengthwise and sidewise are more satisfactory

274 A HANDBOOK ON PIPING

in preventing vibration. Rollers may be provided for the pipe to rest upon, either with hanging or bracket supports. Several forms of supports are shown in Figs. 294 and 295.

Expansion. — The expansion and contraction of pipe under changes in temperature produce severe stresses unless amply

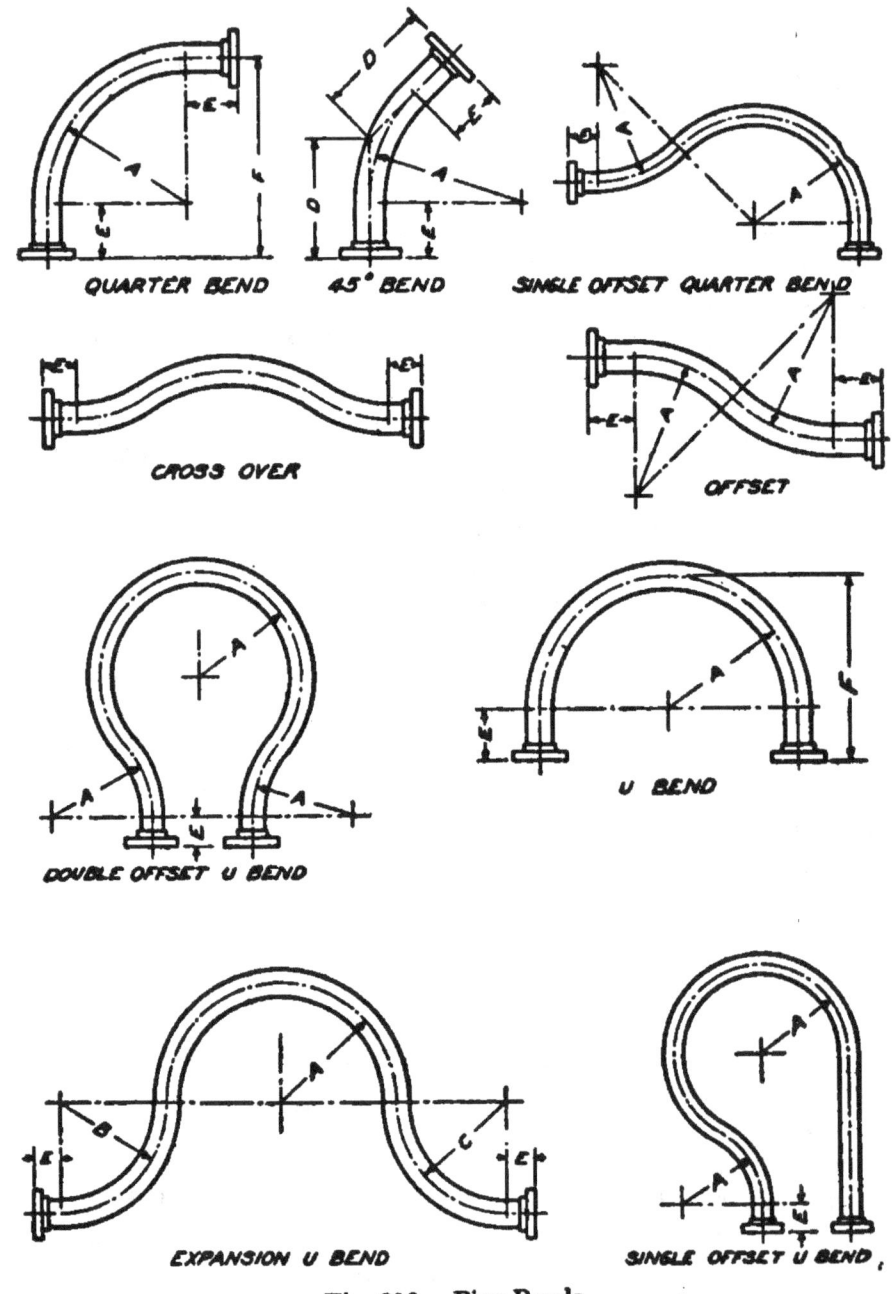

Fig. 296. Pipe Bends.

provided for. There are several ways of doing this. The general method when the line is not too long is by means of expansion bends, depending upon the elasticity of the pipe for the necessary movement. Several forms of bends are shown in Fig. 296 and the ordinary dimensions are given in Table 89.

TABLE 88 (FIG. 295)
CRANE PIPE BRACKETS

Size of Pipe will Support Inches	Safe Load Tons	A Inches	B Inches	C Inches	D Inches	E Inches	F Inches	G Inches
5 to 8	1	25	12	34	5$\frac{1}{2}$	6	1$\frac{1}{4}$	8$\frac{1}{2}$
9 to 14	2	30	14	40	6	6	1$\frac{3}{8}$	9
15 to 18	3	34	16	45	6$\frac{1}{2}$	6	1$\frac{1}{2}$	9$\frac{1}{4}$
20 to 24	4	40	19	51$\frac{1}{2}$	7	6	1$\frac{5}{8}$	11$\frac{1}{2}$
20 to 30	Special	44$\frac{1}{2}$	19	64	7	6	1$\frac{5}{8}$	12$\frac{1}{4}$

TABLE 89 (FIG. 296)
LAP-WELDED STEEL PIPE BENDS

Size of Pipe Inches	A-B-C Advisable Radius of Bends Inches	D Centre to End or Face of Flanges Ft. — Ins.	E Length of Straight Pipe on each Bend Inches	F Centre of Bends to Face of Flanges Ft. — Ins.	Minimum Radius to which Bends can be made from Extra Strong Pipe Only Inches
2$\frac{1}{2}$	12$\frac{1}{2}$	0–9$\frac{3}{16}$	4	1–4$\frac{1}{2}$	7
3	15	0–10$\frac{1}{4}$	4	1–7	8
3$\frac{1}{2}$	17$\frac{1}{2}$	1–0$\frac{1}{4}$	5	1–10$\frac{1}{2}$	10
4	20	1–1$\frac{1}{4}$	5	2–1	12
4$\frac{1}{2}$	22$\frac{1}{2}$	1–3$\frac{5}{16}$	6	2–4$\frac{1}{2}$	14
5	25	1–4$\frac{3}{8}$	6	2–7	15
6	30	1–7$\frac{7}{16}$	7	3–1	20
7	35	1–10$\frac{1}{2}$	8	3–7	24
8	40	2–1$\frac{9}{16}$	9	4–1	28
9	45	2–5$\frac{5}{8}$	11	4–8	35
10	50	2–8$\frac{3}{4}$	12	5–2	40
12	60	3–2$\frac{7}{8}$	14	6–2	50
14	70	3–9	16	7–2	65
15	75	3–11$\frac{1}{16}$	16	7–7	70
16	80	4–3$\frac{1}{8}$	18	8–2	78
18	108	5–2$\frac{3}{4}$	18	10–6	88
20	120	5–7$\frac{3}{4}$	18	11–6	104
22	132	6–0$\frac{3}{8}$	18	12–6	132
24	144	6–5$\frac{5}{8}$	18	13–6	144

276 A HANDBOOK ON PIPING

When screwed fittings are used and the pressures are not too high, expansion may be taken care of by allowing the pipe to turn on the threads as shown in Fig. 297. A swivel joint, Fig. 298, may be used with flanged fittings to allow for expansion. The change in length is allowed for by a turning movement at the two swivels which are packed the same as any gland stuffing box. This turning movement is easier to keep tight than a sliding move-

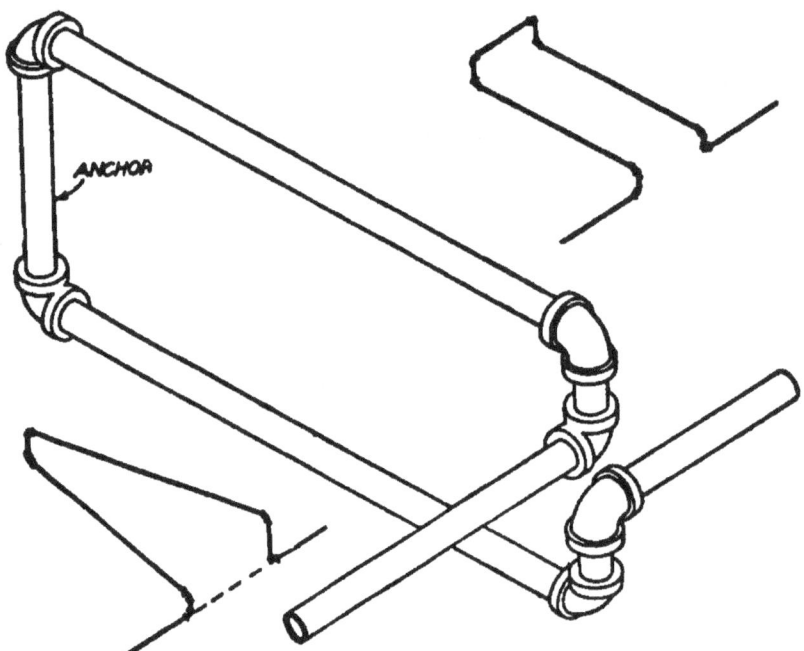

Fig. 297. Expansion Bends, with Screwed Fittings.

ment. They are made by the Walworth Company of cast iron with brass bearings for steam pressures up to 250 pounds, and of cast steel with Monel metal bearings for 350 pounds pressure and 800 degrees F. total temperature. Another method, when bends or angles cannot be used, is to provide an expansion joint. One form is shown in Fig. 299. Tie bolts should always be provided so that the joint cannot pull apart from any cause. The body is usually made of cast iron and the sleeve of brass. Some dimensions of expansion joints are given in Table 90. Diaphragm joints are sometimes used for low pressures. A joint using a copper shell and made for pressures up to 25 pounds is shown in

ERECTION — WORKMANSHIP — MISCELLANEOUS 277

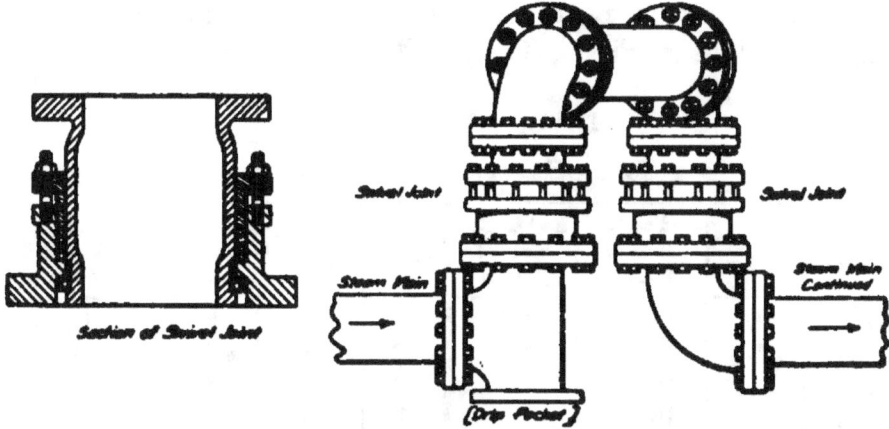

Fig. 298. Swivel Joint.

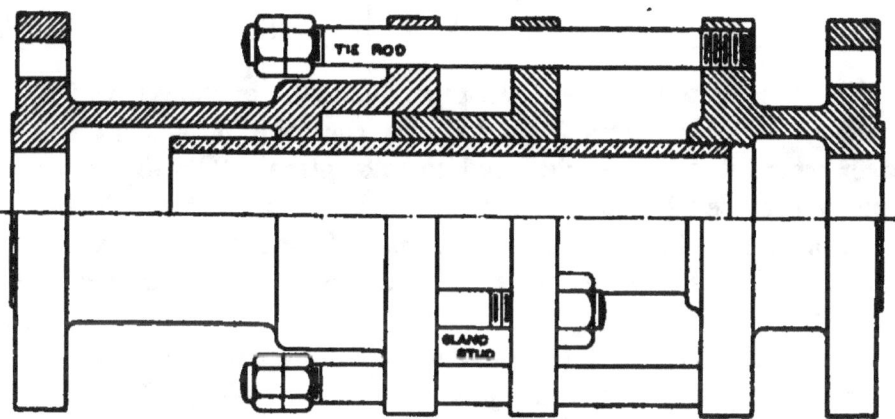

Fig. 299. Expansion Joint.

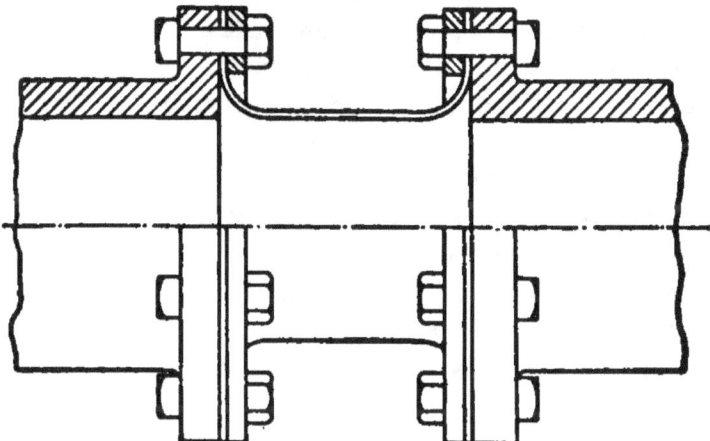

Fig. 300. Low Pressure Expansion Joint.

Fig. 300. Other forms of expansion joints are shown in the chapter on Exhaust Piping (Chapter X).

TABLE 90 (FIG. 300)

Extra Heavy Expansion Joints

Size Inches	Traverse Inches	End to End Screwed Inches	End to End Flanged Inches	Size Inches	Traverse Inches	End to End Screwed Inches	End to End Flanged Inches
2	2¹/₂	15¹/₈	15¹/₂	8	7	30³/₄	31¹/₂
2¹/₂	2¹/₂	15¹³/₁₆	16	9	7	31³/₄	31⁵/₈
3	2³/₄	16⁷/₈	17⁵/₈	10	7	32⁵/₈	33⁵/₈
3¹/₂	3	18	18¹¹/₁₆	12	8	36¹⁵/₁₆	37¹¹/₁₆
4	3¹/₄	18⁷/₈	19¹/₂	14	10	...	43
5	4	21⁵/₈	22⁵/₈	15	10	...	43¹/₈
6	5	24⁷/₈	25³/₄	16	10	...	45
7	6	27⁷/₈	28¹/₂	18	10	...	46¹/₂

Whatever method is used the pipe should be anchored or fastened at suitable places to make sure that the movement will occur where it has been designed to take place. The expansion usually provided for saturated steam is from two to three inches per 100 feet of length. The increase in length for 100 feet of steel pipe for various ranges in temperature may be found from Fig. 301.

Pipe Bends. — For high pressure steam plants long radius bends made of steel pipe are generally used in place of elbows. These bends reduce friction and allow the pipe to expand and contract.

As will be seen by reference to Fig. 296, bends are made purposely for expansion and other requirements.

Extensive tests with various types of bends in several sizes and weights of pipe to determine their relative value have been made by Crane Company and are fully reported in *The Valve World*, October, 1915, from which the following is quoted:

"These tests were made with Full Weight and Extra Strong Quarter Bends, 'U' Bends, Expansion 'U' Bends, and Built-up Bends placed in lines representing the ordinary installation, being anchored at one end and carried on roller supports so that the strains due to expansion and contraction were properly directed to the bend.

ERECTION — WORKMANSHIP — MISCELLANEOUS

"The bends were then extended and compressed repeatedly until something failed. These tests were made with the line cold and also under steam pressure. In this manner the safe allowable movements of bends were determined.

"Combining practical experience, tests, and the formula, it is found that a 180° or 'U' Bend has twice the expansive value of

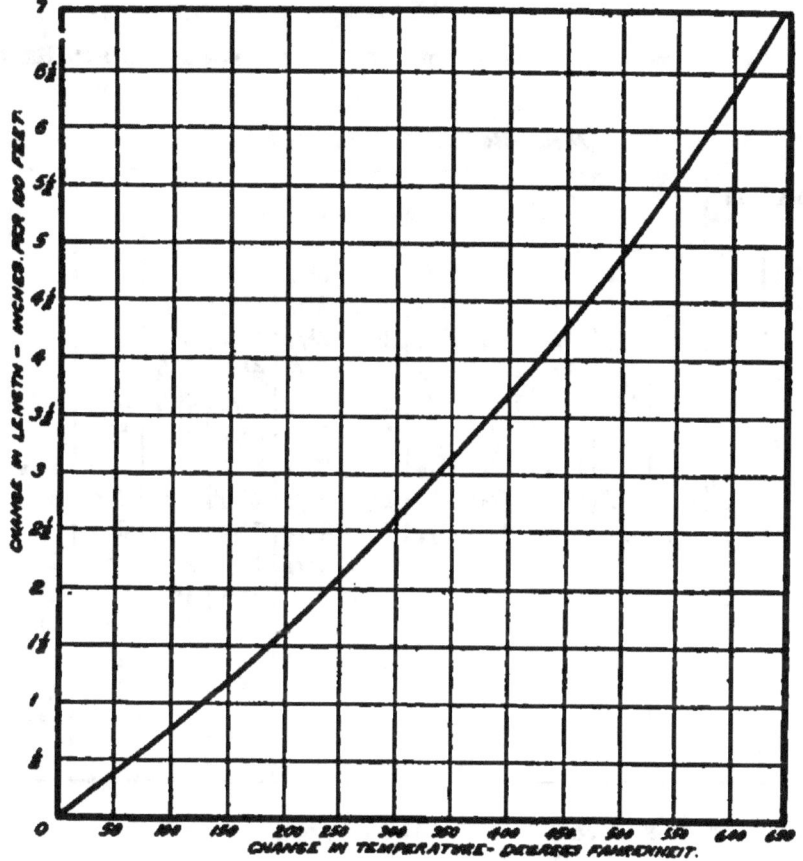

Fig. 301. Curve Showing Expansion of Pipe for Variation in Temperature.

a 90° or Quarter Bend of the same size and radius, and an Expansion 'U' Bend four times the expansive value of a Quarter Bend or twice that of a 'U' Bend. A Double Offset Expansion 'U' Bend has five times the expansive value of a Quarter Bend, two and one half times that of a 'U' Bend and one and one-fourth times that of an Expansion 'U' Bend.

"A battery of Expansion 'U' Bends connected to large headers or manifolds is often used. This method occupies less space and allows of a greater movement than with a single pipe bend of ordinary construction. However, care must be exercised in the design of this type to provide sufficient area.

"We present herewith the expansive value of Quarter Bends of various pipe sizes and radii, Table 91.

TABLE 91

SAFE EXPANSION VALUES OF 90° OR QUARTERS WROUGHT STEEL BENDS IN INCHES

Mean Radius of Bend (in inches)

Sizes	12	15	20	30	40	50	60	70	80	90	100	110	120
1	$1/4$	$3/8$	$3/4$	$1^3/4$	$3^1/8$								
2	$1/8$	$3/8$	$1/2$	1	$1^3/4$	$2^3/4$	$3^7/8$	$5^1/3$					
$2^1/2$	..	$1/4$	$3/8$	$7/8$	$1^1/2$	$2^1/4$	$3^1/4$	$4^1/2$	$5^3/4$				
3	..	$1/8$	$3/8$	$5/8$	$1^1/8$	$1^7/8$	$3^5/8$	$3^5/8$	$4^3/4$	6			
$3^1/2$	..	..	$1/4$	$5/8$	1	$1^5/8$	$2^3/8$	$3^1/8$	$4^1/8$	$5^1/4$			
4	..	..	$1/4$	$1/2$	1	$1^1/2$	2	$2^7/8$	$3^3/4$	$4^3/4$	$5^3/4$		
$4^1/2$	..	..	..	$1/2$	$7/8$	$1^3/8$	$1^7/8$	$2^1/2$	$3^3/8$	$4^1/4$	$5^1/4$		
5	..	..	..	$3/8$	$3/4$	$1^1/8$	$1^5/8$	$2^1/4$	3	$3^3/4$	$4^5/8$	$5^5/8$	
6	..	..	..	$3/8$	$5/8$	1	$1^3/8$	$1^7/8$	$2^1/2$	$3^1/8$	$3^7/8$	$4^3/4$	$5^5/8$
8	..	..	..	...	$1/2$	$3/4$	1	$1^1/2$	$1^7/8$	$2^1/2$	3	$3^5/8$	$4^3/8$
10	..	..	..	...	...	$5/8$	$7/8$	$1^1/8$	$1^1/2$	2	$2^3/8$	$2^7/8$	$3^1/2$
12	..	..	..	...	...	...	$3/4$	1	$1^3/8$	$1^5/8$	2	$2^1/2$	$2^7/8$
14	..	..	..	...	...	...	...	$7/8$	$1^1/2$	$1^5/8$	$1^3/4$	$2^1/8$	$2^1/2$
15	..	..	..	...	...	...	...	...	1	$1^3/8$	$1^5/8$	2	$2^3/8$
16	..	..	..	...	...	...	...	...	$7/8$	$1^1/4$	$1^1/2$	$1^7/8$	$2^1/4$
18	..	..	..	...	...	...	...	...	...	...	...	$1^5/8$	$1^7/8$
20	..	..	..	...	...	...	...	...	...	...	...	...	$1^3/4$

"'U' Bends have twice the above expansive value.

"Expansive 'U' Bends have four times the above expansive value.

"Double Off-set Expansion 'U' Bends have five times the above expansive value.

"An important factor to be considered either when laying out or ordering pipe bends is the weight of the pipe to use. After having obtained the required size, radii of bend, and the working pressure the bend is to be subjected to, the weight of the pipe is the next determination. Based on wide experience in bending pipe and elaborate tests, we recommend the thickness of pipe as follows (Table 92)."

TABLE 92
Thickness of Pipe for Various Bends

Up to 125 Pounds Working Pressure		125 to 250 Pounds Working Pressure	
Diam. of Pipe	Thick. of Pipe	Diam. of Pipe	Thick. of Pipe
Radius 4 to 5 Diameters		Radius 4 to 5 and 6 Diameters	
7″ and smaller	Extra strong	7″ and smaller	Extra strong
8″ and larger	½″ thick	8″ and larger	½″ thick
Radius over 5 Diameters		Radius over 6 Diameters	
7″ and smaller	Full weight	7″ and smaller	Full weight
8″	28.55 lbs. per ft.	8″	28.55 lbs. per ft.
10″	48.48 " " "	10″	40.48 " " "
12″	49.56 " " "	12″	49.56 " " "
14″ to 16″ inclusive	5/16″ thick	14″ to 16″ inclusive	3/8″ thick
18″ to 22″ "	3/8″ "	18″ to 22″ "	7/16″ "
24″ to 30″ "	7/16″ "	24″ to 30″ "	½″ "
250 to 350 Pounds, Working Pressure			
Radius 4 Diameters and over			
7″ and smaller		Extra strong	
8″ and larger		½″ thick	

Bending Pipe. — In the factory pipe is heated and bent to the desired form on a bending floor. Small piping can be bent without crushing by screwing a cap on one end and filling with melted

Fig. 302. Pipe Bending Form.

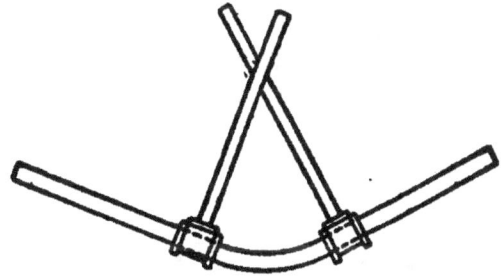

Fig. 303. Bending Small Pipe.

rosin, allowing the rosin to cool and then bending, after which the rosin may be melted out. Sand is sometimes used for the same purpose. There are a number of devices made to assist in bending pipe, one form being shown in Fig. 302. Small pipe can often be bent by using two tees as shown in Fig. 303.

A pipe bending machine for use where a large amount of pipe is to be bent is shown in Fig. 304. With this machine iron or brass pipe up to two inches diameter can be bent cold. The

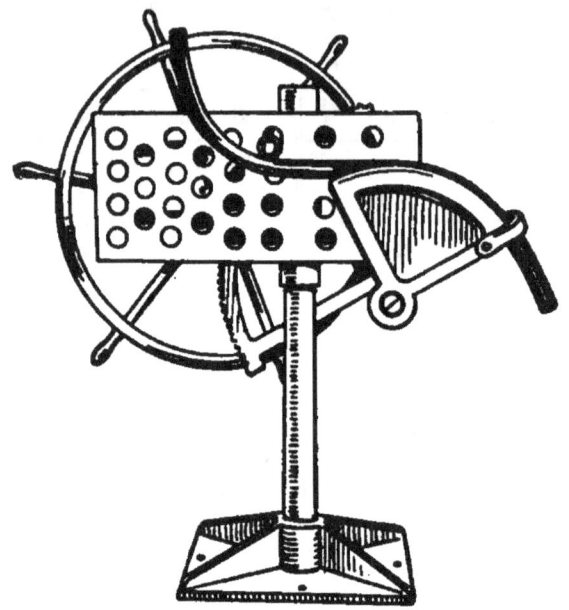

Fig. 304. Pipe Bending Machine.

geared sector which moves the quadrant is operated by a pinion. This pinion is turned by a pilot wheel, 50 inches in diameter. Quadrants are regularly made as follows:

Size of pipe, inches	1/2	3/4	1	1 1/4	1 1/2	2
Radius of bend, inches	4	5	6	9	12	14

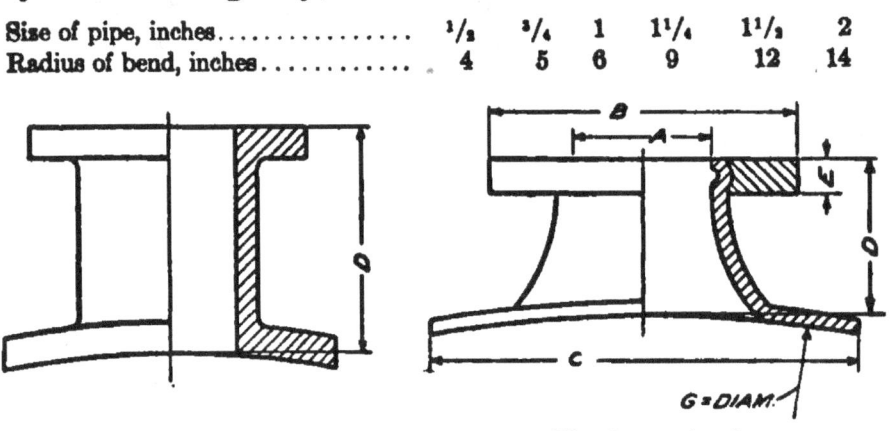

Fig. 305. Nozzles.　　　　Fig. 306. Nozzles.

Nozzles. — Nozzles are used to make the connection between the pipe line and the boiler or for connecting a steam drum to the boiler, Figs. 305 and 306. When made of cast iron or cast steel the dimensions of the upper flange, bolts, thickness of walls, etc., may be made the same as the American Standard. The height D varies from 5 to 16 inches depending upon the size of the outlet. Pressed steel nozzles are stronger and lighter than cast nozzles.

As shown in Fig. 306 the body is pressed out of ⅝ inch flange steel and the upper flange from 1¼ inch flange steel. The flange is connected to the body by expanding the metal of the body under hydraulic pressure into a groove turned in the flange. The joint is tight under a pressure of 1500 pounds per square inch.

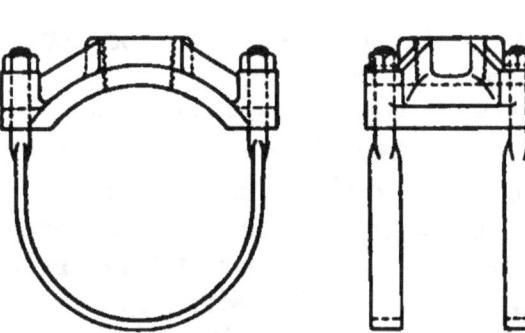

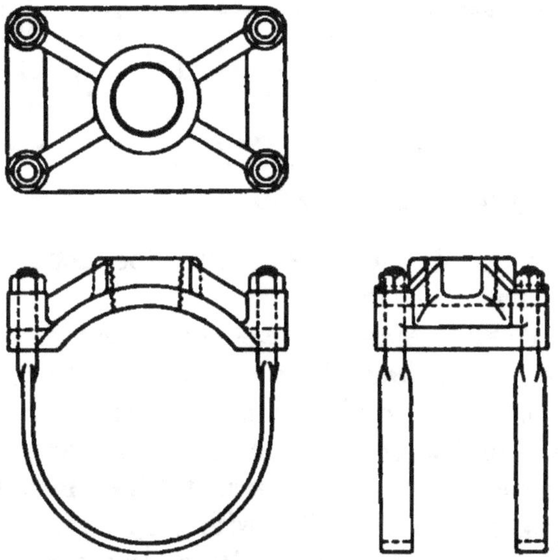

Fig. 307. Pipe Saddle.

Pipe Saddles. — Steam pipe saddles for making connections to wrought iron pipe are made as shown in Fig. 307. These are convenient for use in adding to existing pipe lines, and may be arranged so that they can be put in place upon pipes under pressure. The boss is made of malleable iron and the straps of wrought iron. The combinations of pipe and branches are shown in Table 93.

TABLE 93 (Fig. 307)

PIPE SADDLES

Size of Pipe. Inches	Tapped for Pipe. Inches	Size of Pipe. Inches	Tapped for Pipe. Inches
1½	½ and ¾	6	2½ to 4
2	½ to 1½	7	1 to 4
2½	¾ to 1½	8	1 to 4
3	¾ to 2	9	1½ to 4
3½	¾ to 2	10	1½ to 4
4	¾ to 2	10	4½ to 6
4½	¾ to 2	12	1½ to 4
5	¾ to 2	12	4½ to 6
5	2½ and 3	15	3 to 6
6	¾ to 2	16	3 to 6

Supporting Large Thin Pipe. — Large lead pipe and fittings for acid and other work may be made up from sheets of lead by forming from developed patterns, and burning the edges together. The supports for such piping should be arranged to carry the upper as well as the lower half of the pipe. Thinness of material makes this necessary. Fig. 308 shows such a support with the two halves of the iron ring bolted together and a strip of lead burned over the upper half, thus holding the shape of the pipe.

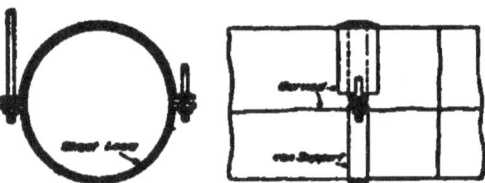

Fig. 308. Supporting Large Lead Pipe.

Flexible Metal Hose. — For many purposes a flexible pipe connection is desirable, such as for blowing boiler tubes, operating steam or air drills, temporary steam, air, oil, or gas lines, for oil feed piping, connections to moving parts of machines and similar services. For such uses metal hose may be had which will give good results if handled with proper care. A section of hose made by the American Metal Hose Company is shown in Fig. 309. It is made from a continuous strip of high tensile strength phosphor bronze, which is wound spirally over itself and made pressure tight by means of a special prepared asbestos cord that is fed into place between the metal surfaces during the winding operation. This hose is also made of steel which is somewhat stronger than bronze and is preferred for superheated steam and where subject to hard usage. Information concerning "American" bronze metal hose is given in Table 94. Sizes are specified by the inside diameter.

Fig. 309. Metal Hose.

Aluminum Piping and Tubing. — Aluminum tubing is specified by outside diameter and thickness of wall. The tables and information in this article are from the catalog of the Aluminum

TABLE 94

Sizes and Dimensions of Metal Hose

Bronze Hose Diam.	Approx. bending diameter	Weight per foot in lbs.	Bronze Hose Diam.	Approx. bending diameter	Weight per foot, in lbs.
1/4"	4"	.11	1 1/2"	22"	1.75
3/8	5"	.25	2"	26"	2.65
1/2"	7"	.35	2 1/2"	32"	3.15
3/4"	12"	.80	3"	38"	4.50
1"	14"	1.00	4"	44"	5.70
1 1/4"	18"	1.50	6"	56"	9.00

Steel hose is approximately 10% lighter than bronze

Company of America. Table 95 gives the outside diameter and several wall thicknesses for aluminum tubing. The safe pressure may be figured by formula 2 Chapter II. The allowable unit stress when the temperature is less than 100 degrees Centigrade is given as follows:

```
Pure aluminum (cast)......................5000 lbs. per sq. inch
Special casting alloy (cast).................5000 to 6000  lbs. per sq. inch
No. 12 casting alloy (cast).................6000 to 8000   "    "   "    "
Pure aluminum tubing (made from sheet).....6000 to 8000   "    "   "    "
3S aluminum tubing (made from sheet).......8000 to 10000  "    "   "    "
```

When the temperature is more than 100 degrees Centigrade the above values should be halved and when more than 200 degrees Centigrade aluminum should not be used under pressure.

Aluminum tubing up to 1 1/2 inches outside diameter can be made by extrusion in almost any desired length. Such continuous lengths have an especial advantage for condensing coils for chemical works as the entire coil can be made from a single piece without joints.

Seamless drawn aluminum tubes are also made to the same dimensions as standard wrought pipe. The weight of aluminum pipe when made to iron pipe sizes is given in Table 96. Such piping can be used with iron fittings, but aluminum fittings can be had in most pipe sizes and are preferable as being less liable to induce galvanic action, than when the fitting is made of another metal.

Brass and Copper Tubing. — The outside diameter is generally used in specifying brass or copper tubing. The thickness may be

TABLE 95.—WEIGHTS IN POUNDS PER FOOT OF ALUMINUM TUBING OUTSIDE MEASUREMENT (Stubs' Gauge)

Nos. of Gauge. Thickness in Thousandths of an Inch	1	2	3	4	5	6	7	8	9	10	11	12	13	14	15	16	17	18	19	20	21	22	23	24	Nos. of Gauge. Thickness in Thousandths of an Inch
Dia.	.300	.284	.259	.238	.220	.203	.180	.165	.148	.134	.120	.109	.095	.083	.072	.065	.058	.049	.042	.035	.032	.028	.025	.022	Dia.
1/8 in.																									1/8 in.
3/16 in.																	.044	.038	.033	.030	.027	.025	.023	.020	3/16 in.
1/4 in.														.053	.050	.047	.069	.063	.053	.046	.043	.037	.033	.030	1/4 in.
5/16 in.											.12	.11	.060	.093	.083	.076	.100	.086	.073	.063	.056	.050	.046	.040	5/16 in.
3/8 in.							.33	.22	.20	.19	.18	.17	.100	.133	.120	.110	.130	.110	.093	.080	.073	.063	.056	.050	3/8 in.
7/16 in.						.43	.31	.30	.27	.25	.23	.22	.147	.170	.150	.140	.150	.130	.110	.096	.090	.070	.070	.060	7/16 in.
1/2 in.					.56	.52	.39	.37	.34	.32	.29	.27	.190	.210	.190	.170	.180	.160	.130	.110	.100	.090	.083	.073	1/2 in.
9/16 in.			.61	.58	.66	.62	.48	.45	.42	.38	.35	.32	.240	.250	.220	.200	.210	.180	.160	.130	.100	.100	.093	.083	9/16 in.
1 in.	.80	.78	.74	.70	.76	.72	.57	.53	.49	.45	.41	.37	.290	.290	.260	.230	.240	.200	.180	.150	.120	.100	.110	.093	1 in.
1 1/8 in.	.94	.91	.86	.81	.87	.81	.65	.61	.56	.51	.46	.43	.330	.330	.290	.270	.240	.200	.180	.150	.140	.120	.110	.10	1 1/8 in.
1 1/4 in.	1.09	1.05	.98	.92	.97	.91	.74	.69	.63	.58	.52	.48	.38	.37	.33	.30	.27	.23	.20	.16	.15	.13	.12	.11	1 1/4 in.
1 3/8 in.	1.23	1.18	1.11	1.03	1.08	1.00	.83	.77	.70	.64	.58	.53	.41	.41	.36	.33	.29	.25	.22	.18	.17	.15	.13	.12	1 3/8 in.
1 1/2 in.	1.36	1.33	1.23	1.15	1.18	1.11	.92	.85	.77	.70	.64	.56	.48	.45	.39	.36	.32	.27	.24	.19	.18	.16	.14	.12	1 1/2 in.
1 5/8 in.	1.52	1.45	1.36	1.26	1.26	1.18	.99	.93	.84	.77	.69	.63	.51	.49	.43	.39	.35	.29	.26	.21	.19	.17	.15	.13	1 5/8 in.
1 3/4 in.	1.66	1.59	1.48	1.38	1.38	1.29	1.09	1.01	.91	.83	.75	.69	.56	.53	.47	.42	.38	.32	.27	.23	.21	.18	.16	.14	1 3/4 in.
1 7/8 in.	1.81	1.73	1.60	1.49	1.49	1.39	1.17	1.09	.98	.90	.81	.74	.60	.57	.50	.45	.41	.34	.29	.24	.23	.20	.18	.15	1 7/8 in.
2 in.	1.94	1.84	1.73	1.61	1.60	1.50	1.25	1.17	1.05	.96	.87	.79	.65	.61	.53	.48	.43	.36	.31	.26	.24	.21	.19	.16	2 in.
2 1/8 in.	2.23	2.13	1.97	1.77	1.71	1.59	1.43	1.32	1.20	1.09	.98	.90	.70	.69	.60	.54	.49	.41	.36	.30	.27	.24	.21	.18	2 1/8 in.
2 1/4 in.	2.52	2.40	2.22	2.06	1.92	1.78	1.60	1.48	1.34	1.22	1.10	1.00	.79	.77	.69	.61	.54	.46	.39	.33	.30	.26	.24	.21	2 1/4 in.
2 3/8 in.	2.67	2.47	2.28	2.12	2.12	1.98	1.78	1.64	1.48	1.35	1.21	1.11	.88	.85	.74	.67	.60	.51	.43	.36	.33	.29	.26	.23	2 3/8 in.
3 in.	2.80	2.67	2.47	2.28	2.34	2.17	1.95	1.82	1.62	1.48	1.33	1.21	.97	.93	.81	.73	.65	.55	.48	.40	.36	.32	.28	.25	3 in.
3 1/8 in.	3.10	2.95	2.71	2.51	2.52	2.36	2.12	1.96	1.76	1.60	1.44	1.32	1.07	1.01	.88	.80	.71	.60	.52	.43	.39	.34	.31	.27	3 1/8 in.
3 1/4 in.	3.37	3.21	2.96	2.74	2.74	2.56	2.29	2.11	1.90	1.73	1.56	1.42	1.15	1.09	.95	.86	.77	.65	.56	.46	.42	.37	.33	.29	3 1/4 in.
3 3/8 in.	3.65	3.48	3.21	2.97	2.96	2.75	2.46	2.27	2.05	1.86	1.67	1.52	1.24	1.17	1.02	.92	.82	.70	.60	.50	.46	.40	.36	.31	3 3/8 in.
3 1/2 in.	3.81	3.47	3.19	2.96	2.75	2.46	2.29	2.11	2.05	1.99	1.79	1.63	1.34	1.25	1.09	.98	.88	.74	.64	.53	.49	.42	.38	.33	3 1/2 in.
4 in.	4.24	4.03	3.70	3.42	3.18	2.90	2.64	2.43	2.19	1.99	1.79	1.63	1.43	1.33	1.16	1.05	.93	.79	.68	.56	.52	.45	.40	.35	4 in.
4 1/8 in.	4.51	4.30	3.71	3.65	3.39	3.14	2.81	2.59	2.33	2.12	1.90	1.73	1.52	1.41	1.23	1.11	.99	.84	.72	.60	.55	.48	.43	.36	4 1/8 in.
4 1/4 in.	4.80	4.57	4.20	3.88	3.61	3.33	2.98	2.75	2.47	2.24	2.02	1.83	1.61	1.49	1.30	1.18	1.05	.88	.76	.63	.58	.51	.45	.38	4 1/4 in.
4 3/8 in.	5.10	4.84	4.45	4.11	3.81	3.53	3.15	2.91	2.61	2.37	2.13	1.94	1.70	1.57	1.36	1.23	1.07	.93	.80	.67	.61	.53	.48	.40	4 3/8 in.
5 in.	5.40	5.12	4.70	4.33	4.02	3.72	3.32	3.06	2.76	2.50	2.25	2.05	1.79	1.65	1.43	1.30	1.16	.98	.84	.70	.64	.56	.50	.43	5 in.
5 1/8 in.	5.67	5.40	4.94	4.54	4.24	3.91	3.49	3.22	2.89	2.62	2.36	2.15	1.88	1.73	1.50	1.36	1.21	1.03	.88	.73	.67	.59	.53	.44	5 1/8 in.
5 1/4 in.	5.96	5.66	5.19	4.79	4.44	4.07	3.67	3.38	3.04	2.76	2.48	2.26	1.96	1.81	1.57	1.42	1.27	1.07	.92	.77	.70	.61	.55	.46	5 1/4 in.
5 3/8 in.	6.20	5.93	5.44	5.02	4.55	4.30	3.84	3.54	3.18	2.89	2.59	2.36	2.07	1.88	1.64	1.48	1.33	1.08	.93	.77	.70	.61	.55	.46	5 3/8 in.
6 in.	6.55	6.20	5.68	5.24	4.86	4.49	4.01	3.70	3.32	3.01	2.71	2.47	2.16	1.99	1.64	1.48	1.33	1.12	.96	.80	.73	.64	.57	.50	6 in.

TABLE 96

Weight of Aluminum Pipe for Iron Pipe Sizes

Same as Iron, Size	Outside Diameter	Inside Diameter	Weights per foot Aluminum lbs.
1/8	.405	.270	.083
1/4	.540	.364	.145
3/8	.675	.494	.193
1/2	.840	.623	.290
3/4	1.050	.824	.387
1	1.315	1.048	.577
1 1/4	1.660	1.380	.777
1 1/2	1.900	1.611	.928
2	2.375	2.067	1.24
2 1/2	2.875	2.468	1.98
3	3.500	3.067	2.59
3 1/2	4.000	3.548	3.11
4	4.500	4.026	3.69

given in Stubs' gauge or B. and S. gauge, but more commonly the former is used. Almost any combination of diameter and thickness may be obtained. The Handbook of Seamless Tubing of the Bridgeport Brass Company gives very complete tables and information.

Boiler Tubes. — The dimensions of standard lap-welded steel or charcoal iron boiler tubes are given in Table 97. The size of tube is specified by the outside diameter.

TABLE 97. — Standard Boiler Tubes

Outside Diameter. Inches	Thickness. Inches	Thickness Nearest B.W.G.	Nominal Weight per Foot	Outside Diameter. Inches	Thickness. Inches	Thickness Nearest B.W.G.	Nominal Weight per Foot
1 1/4	.095	13	1.16	4	.134	10	5.53
1 1/2	.095	13	1.42	4 1/2	.134	10	6.25
1 3/4	.095	13	1.68	5	.148	9	7.67
2	.095	13	1.93	6	.165	8	10.28
2 1/4	.095	13	2.18	7	.165	8	12.04
2 1/2	.109	12	2.78	8	.165	8	13.81
2 3/4	.109	12	3.07	9	.180	7	16.95
3	.109	12	3.36	10	.203	6	21.24
3 1/4	.120	11	4.01	11	.220	5	25.33
3 1/2	.120	11	4.33	12	.229	4 1/2	28.79
3 3/4	.120	11	4.65				

Color System to Designate Piping. — For convenience in distinguishing pipe systems various methods have been devised, for using different colors on the pipes. The A. S. M. E. standard markings are given in Vol. 33 of the Transactions, from which the following is abstracted: "In the main engine rooms of plants which are well lighted and where the functions of the exposed pipes are obvious, all pipes shall be painted to conform to the color scheme of the room, and if it is desirable to distinguish pipe systems, colors shall be used only on flanges and on valve fitting flanges.

In all other parts of the plant, such as boiler house, basements, etc., all pipes (exclusive of valves, flanges and fittings) except the fire system, shall be painted black, or some other single, plain, durable, inexpensive color.

All fire lines (suction and discharge) including pipe lines, valve flanges and fittings, shall be painted red throughout.

The edges of all flanges, fittings or valve flanges on pipe lines, larger than 4 inches, inside diameter, and the entire fittings valves and flanges of lines 4 inches inside diameter and smaller, shall be painted the following distinguishing colors:

DISTINGUISHING COLORS TO BE USED ON VALVES, FLANGES AND FITTINGS

Steam Division
   High pressure — white
   Exhaust steam — buff
Water Division
   Fresh water, low pressure — blue
   Fresh water, high pressure, boiler feed lines — blue and white
   Salt water piping — green
Oil Division
   Delivery and discharge — brass or bronze yellow
Pneumatic Division — all pipe gray
Gas Division
   City Lighting Service — aluminum
   Gas Engine Service — black, with red flanges
Fuel Oil Division — all piping black
Refrigerating System
   Flanges and fittings — white and green stripes, alternately
   Body of pipe — black
Electric Lines and Feeders
   Flanges and fittings — black and red stripes, alternately
   Body of pipe — black

CHAPTER XVI

PIPING INSULATION

Pipe Coverings. — The importance of providing suitable insulation or covering for steam pipes is well known. The loss due to radiation with bare pipes is about 3 B.t.u. per square foot of surface, per degree difference in temperature between steam and air, per hour. With a good covering about one inch thick from 80 to 90 per cent. of this loss can be saved. Some points to be considered in the selection of a pipe covering are as follows: the material should not carbonize after being in contact with a hot surface; the material should be fireproof; the material should not lose its shape after being in use; the material should not contain sulphate of lime or any other substance which might corrode the pipe; the life of the material; the thickness of the material; the value of the coal saved by use of the material; the cost of the material; with superheated steam it is especially necessary that the material contain no organic substances — magnesia and similar materials are desirable; the material should not loosen or disintegrate under vibration. The losses with small pipes are greater than with large ones (relatively). The thickness of material should be between one and two inches. Flanges, valves, etc., should be covered as well as pipes.

Tests on Pipe Coverings. — A valuable series of tests on 26 coverings by L. B. McMillan is described in the Journal of the A. S. M. E., January, 1916. Very complete data is given, and the interested reader is advised to secure the complete paper. The following is abstracted. The tests were made on a 16-foot section of 5-inch pipe. Table 98 gives the B.t.u. losses for the bare pipe and for various kinds of coverings.

Sectional moulded coverings can be obtained for flanges and valves and are especially advisable when the coverings may have to be removed. The material to be used and the exterior covering or casing will be influenced by the location of the piping. Low pressure steam, hot and cold pipes all require separate consideration.

TABLE 98

Data on Efficiencies for Single Thickness Coverings

Covering No.	Kind of Covering	Temperature Difference (Pipe and Room)	Actual Temperature (Room = 80 deg. Fahr.)	B.t.u. Loss /Sq. ft./ Deg. Temperature Difference/Hr.		B.t.u. Saving Due to Covering /Deg./ Sq. Ft./ Hr.	Efficiency of Covering — Per Cent.
				Bare Pipe	Covered Pipe		
I	J-M 85% Magnesia	50	130	1.950	0.435	1.515	77.7
		100	180	2.152	0.438	1.714	79.6
		200	280	2.665	0.446	2.219	83.3
		300	380	3.260	0.455	2.805	86.1
		400	480	4.035	0.469	3.566	88.4
		500	580	5.180	0.488	4.692	90.6
II	J-M Indented	50	130	1.950	0.472	1.478	75.6
		100	180	2.152	0.483	1.669	77.6
		200	280	2.665	0.309	2.156	80.9
		300	380	3.260	0.549	2.711	83.2
		400	480	4.035	0.603	3.432	85.1
		500	580	5.180	0.666	4.514	87.1
III	J-M Vitribestos	50	130	1.950	0.626	1.324	67.9
		100	180	2.152	0.654	1.498	69.6
		200	280	2.665	0.715	1.950	73.2
		300	380	3.260	0.781	2.481	76.0
		400	480	4.035	0.856	3.177	78.8
		500	580	5.180	0.967	4.213	81.4
IV	J-M Eureka	50	130	1.950	0.440	1.510	77.4
		100	180	2.152	0.451	1.701	79.0
		200	280	2.665	0.464	2.201	82.6
		300	380	3.260	0.478	2.782	85.4
		350	430	3.627	0.487	3.140	86.6
V	J-M Molded	50	180	1.950	0.517	1.433	73.4
		100	180	2.152	0.522	1.630	75.8
		200	280	2.665	0.539	2.126	79.8
		300	380	3.260	0.561	2.699	82.8
		400	480	4.035	0.596	3.439	85.2
VI	J-M Wool-Felt	50	130	1.952	0.386	1.564	80.2
		100	180	2.152	0.400	1.752	81.4
		200	280	2.665	0.421	2.244	84.2
		300	380	3.260	0.442	2.818	86.4
		350	430	3.627	0.453	3.174	87.6

TABLE 98 (*Continued*)

Covering No.	Kind of Covering	Temperature Difference (Pipe and Room)	Actual Temperature (Room = 80 deg. Fahr.)	B.t.u. Loss/Sq. Ft./Deg. Temperature Difference/Hr.		B.t.u. Saving Due to Covering /Deg./Sq. Ft./Hr.	Efficiency of Covering — Per Cent.
				Bare Pipe	Covered Pipe		
VII	Sall-Mo Expanded	50	130	1.950	0.409	1.541	79.0
		100	180	2.152	0.427	1.725	80.2
		200	280	2.665	0.464	2.201	82.6
		300	380	3.260	0.503	2.757	84.6
		400	480	4.035	0.541	3.494	86.6
		500	580	5.180	0.581	4.599	88.8
VIII	Carey Carocel	50	130	1.950	0.358	1.592	81.6
		100	180	2.152	0.378	1.774	82.4
		200	280	2.665	0.421	2.244	84.2
		300	380	3.260	0.466	2.794	85.7
		400	480	4.035	0.510	3.525	87.4
		500	580	5.180	0.562	4.618	89.2
IX	Carey Serrated	50	130	1.950	0.454	1.496	76.7
		100	180	2.152	0.468	1.684	78.2
		200	280	2.665	0.506	2.159	81.0
		300	380	3.260	0.546	2.714	83.3
		400	480	4.035	0.587	3.448	85.4
		500	580	5.180	0.634	4.546	87.8
X	Carey Duplex	50	130	1.950	0.423	1.527	78.3
		100	180	2.152	0.447	1.705	79.2
		200	280	2.665	0.498	2.167	81.3
		300	380	3.260	0.548	2.712	83.2
		350	430	3.627	0.574	3.053	84.2
XI	Carey 85% Magnesia	50	130	1.950	0.413	1.537	78.8
		100	180	2.152	0.418	1.734	80.5
		200	280	2.665	0.424	2.241	84.1
		300	380	3.260	0.436	2.824	86.6
		400	480	4.035	0.454	3.581	88.8
		500	580	5.180	0.472	4.708	90.9
XII	Sall-Mo Wool-Felt	50	130	1.950	0.395	1.555	79.8
		100	180	2.152	0.401	1.751	81.4
		150	230	2.400	0.421	1.979	82.5
		200	280	2.665	0.433	2.232	83.8
		250	330	2.951	0.455	2.506	84.9
		300	380	3.260	0.459	2.801	85.9

TABLE 98 (*Continued*)

Covering No.	Kind of Covering	Temperature Difference (Pipe and Room)	Actual Temperature (Room = 80 deg. Fahr.)	B.t.u. Loss/Sq. Ft./Deg. Temperature Difference/Hr.		B.t.u. Saving Due to Covering /Deg./Sq. Ft./Hr.	Efficiency of Covering — Per Cent.
				Bare Pipe	Covered Pipe		
XIII	Nonpareil High Pressure	50	130	1.950	0.399	1.551	79.5
		100	180	2.152	0.402	1.750	81.3
		200	280	2.665	0.412	2.253	84.6
		300	380	3.260	0.426	2.834	68.9
		400	480	4.035	0.444	3.591	89.0
		500	580	5.180	0.465	4.715	91.0
XIV	J-M Fire Felt	50	130	1.950	0.694	1.256	64.4
		100	180	2.152	0.711	1.441	67.0
		200	280	2.665	0.749	1.916	71.9
		300	380	3.260	0.795	2.465	75.6
		400	480	4.035	0.845	3.190	79.0
		500	580	5.180	0.901	4.279	82.6
XV	J-M Sponge Felted	50	130	1.950	0.336	1.614	82.7
		100	180	2.152	0.347	1.805	83.8
		200	280	2.665	0.369	2.296	86.2
		300	380	3.260	0.391	2.869	88.0
		400	480	4.035	0.414	3.621	89.8
		500	580	5.180	0.439	4.741	91.5
XVI	J-M Asbestocel	50	130	1.950	0.418	1.532	78.5
		100	180	2.152	0.429	1.723	80.0
		200	280	2.665	0.454	2.211	83.0
		300	380	3.260	0.493	2.767	84.8
		400	480	4.035	0.544	3.491	86.5
		500	580	5.180	0.609	4.571	88.2
XVII	J-M Air Cell	50	130	1.950	0.459	1.491	76.4
		100	180	2.152	0.475	1.677	77.9
		200	280	2.665	0.515	2.150	80.7
		300	380	3.260	0.571	2.689	82.7
		400	480	4.035	0.643	3.392	84.1
		500	580	5.180	0.733	4.447	85.8

PIPING INSULATION

Table 99 gives data for various thicknesses of 85 per cent. magnesia.

TABLE 99

Data on Efficiencies for Various Thicknesses of 85 Per Cent. Magnesia Covering

Temperature Difference	Thickness	B.t.u./Sq. ft./Deg. Dif./Hr.			Saving	Efficiency
		Bare Pipe	Plastic 85 Per Cent. Magnesia	Sectional 85 Per Cent. Magnesia		
100	0.5	2.152	0.735	0.691	1.461	67.8
100	1.0		0.492	0.462	1.690	78.4
100	2.0		0.319	0.300	1.852	85.5
100	3.0		0.248	0.233	1.919	89.1
100	4.0		0.209	0.196	1.956	90.8
100	5.0		1.185	0.174	1.978	91.9
300	0.5	3.260	0.805	0.757	2.503	76.8
300	1.0		0.524	0.493	2.767	84.9
300	2.0		0.335	0.315	2.945	90.4
300	3.0		0.260	0.244	3.016	92.5
300	4.0		0.219	0.206	3.054	93.7
300	5.0		0.192	0.181	3.079	94.4
500	0.5	5.180	0.895	0.842	4.338	83.7
500	1.0		0.557	0.524	4.656	89.9
500	2.0		0.350	0.329	4.851	93.6
500	3.0		0.273	0.257	4.923	95.0
500	4.0		0.229	0.215	4.965	95.8
500	5.0		0.199	0.187	4.993	96.4

The seventeen coverings listed in Table 83 are described as follows:

"I. *J-M 85 Per Cent. Magnesia.* A moulded sectional covering for use on high pressure steam pipes. Contains 85 per cent. by weight of magnesium carbonate and the remainder is principally asbestos fibre. Weight per foot is 2.92 lbs. and the thickness 1.08 in.

II. *J-M Indented.* Made up of layers of asbestos felt which has in it indentations, about 1 1/4 in. in diameter and 1/8 in. deep, spaced very close to each other in staggered rows. Suitable for use on pipes containing high pressure steam. Weight per foot 3.46 lbs. and thickness 1.12 in.

III. *J-M Vitrebestos.* An asbestos air cell covering made of alternate layers of smooth and corrugated vitrified asbestos sheets. Corrugations are about 1/4 in. deep and run lengthwise of the pipe. Recommended for use on high pressure and superheated steam pipes and for stack linings, etc. Weight per foot 4.05 lbs. and thickness 0.96 in.

IV. *J-M Eureka.* For use on low pressure steam and hot water pipes. Made of $1/4$ in. of asbestos felt on the inside of the section and the balance of alternate layers of asbestos and wool felt. Weight 4.60 lbs. per ft. and 1.04 in. thick.

V. *J-M Molded Asbestos.* A molded sectional covering for use on low and medium pressure steam pipes. Made of asbestos fiber and other fireproof material. Weight per ft. 5.53 lbs. and thickness is 1.25 in.

VI. *J-M Wool Felt.* A sectional covering made of layers of wool felt with an interlining of two layers of asbestos paper. May be used on low pressure steam and hot water pipes. Weight per ft. 2.59 lbs. and thickness 1.10 in.

VII. *Sall-Mo Expanded.* A covering for use in high and low pressure steam pipes. Made of eight layers of material, each consisting of a smooth and a corrugated piece of asbestos paper, the corrugations being so crushed down to form small longitudinal air spaces. Weight 3.47 lbs. per ft., and thickness 1.07 in.

VIII. *Carey Carocel.* Composed of plain and corrugated asbestos paper firmly bound together. Corrugations are approximately $1/8$ in. deep and run lengthwise of the pipe. For use on medium and low pressure steam pipes. Weight 3.06 lbs. per ft. and thickness 0.99 in.

IX. *Carey Serrated.* A covering for use on high pressure steam pipes. Composed of successive layers of heavy asbestos felt having closely spaced indentations in it. Weight 5.66 lbs. per ft., and thickness 1.00 in.

X. *Carey Duplex.* For use on low pressure steam and hot water pipes. Made of alternate layers of plain wool felt and corrugated asbestos paper firmly bound together. Corrugations run lengthwise of the pipe and make air cells approximately $1/4$ in. deep. Weight 1.79 lbs. per ft. and 0.96 in. thick.

XI. *Carey 85 Per Cent. Magnesia.* A covering for high pressure steam and similar in composition to No. 1. Weight per foot 2.75 lbs. and thickness is 1.10 in.

XII. *Sall-Mo Wool Felt.* Similar to No. VI except that it has no interlining of asbestos paper. For use on low pressure steam and hot water pipes. Weight per foot 3.73 lbs. and thickness is 1.01 in.

XIII. *Nonpareil High Pressure.* A molded sectional covering consisting mainly of silica in the form of diatomaceous earth — the skeletons of microscopic organisms. For use on high pressure and superheated steam pipes. Weight 2.96 lbs. per ft., and is 1.16 in. thick.

XIV. *J-M Asbestos Fire Felt.* Consists of asbestos fiber loosely felted together, forming a large number of small air spaces. For use on high pressure and superheated steam pipes. Weight per ft. is 3.75 lbs., and thickness 0.99 in.

XV. *J-M Asbestos Sponge Felted.* Covering is made from a thin felt asbestos fiber and finely ground sponge forming a very cellular fabric. Made up of 41 of these sheets per inch thickness and air spaces are formed between the sheets in addition to those in the felt itself. Specially recommended for high pressure and superheated steam pipes. Weight per foot 4.04 lbs. and thickness 1.16 in.

XVI. *J-M Asbestocel.* For use on medium pressure steam and heating pipes. Consists of alternate sheets of corrugated and plain asbestos paper

PIPING INSULATION

forming air cells about $1/8$ in. deep that run around the pipe. Weight per foot 1.94 lbs., and thickness 1.10 in.

XVII. *J-M Air Cell.* Made of corrugated and plain sheets of asbestos paper arranged alternately so as to form air cells about $1/4$ in. deep running lengthwise of the pipe. For use on medium pressure steam and heating pipes. Its weight per foot is 1.55 lbs., and thickness is 1.00 in."

The results of exhaustive tests made on Nonpareil coverings are given in very complete form in a book published by the Armstrong Cork and Insulation Company. This covering is composed of diatomaceous earth (kieselguhr) and asbestos fibre. These tests showed the conductivity of Nonpareil High Pressure Covering per square foot at the mean circumference per one inch thickness per degree difference in temperature to be 7.363 B.t.u. and the transmission through bare pipe 51.07 B.t.u. per square foot of pipe surface per degree difference in temperature for 24 hours. These transmissions were measured in still air and consequently are less than would obtain under operating conditions. The following thicknesses of Nonpareil High Pressure covering are considered economical for the purposes listed under average conditions. Standard thickness ranges from one inch for the small sizes to $1\frac{1}{2}$ inches for the large sizes of pipe. For high pressure piping, inside of buildings.

Cost of Steam per 1000 Pounds	Saturated Steam	Superheated Steam
Less than 10 cents	Standard thickness	$1\frac{1}{2}''$ thick
10 cents to 15 cents	Standard thickness	$2''$ thick
15 cents to 20 cents	$1\frac{1}{2}''$ thick	Double layer $1\frac{1}{2}''$
20 cents and over	$2''$ thick	Double layer $1\frac{1}{2}''$

For exhaust feed and hot well, high pressure drip piping, etc., under all conditions listed above — standard thickness. For high pressure steam outside of buildings under all conditions listed above — double layer of $1\frac{1}{2}$ inch thickness.

Low Pressure Steam, Hot and Cold Water Pipes. — All heating piping, either steam or hot water, should be fully covered where radiation losses are to be avoided. Cold water pipes are frequently insulated in order to prevent "sweating" and dripping. For the above conditions, and where exhaust pipes are to be insulated, the low temperatures do not require thick coverings and wool felt or air cell coverings $1/2$ inch to one inch thickness may be used.

Cold Pipes. — It is important to consider the question of insulation of pipe used to convey ammonia or brine for refrigeration purposes, if serious losses are to be prevented. The problem is not very different from insulation of hot pipes, but it is very essential that the material used is not easily injured by moisture. Hair, felt and paper in alternate layers has been used as a protection for cold pipes. Hair felt soaked in boiling resin and applied to the pipes while hot is also used. Sectional coverings composed of granulated cork may be obtained ready for use on brine or ammonia pipes and fittings.

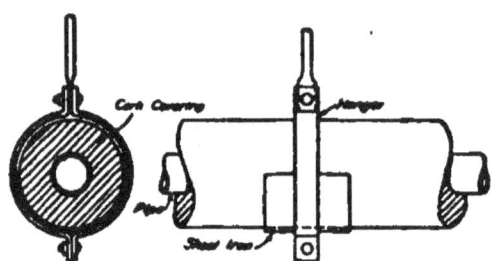

Fig. 310. Support for Pipe with Cork Insulation.

Nonpareil cork covering is made by the Armstrong Cork and Insulation Company by compressing and then baking pure granulated cork in metal moulds. After this the covering is coated inside and out with a waterproof mineral rubber finish, ironed on hot. Tests by the above company gave an average transmission per square foot at mean circumference, per one inch thickness per degree difference in temperature per 24 hours of 8.6 B.t.u. for cork covering and of 43.2 B.t.u. for bare pipe. Four grades of this covering are made. Standard brine covering, from two to three inches thick for temperatures of 0 to 25 degrees F.; special thick brine covering, from three to four inches thick for temperatures below zero degrees F.; ice water covering, about $1\frac{1}{2}$ inches thick for temperatures of 25 to 45 degrees F.; and cold water covering for use on cold water piping to prevent sweating. The method of supporting the pipe is shown in Fig. 310 where a hanger is on the outside with a piece of sheet iron protecting the covering.

Forms of Pipe Coverings. — The materials for pipe coverings may be had in a variety of forms. For covering pipe, sheets of material may be wrapped around the pipe and fastened with wire or heavy twine; the material may be in plastic form and applied in the shape of a mortar; or any of the large variety of moulded or sectional coverings, Fig. 311, may be used. Sectional coverings are made in lengths of three feet, and are split lengthwise into halves. When applied to the pipe they are wrapped with canvas and then held on with iron or brass bands spaced from one to two feet apart. Fittings and valves may be insulated with a plastic coating or with moulded covers made in sections to fit over them.

Fig. 311. Sectional Pipe Covering.

Hair felting comes in rolls six feet wide and in thicknesses of $1/4$ to $1\frac{1}{2}$ inches. Asbestos paper is made in varying thicknesses and in rolls 36 inches wide.

Underground Piping. — Two methods of insulating underground piping are described in Chapter XIII. Careful underdrainage is essential to any system.

Forms of wood casing for underground steam and hot water piping made by A. Wyckoff & Son Company are shown in Figs.

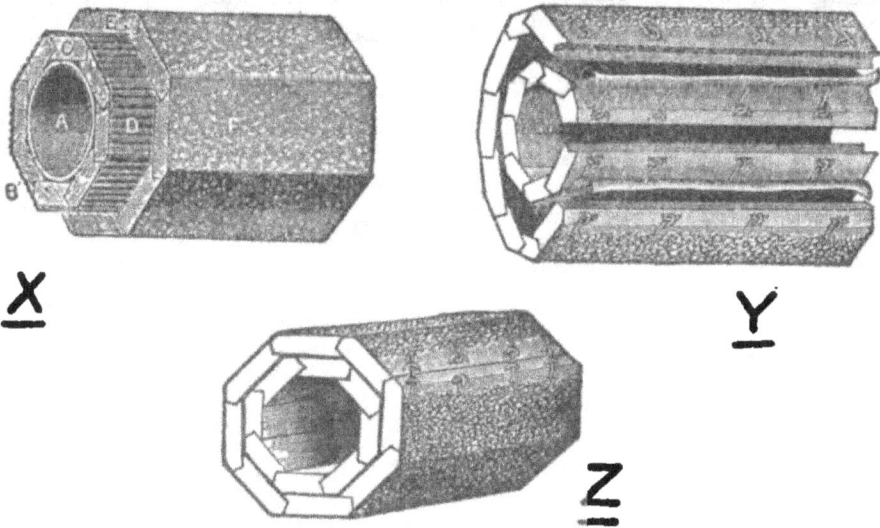

Fig. 312. Wood Casing — Split Form.

312 and 313. The form shown at X, Y and Z is made of thoroughly seasoned gulf cypress staves, one inch thick, closely jointed together, wound with heavy galvanized steel wire, and then

wrapped with two layers of heavy corrugated paper. Another casing of one inch cypress staves is put on the outside and wound with galvanized wire. For use with high pressure steam pipe the

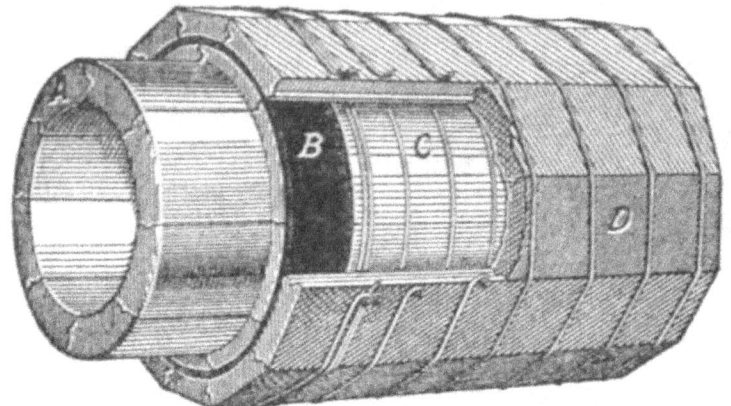

Fig. 313. Improved Wood Casing.

casing is lined with tin and two layers of asbestos paper to prevent the wood from charring. The casing is made in lengths of from four to eight feet which are connected by tenon and socket joints X, Fig. 312. For use on pipes which are already in place the casing may be had split in the form shown at Y and Z, Fig. 312. The casing shown in Fig. 313 is an improved form in which A is a two inch inner shell, B is asphaltum packing, C is a $1/4$ inch air

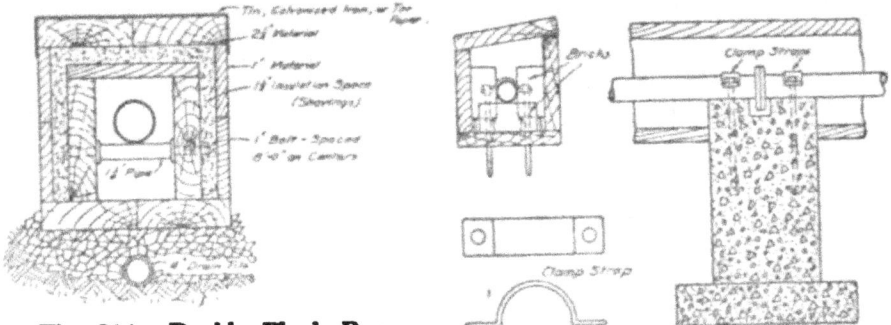

Fig. 314. Double Plank Box Insulation.

Fig. 315. Plank Box Insulation.

space and D is a one inch outer shell. The casing is afterwards coated with Hydolene-B and rolled in sawdust. This form is made in lengths of from four to twelve feet, with tenon and socket joints. It cannot be split, but must be slipped over the pipes, while they are being connected up.

PIPING INSULATION

Two forms of plank box insulation for underground piping are shown in Figs. 314 and 315, which have appeared in *Power*, and are described as being in successful use. Fig. 314 is by W. H. Wolfang, and shows double planking with shavings filled in be-

Fig. 316. Split Tile Conduit.

tween. The supports are rollers made from $1\frac{1}{4}$ inch pipe and one inch rods. The side dimensions for four inch pipe are eight by twelve inches. Fig. 315 is by Henry G. Pope, and is composed of rough two inch plank. As noted, the top plank slopes to one side to shed water. Waterproofed building paper was tacked over each joint. Bricks were used for supporting the pipe. The method of anchoring is also shown in the figure.

A method of constructing underground mains up to 20 inch pipe using split tile is illustrated in Figs. 316 and 317, and described by the Armstrong Cork and Insulation Company.

TABLE 100

Sizes of Steam Lines and Protecting Tile

Steam Line Size Inches	Protecting Tile Size, Inches	Steam Line Size, Inches	Protecting Tile Size, Inches
1	8	7	15
1¼	8	8	15
1½	8	9	18
2	10	10	18
2½	10	12	20
3	10	14	21
3½	12	16	24
4	12	18	27
4½	12	20	27
5	12	24	30
6	15	30	36

"For underground lines excellent results can be secured by using two inch thick, nonpareil, high pressure covering, protected with a good grade of hard-glazed, split tile, although for lines larger

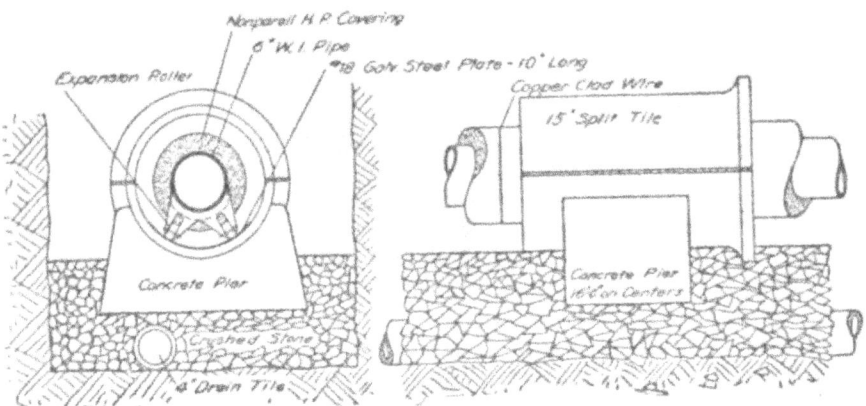

Fig. 317. Split Tile Conduit.

than twenty inches it is often advisable to use regular tunnel construction. A four inch drain is laid in the bottom of the trench to carry off seepage water and concrete supporting piers are installed on sixteen-foot centres. A bed of crushed stone or coarse gravel is then put down to grade, and upon this the lower half of the tile is laid. The expansion rollers are strapped to the steam pipe so that they will rest directly over the concrete supporting piers. To prevent abrasion of the tile, No. 18 gauge galvanized

PIPING INSULATION

steel plates are inserted between the expansion rollers and the tile. After the pipe is in position, the covering is applied and held in place by copper-clad steel wire, canvas on the outside being usually dispensed with. The joints are pointed up with nonpareil high pressure cement, and the top of the tile is then cemented in place with Portland cement mortar."

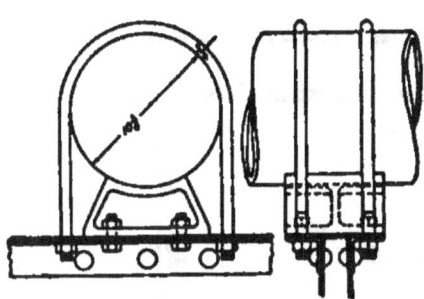

Fig. 318. Method of Anchoring.

The sizes of protecting tile are given in Table 100.

Where a number of pipes, electric wires, etc., are to be carried underground, some form of tunnel is about the best arrangement. Such tunnels can be built up of brick or can be made of concrete. Electric wires may be run in tile set in the walls or roof of the tunnel. Pipe lines can be carried on brackets or supports at the sides, with provision for expansion and drainage and regular methods of insulation. The floor of the tunnel should be arranged with drain connections to take care of any water that may accumulate from leaks in the piping or other causes.

Out-of-Doors Piping. — The methods of insulation shown in Figs. 312 and 313 are well adapted for use on steam pipes running out of doors and exposed to the weather. For such purposes the outer wooden casing is painted with black asphaltum paint.

Very often the regular method of insulation as used on in-door lines are employed, making the covering somewhat thicker and enclosing it in waterproof paper,

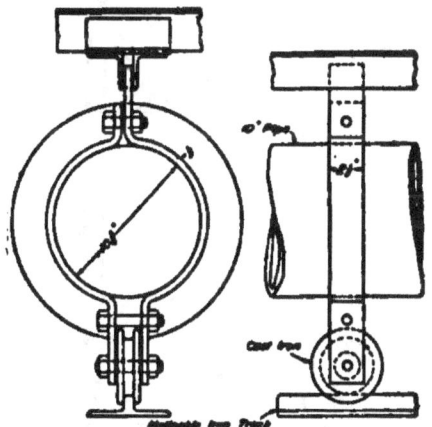

Fig. 319. Roller Support.

or wooden or steel plate boxing may be constructed for a protection from the weather.

Some details of an interesting out-door pipe line forming part of the River Power Plant of the Victor Talking Machine Company are shown in Figs. 318, 319, 320, 321 and 322. This plant

302   A HANDBOOK ON PIPING

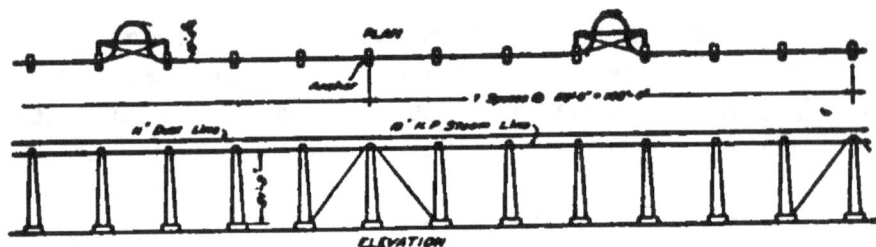

Fig. 320. Part Plan and Elevation of Outdoor Steam Line.

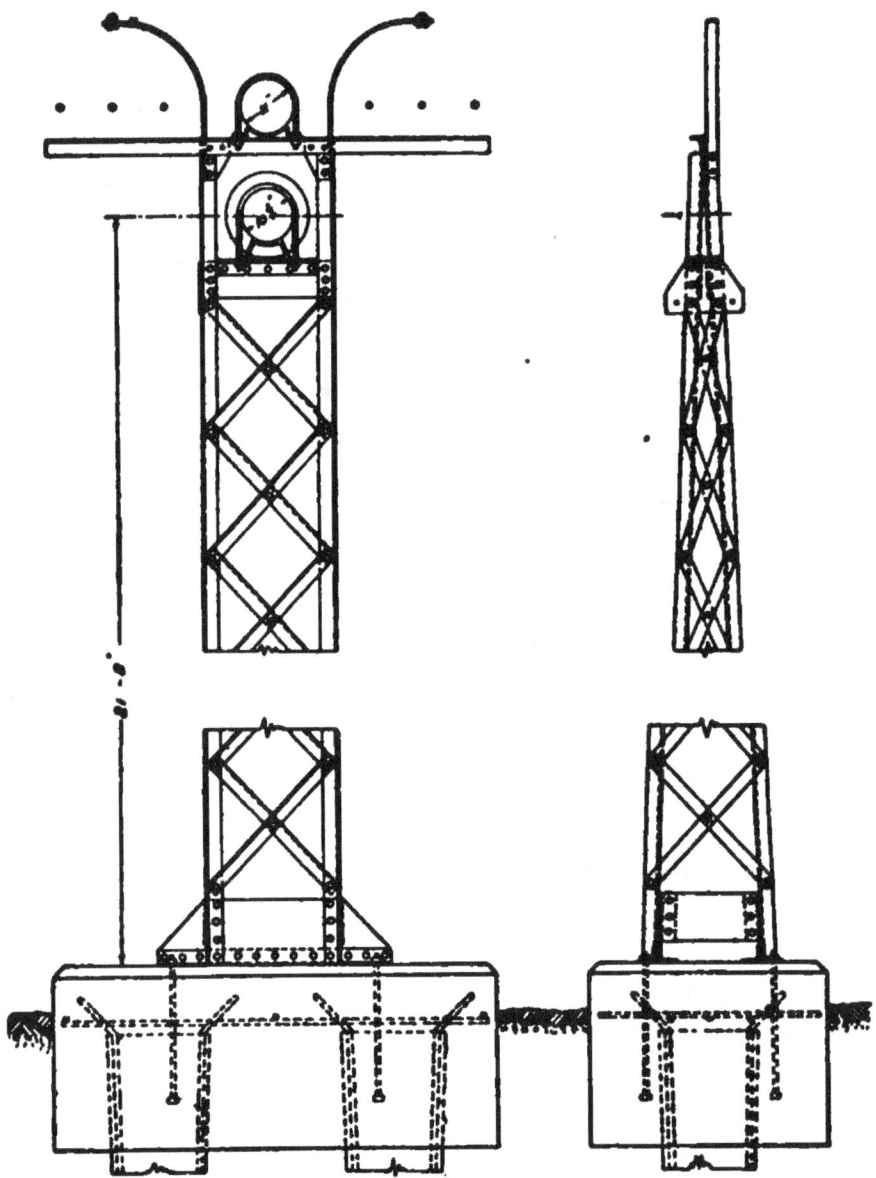

Fig. 321. Drawing of Supporting Structure for Outdoor Steam Line.

PIPING INSULATION

is the design of Mr. Albert C. Wood, consulting engineer, who has furnished the information concerning it. A part plan and elevation of the line which is several hundred feet long is show in Fig. 320. One of the supporting structures is shown in Fig. 321 with its foundation resting upon two concrete piles, which were necessary because the ground is made and is underlaid with

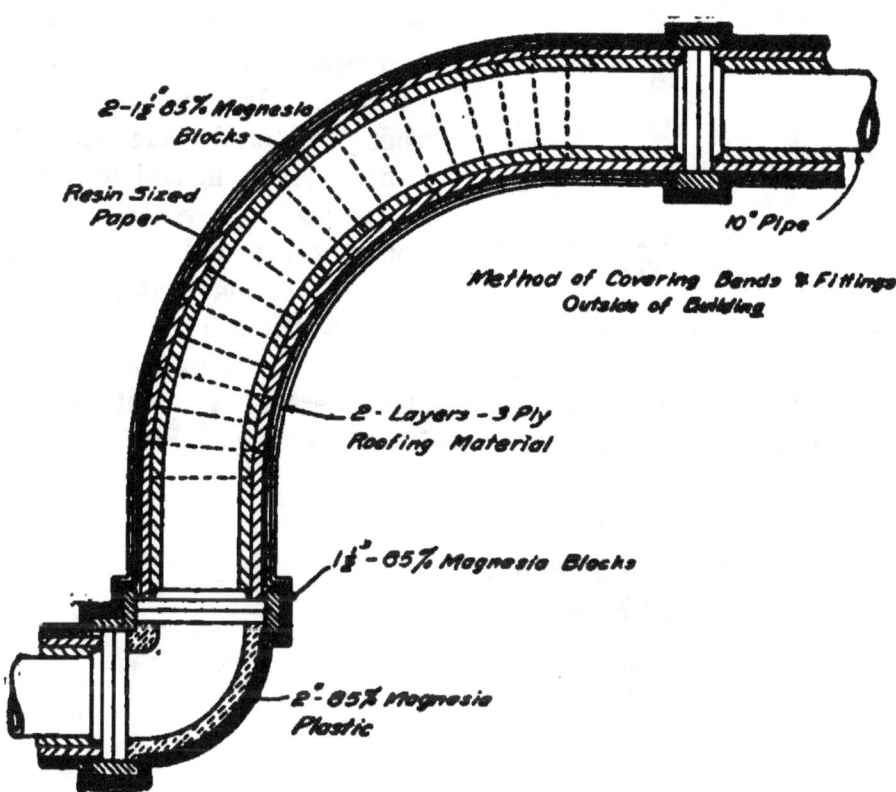

Fig. 322. Method of Covering Bends and Fittings.

river mud. The supports were made very heavy in order to provide for the possibility of lumber stacks falling against them and also that the high pressure steam line might be substantially supported. These supports are placed about 20 feet apart. They carry a ten inch high pressure steam line, 160 pounds per square inch (150 degrees superheat) and an eleven inch sawdust line, as well as brackets for 500,000 C.N., 250 volt D.C. cables. The method of anchoring is shown in Fig. 320. The roller support, which allows freedom for movement due to expansion, is clearly indicated in Fig. 319.

The insulation of the high pressure steam pipe consists of two layers, 1½ inches thick, 85 per cent. magnesia blocks, moulded to proper radius to suit the pipe with the joints broken both longitudinally and circumferentially. The joints and interstices were filled with 85 per cent. magnesia plastic. Over this resin sized paper was applied and wired every twelve inches with two turns of No. 16 copper wire. Then two layers of roofing material were applied with all joints lapped at least two inches and wrapped with roofing compound. The first layer of roofing material was secured at the joints and at intervals of about 18 inches with three turns of No. 16 copper wire, while the second layer was secured at the joints and at regular intervals of about twelve inches with three turns of No. 14 copper wire. Fittings and

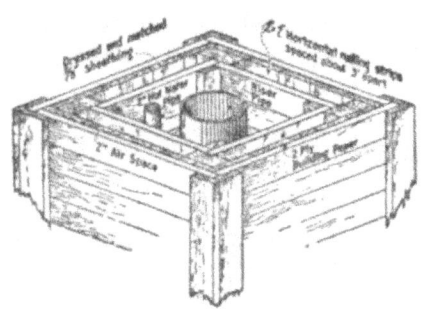

Fig. 323. Frost Boxing for Water Stand Pipe.

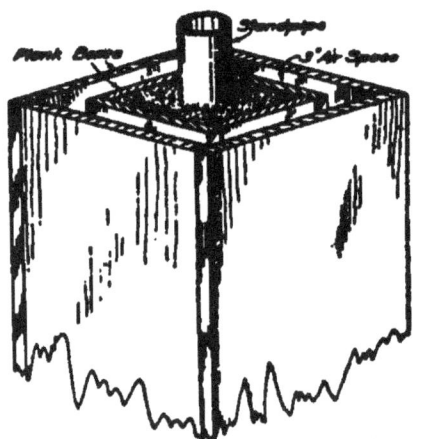

Fig. 324. Square Boxing for Water Pipe.

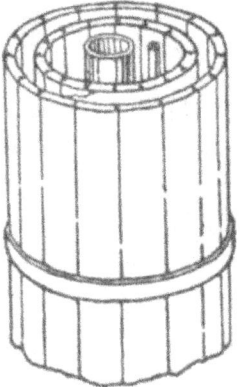

Fig. 325. Circular Boxing for Water Pipe.

valves were covered as indicated in Fig. 322, blocks being used, together with 85 per cent. magnesia plastic.

Air piping may be run on the surface of the ground or carried on trussed poles or towers. Proper care must be taken to provide for drainage and necessary expansion.

PIPING INSULATION

The protection of water standpipes from freezing is an important matter. In Fig. 323 is shown a tightly constructed frost boxing described by Mr. W. C. Teague in the A. S. M. E. Journal, April, 1914. Arrangements should be made for keeping the water heated by a hot water heater or steam coil placed in the bottom of the tank.

A simple form of protection is shown in Fig. 324, composed of two plank boxes with an air space between them. The joints should be made very tight and the outside painted with asphaltum paint or be otherwise protected. A circular form of protection is shown in Fig. 325.

CHAPTER XVII

PIPING DRAWINGS

The underlying principles are the same for all classes of drawings, but for each branch there are certain conventions and general methods of representation. It is the purpose of this chapter to deal with some of these general customs and details rather than to present a collection of complicated drawings.

Classification of Piping Drawings.—There are several kinds of piping drawings depending upon the purpose and requirements of the work. Sometimes a freehand sketch is sufficient, sometimes a line diagram, and sometimes a large scale drawing, consisting of several views of the entire system, together with working drawings of details is necessary. A drawing for construction purposes must give complete information as to sizes, position of valves, branches and outlets. A drawing to show the layout of existing pipe lines need not be as complete and is often made to small scale, using single lines to represent the pipes, with notes to tell sizes, location and purpose for which the pipe is used. A drawing to show proposed changes should give both existing and proposed piping, using different kinds of lines to distinguish the changes. Dot and dash lines, dash lines, or red or other colored ink may be used for this purpose. A drawing for repairs may consist of simply the part to be repaired, or may show the location or connection between the repairs and apparatus or other parts of the system. Drawings for repairs should be checked very carefully and just what is to be replaced or repaired should be made clear.

Erection Drawings.—Drawings for erection are sometimes made with very few dimensions but with all pieces numbered and accompanied by a list giving complete information concerning each piece. A piping list may be made up in a variety of ways. One method is to list each piece of pipe, fitting and valve in order from one end of the system, and then collect all the pipe of each size, all the ells, tees, unions, valves, etc. A form similar to Fig. 326 is often useful.

PIPING DRAWINGS

Detail drawings should be made in the same manner as for any other purpose. The detail drawing for a special fitting is shown in Fig. 327. All piping drawings should have a title giving the purpose of the piping, scale of drawing, and date, together with provision for changes and date of changes and any other necessary information. It is particularly important that piping drawings be kept up to date. The dimensions for standard flange fittings are given in Chapter IV, and throughout this book will be found tables giving dimensions for various piping fixtures and

Size	Pipe Feet	Number Valves	Number Fittings	Thds.	Mat'l	Make
⅞	365	—	—	R	W.I.	
1¼	185	—	—	R	W.I.	
1¼	—	8 Globe	—	R	Brass	T.Z.Co.
1¼	—	—	21 Ells	R	C.I.	
1¼	—	—	7 Tees	R	C.I.	
1¼	—	—	5 Couplings	R & L	C.I.	

Fig. 326. Form for Listing Fittings.

fittings, etc. When possible it is always well to use the manufacturers' catalogs, provided the makes to be used are known. A steam piping drawing is shown in Fig. 328, in which the dimensions are indicated without the figures, for the sake of clearness.

Conventional Representation. — Fittings and valves may be drawn as in the various figures shown throughout this book. When drawn to a small scale conventional representations are often used. A variety of such conventions are shown in Figs. 329 and 330. It is desirable to add an explanatory list to a drawing when these are used, unless notes make clear the meaning of each one. They are very convenient for sketching and diagrammatic purposes. Several methods of showing pipe are given in Fig. 331. Except in special cases, or for small pieces, it is not necessary to use shading. When a single line is used it should be

308 A HANDBOOK ON PIPING

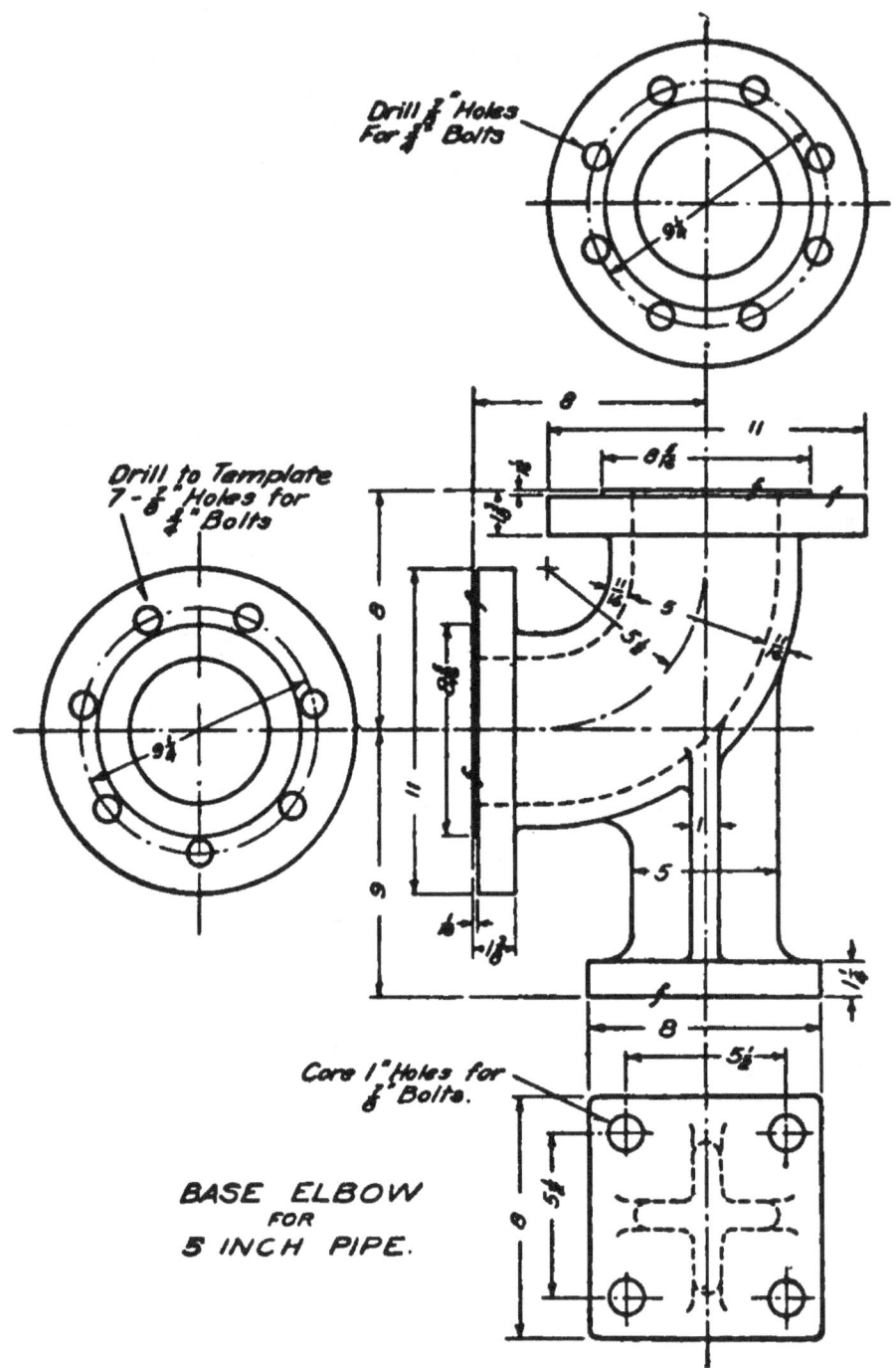

Fig. 327. Detail of Base Elbow.

PIPING DRAWINGS

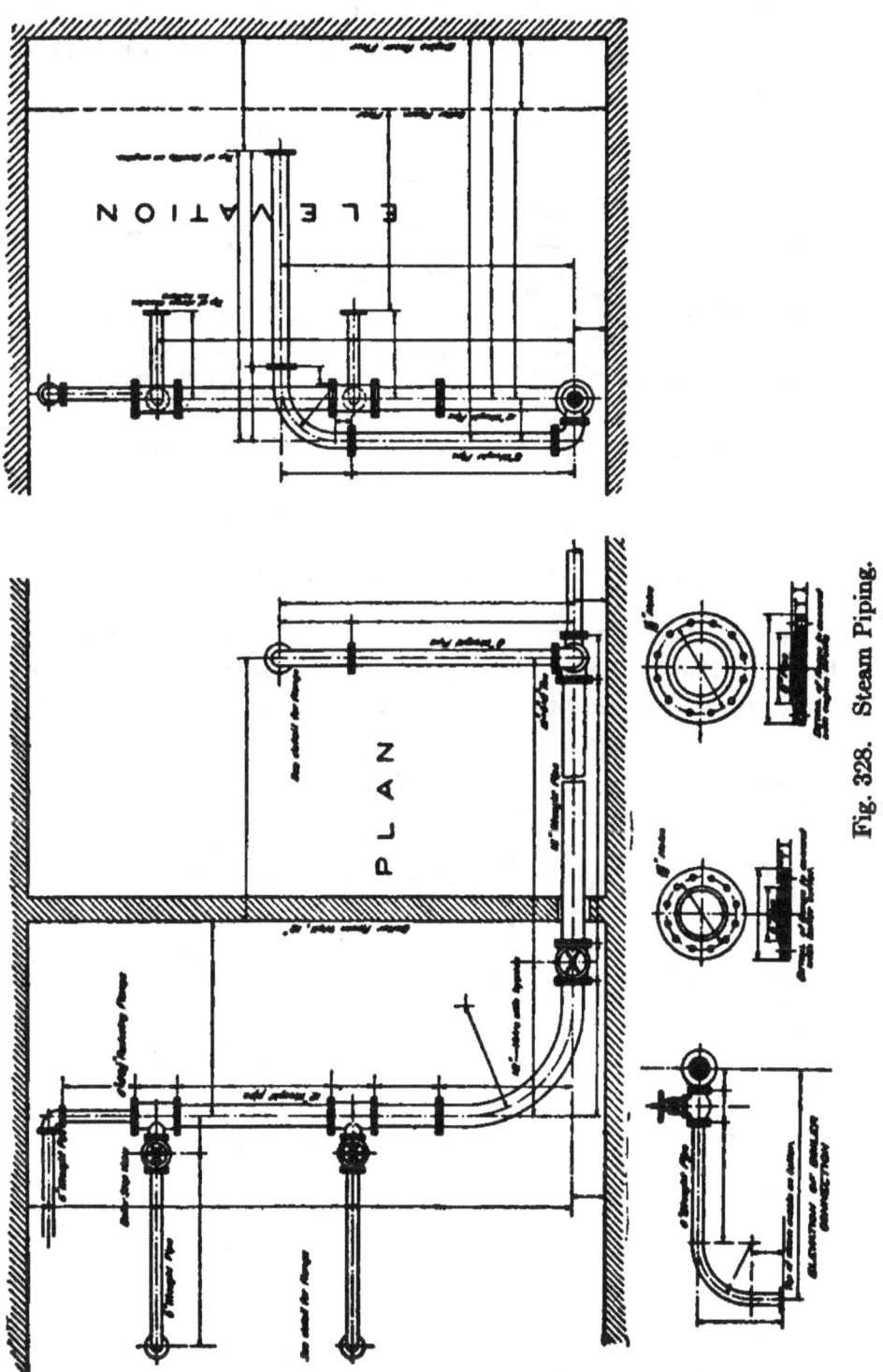

Fig. 328. Steam Piping.

310   A HANDBOOK ON PIPING

enough heavier than the other lines of the drawing to stand out clearly, usually about three times as heavy will be satisfactory.

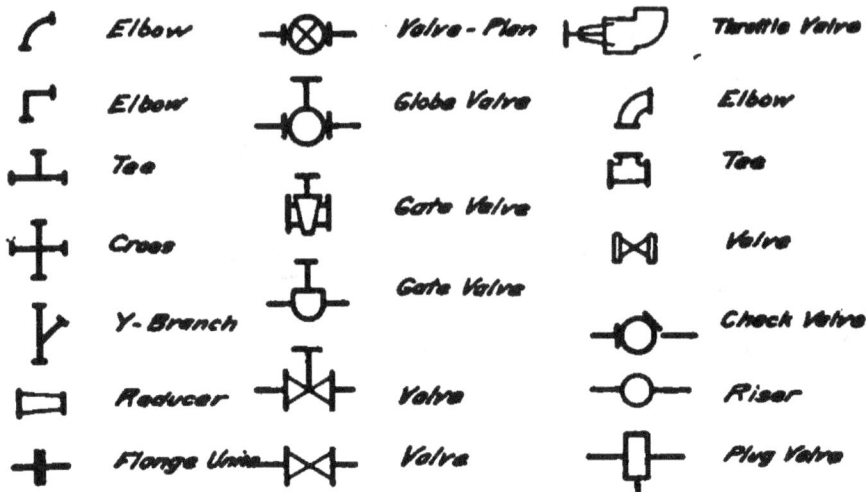

Fig. 329. Conventional Representations for Fittings.

Apparatus used in connection with piping as well as the machines to which it is connected are frequently represented by diagrams, more or less conventional. Several methods in use are shown in

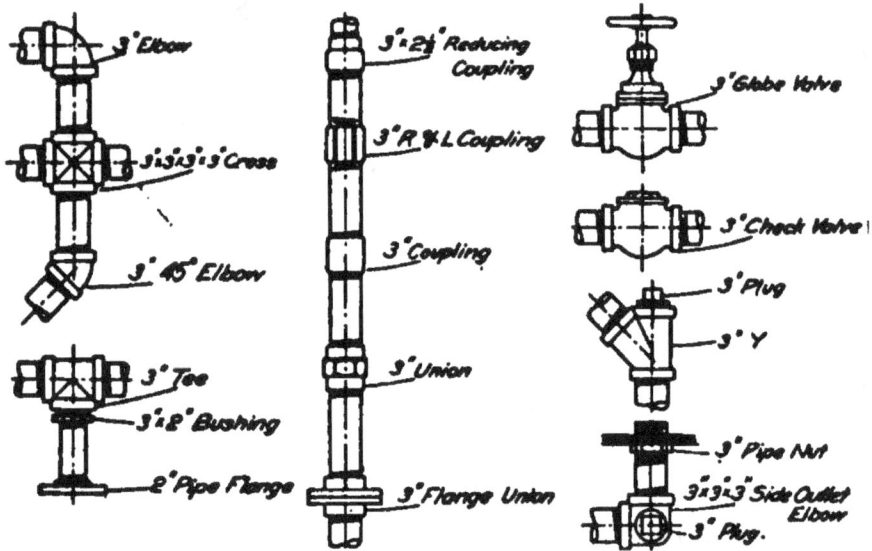

Fig. 330. Conventional Representations for Fittings.

Fig. 332, and these will serve to suggest such others as may be required. The over all dimensions together with notes and loca-

PIPING DRAWINGS

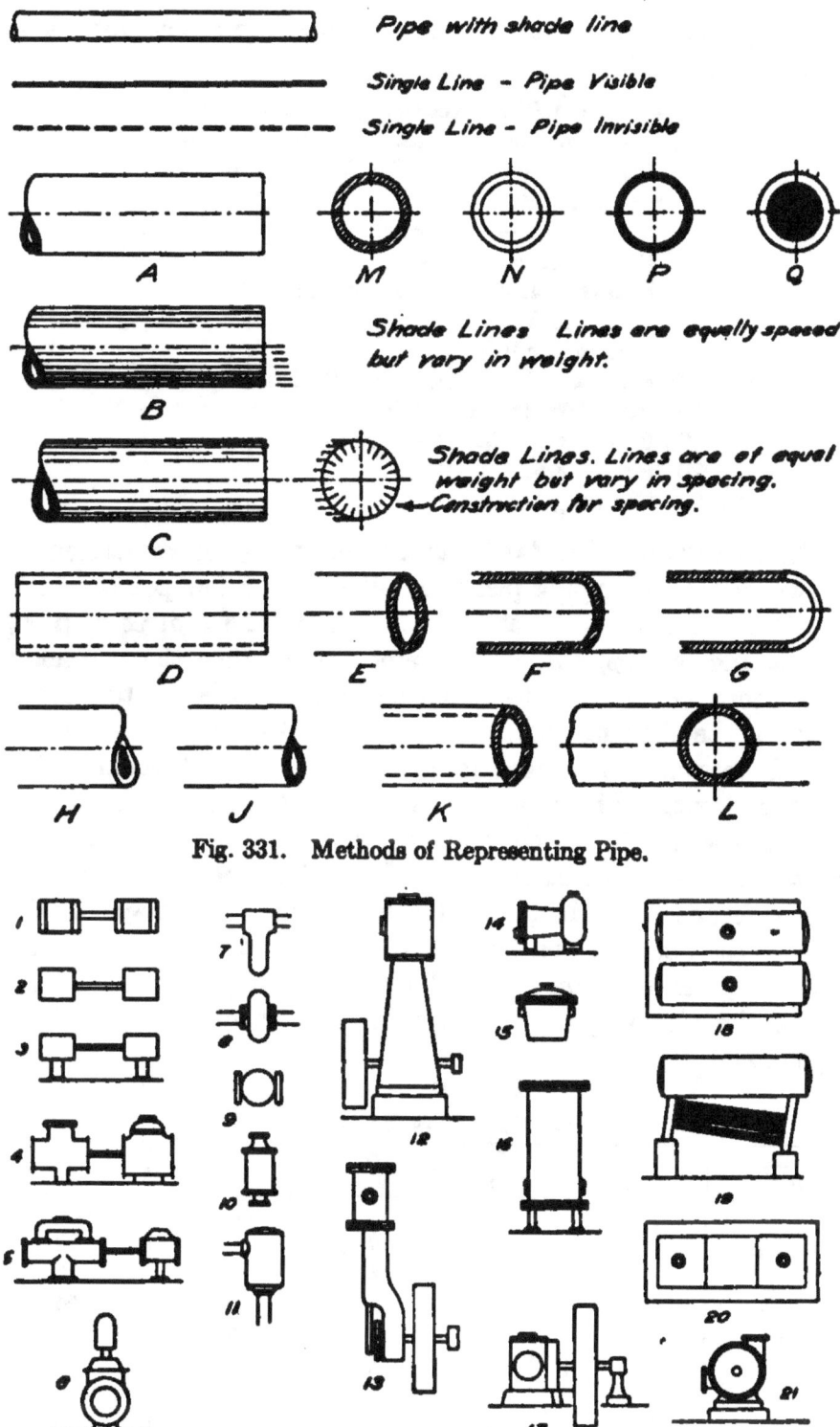

Fig. 331. Methods of Representing Pipe.

Fig. 332. Conventional Representations for Apparatus.

A HANDBOOK ON PIPING

tion of pipe flanges or openings are necessary in many cases, and always desirable.

  1, 2. Plan of Direct Acting Steam Pump.
  3, 4, 5. Elevation of Direct Acting Steam Pump.
  6. End View of Direct Acting Steam Pump.
  7, 8, 9. Separator.
  10, 11. Receiver — or Receiver Separator.
  12. Vertical Steam Engine.
  13. Plan of Horizontal Steam Engine.
  14, 15. Steam Trap.
  16. Feed Water Heater.
  17. End View Horizontal Steam Engine.
  18. Plan of Water Tube Boiler.
  19. Elevation of Water Tube Boiler.
  20. Plan of Fire Tube Boiler.
  21. Centrifugal Pump.

Dimensioning. — Most of the general rules for dimensioning drawings hold for piping plans, but there are a few points which may be mentioned. Always give figures to the centres of pipe, valves and fittings, and let the pipe fitters make the necessary allowances. If a pipe is to be left unthreaded, it is well to place a note on the drawing calling attention to the fact. If left-hand (L.H.) threads are wanted it should be noted. Wrought pipe sizes can generally be given in a note using the nominal sizes.

The bosses into which pipe screws should be located from centre lines of the machines and from the base or foundation. Flange connections should be located in the same way. Satisfactory sizes of cast-iron bosses to be provided for pipe to screw into are given in Table 101. This table also gives the distance which the pipe may be expected to enter in order to obtain a tight joint.

TABLE 101 (Fig. 333)

Cast-Iron Bosses

Size Inches	B Inches	C Inches	Size Inches	B Inches	C Inches
$1/8$	$7/8$	.19	2	$3 1/2$	.58
$1/4$	1	.29	$2 1/2$	$4 1/4$	.89
$3/8$	$1 1/8$	.30	3	5	.95
$1/2$	$1 3/8$	.39	$3 1/2$	$5 1/2$	1.00
$3/4$	$1 3/4$	.40	4	6	1.05
1	$2 1/8$	.51	$4 1/2$	$6 3/4$	1.10
$1 1/4$	$2 1/2$	.54	5	$7 1/2$	1.16
$1 1/2$	$2 3/4$	.55	6	$8 1/2$	1.26

PIPING DRAWINGS

These values may be used where it is necessary to make an allowance for the thread. Crane Company gives the values shown in Table 102 for length of thread on pipe that is screwed into valves or fittings to make a tight joint.

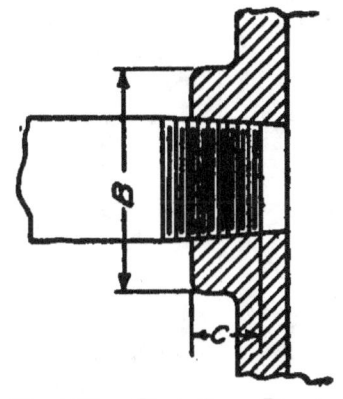

Fig. 333. Cast Iron Bosses.

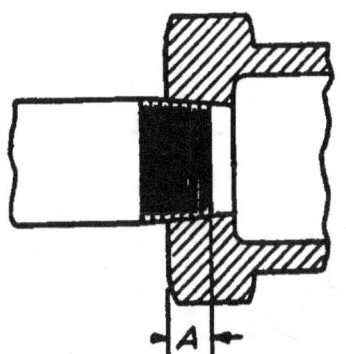

Fig. 334. Distance Pipe Enters Fitting.

TABLE 102 (FIG. 334)

DISTANCE FOR PIPE TO ENTER FITTINGS

Size Inches	A Inches	Size Inches	A Inches	Size Inches	A Inches
1/8	1/4	1 1/2	5/8	5	1 3/16
1/4	3/8	2	11/16	6	1 1/4
3/8	3/8	2 1/2	15/16	7	1 1/4
1/2	1/2	3	1	8	1 5/16
3/4	1/2	3 1/2	1 1/16	9	1 3/8
1	9/16	4	1 1/16	10	1 1/2
1 1/4	5/8	4 1/2	1 1/8	11	1 5/8

Flanged valves when drawn to large scale may have the over all dimensions given, the distance from centre to top of hand wheel or valve stem when open and when closed, diameter of hand wheel, etc., about as shown in Chapter VI. Separate flanges should be completely dimensioned as in Fig. 338, as should all special parts. It is necessary that the location of the piping should be definitely given, which means that the parts of the building containing the piping must be shown and must be accurately dimensioned. The location of apparatus and the pipe connections should be given by measurements from the centre lines of the machines, distances between centres of machines, heights of connections, etc.

In all cases the principal object of dimensioning must be kept in mind, namely, to tell exactly what is wanted in size, location

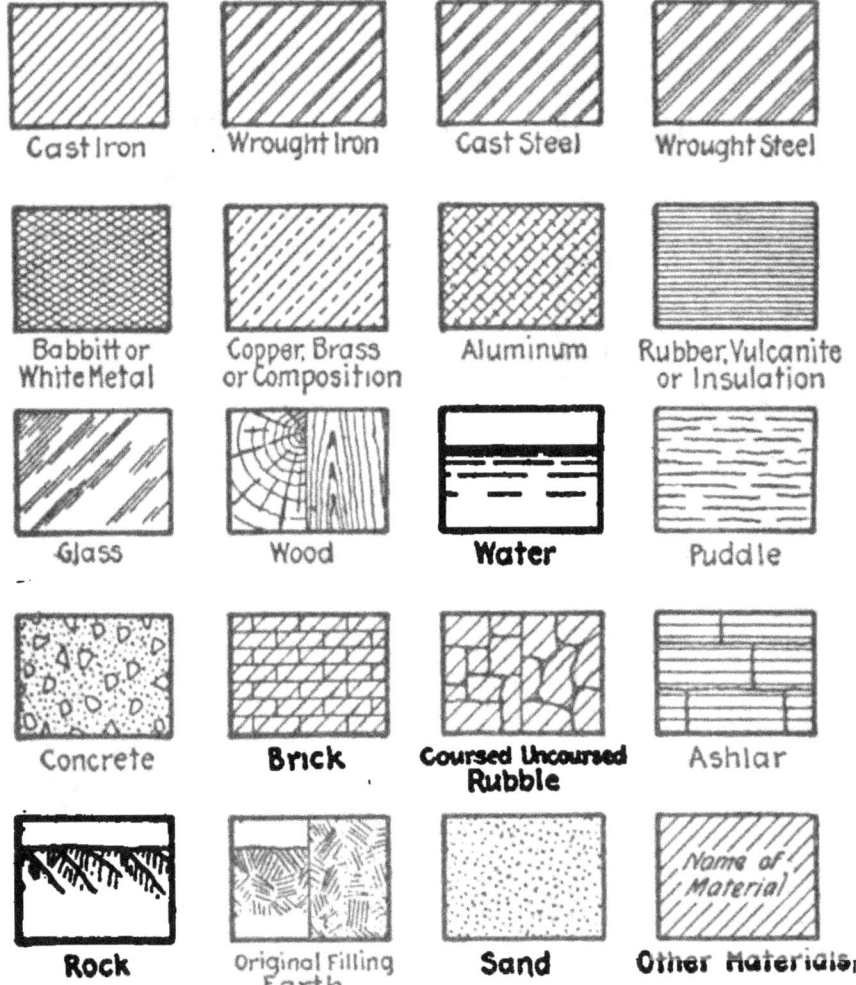

Fig. 335. A. S. M. E. Cross Sections.

and material, in such a way as to leave no room for misunderstanding. To this end clearness and exactness are essential. Several examples of dimensioning are shown in Figs. 327, 328 and 343.

When it is desired to indicate the different materials appearing in cross-section, the standard recommended by a committee of the A. S. M. E. may be used. This standard is shown in Fig.

PIPING DRAWINGS

335. It is not advisable to depend upon such representations, and a note should always be added to tell the material. Their chief value is to make it easier to distinguish different pieces.

Final drawings should be made after the engines, boilers and other machinery have been decided upon, as they can then be

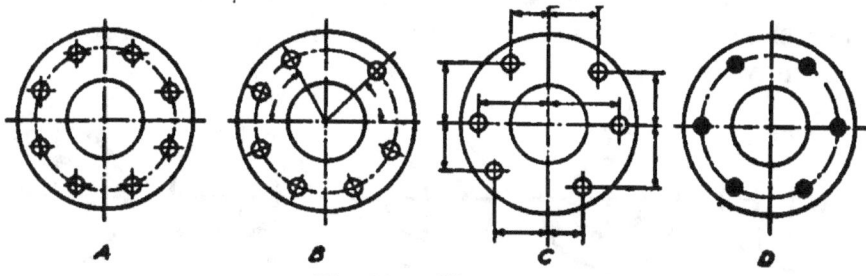

Fig. 336. Flanges.

drawn completely and accurately. At least two views should be drawn, a plan and elevation. Often extra elevations and detail drawings are necessary. Every fitting and valve should be shown. A scale of ³/₈ inches equals 1 foot is desirable for piping drawings when it can be used, as it is large enough to show the system to scale.

Flanges. — The dimensions of the American Standard for flanges are given in Tables 39 and 40, but sometimes special flanges or drilling are required. The number of bolts used for the

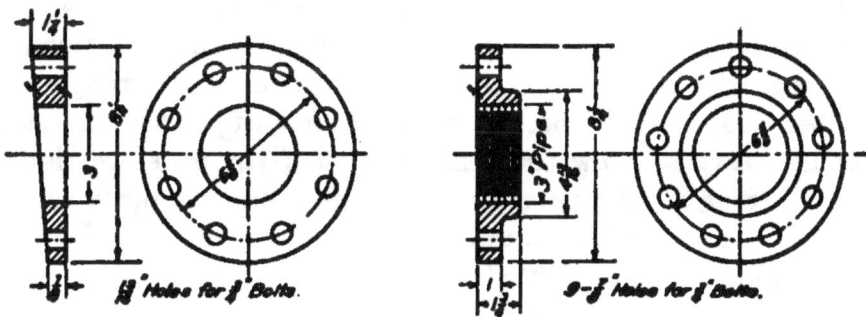

Fig. 337. Tapered Filling-in Piece.   Fig. 338. Flange.

flanges or fittings and valves is generally divisible by four, and placed "two-up" or to "straddle" the centre line. If any other arrangement is required the location of bolt holes should be clearly shown, as in Fig. 336 at *B* and *C*. Regular spacing can be given in a note, as "16 holes equally spaced," etc. The drawing for a tapered filling-in piece is shown in Fig. 337, and for a special flange in Fig. 338. The bolt holes are sometimes blacked in to indicate that the bolts or studs are not required, in which case a note should be added indicating such a meaning. A tapped or threaded hole may be shown by the methods of Fig. 339. The nominal diameter may be used or the actual diameter obtained from Table 4. The taper of the thread is usually exaggerated when shown. A straight hole with ordinary thread representations may be used.

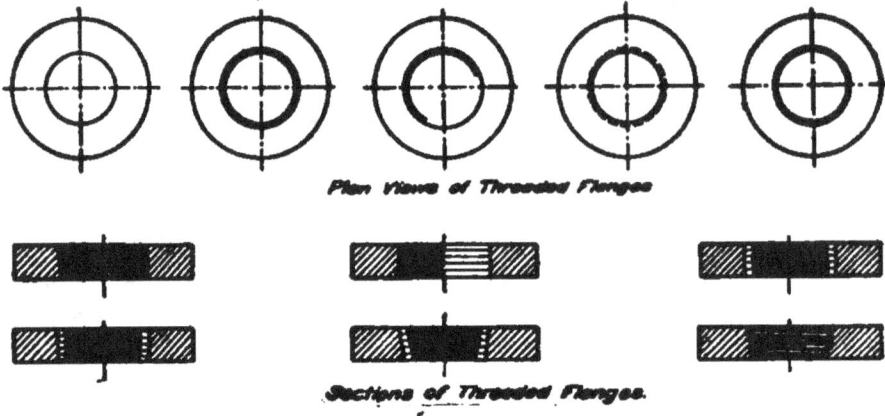

Fig. 339. Threaded Holes.

Coils. — Several drawings for pipe coils are shown in Fig. 340. Such drawings should tell the thickness of the pipe and the materials, the diameter of the coil taken either inside or outside of the pipe as indicated; the length of the pipe or coil; the number of turns; the pitch of the turns; the position and arrangement of the ends, and the method of connection, support, etc. It is not necessary to draw the complete coil if the ends are clearly drawn. Single line representations require explicit notes to tell whether centre line or outside dimensions are meant and otherwise explain what is wanted.

Sketching. — Sketching is an invaluable aid as a preliminary step in any kind of drawing, and a sketch is often the only drawing needed. One's ideas can be made clear and the number and kind of fittings and valves checked up in this way. Where only a small amount of work is to be done, a sketch may be made and fully dimensioned, from which a list of pieces can be made with lengths, sizes, etc. This will avoid mistakes in cutting, and the sketch shows just how the parts go together without depending upon memory. Such a sketch may be used to order with, but

PIPING DRAWING

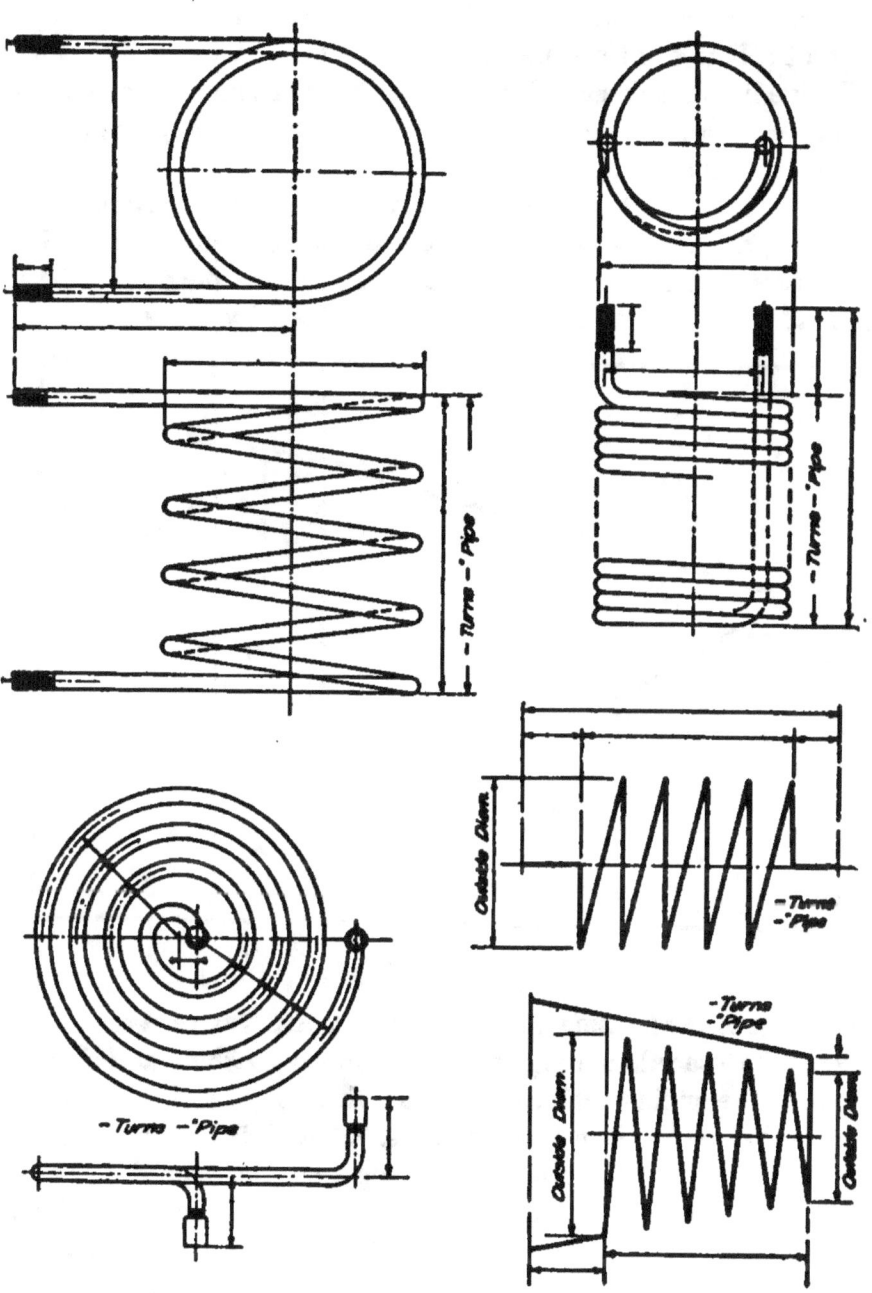

Fig. 340. Pipe Coils.

in such cases it should be made upon tracing cloth or thin paper so that a blue print can be made as a record. An H or 2H pencil will give lines black enough to print if ink is not used. The figures, however, should be put on in ink in all cases. If only one or two copies are wanted carbon paper may be used. Dimensions and notes should be put on as carefully as on a finished drawing. The general procedure is much the same as for all kinds of sketching. First sketch the arrangement using a single line diagram. When satisfactory the real sketch may be started by drawing in the

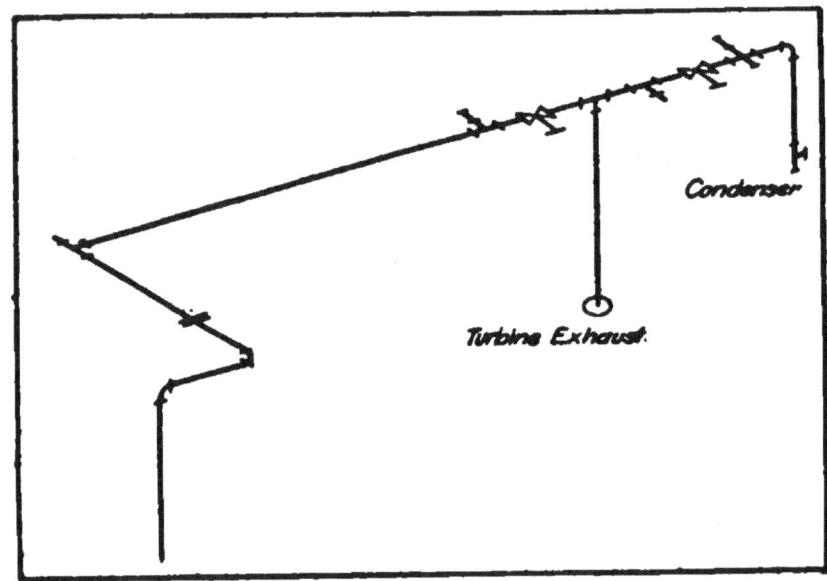

Fig. 341. Pictorial View of Piping.

centre lines, estimating locations of fittings, valves, etc., which should be spaced in roughly in proportion to their actual positions. The piping, valves, etc., can then be sketched in, using any of the conventions shown in Figs. 329 and 330. Finally locate dimension lines, figures and notes, together with the date and a title of some kind. Pictorial methods can be used to great advantage for sketching purposes, especially for preliminary layouts, as the directions and changes in levels can be clearly shown, Fig. 341.

Developed or Single Plane Drawings. — It will often be found convenient to swing the various parts of a piping layout into a single plane in order to show the various lengths and fittings in one view. Different methods of showing the same piping are here

PIPING DRAWING

illustrated. Fig. 341 is a pictorial view using single lines to show the position in space; Fig. 342 is a developed line sketch with the sizes, fittings, etc., written on, and Fig. 343 is a developed drawing with complete dimensions and notes. Such drawings are valuable when listing or making up an order as well as for the pipe fitters to work from. A free-hand line sketch, as a preliminary step in laying out a steam line, can often be made in this way.

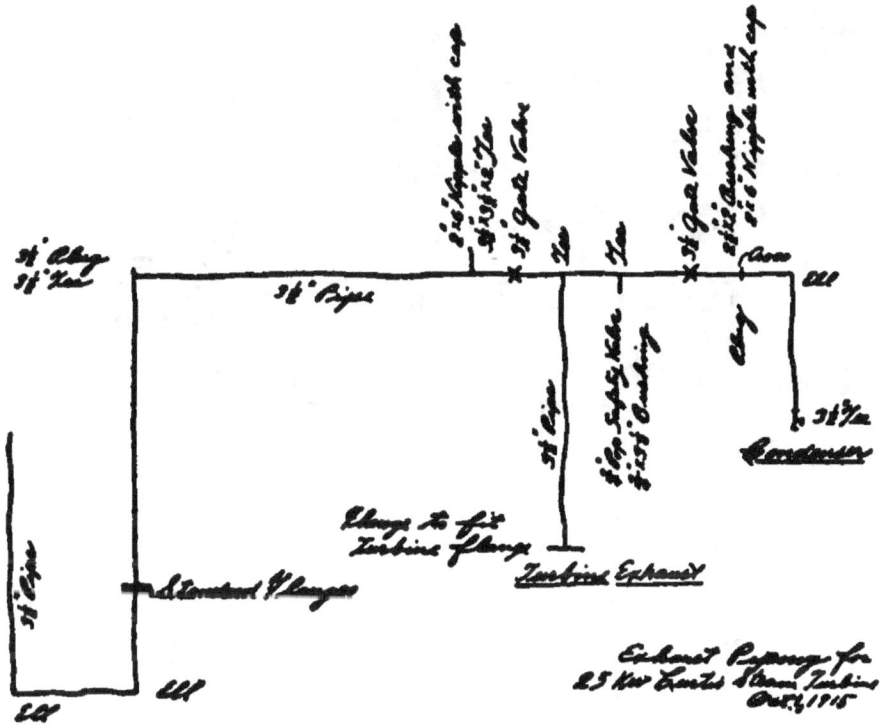

Fig. 342. Developed Sketch.

Isometric Drawing. — Two forms of pictorial drawing lend themselves readily to piping drawings, isometric and oblique. Both show the position of the pipe in space and are easily drawn and easily understood. They are especially valuable for sketching and preliminary layout work. The principles here given will enable anyone to make use of this convenient form of representation. Isometric drawing is based upon the three edges of a cube which come together at a corner. The lines representing these three edges are called isometric axes. One of these axes is vertical and the other two make angles of 30 degrees with the horizontal. See Fig. 344. These three lines represent three directions in space.

320 A HANDBOOK ON PIPING

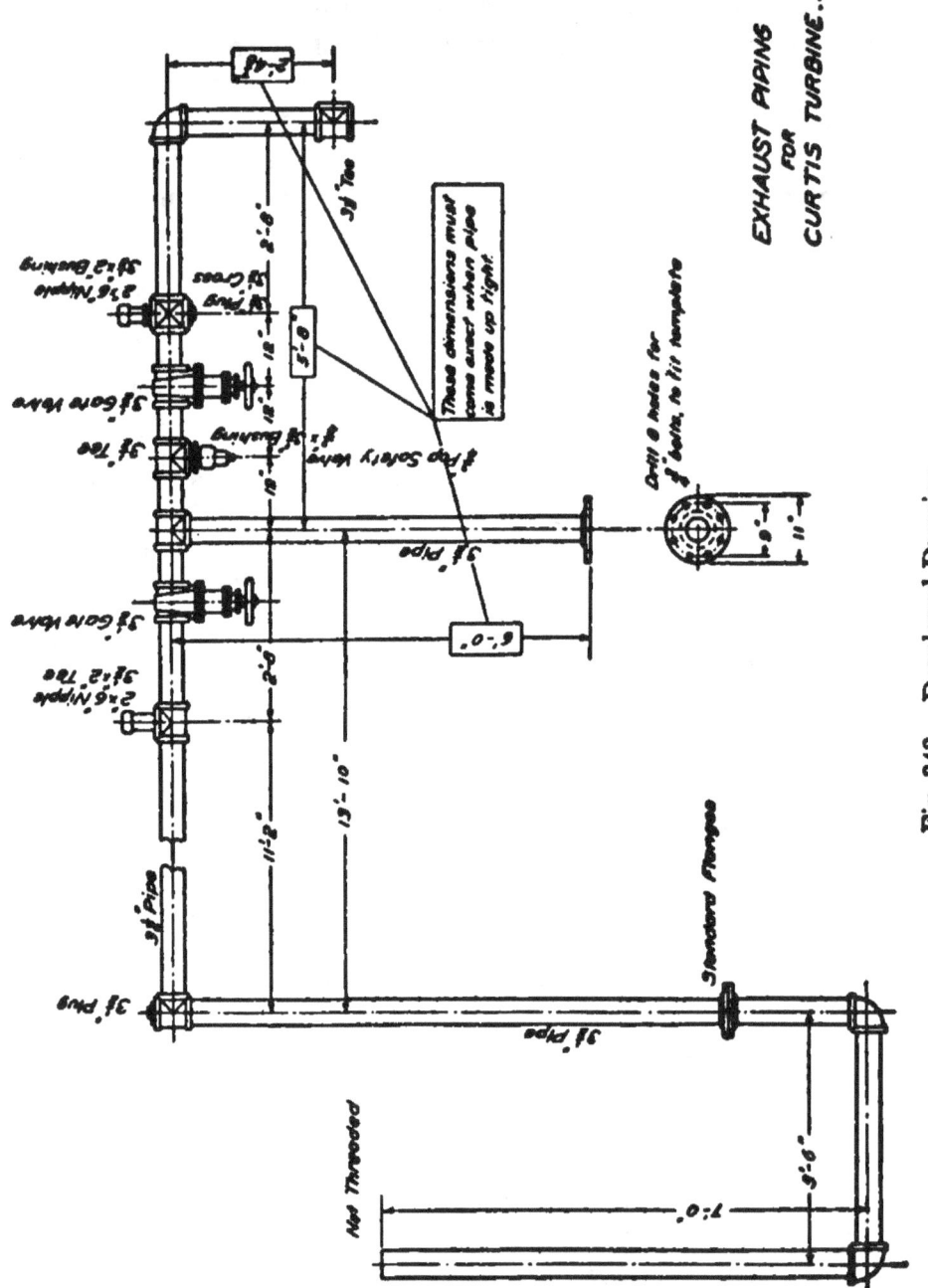

Fig. 343. Developed Drawing.

PIPING DRAWINGS

Lines parallel to the axes are called isometric lines. All other lines are non-isometric lines. All measurements are made along the axes or along isometric lines. Non-isometric lines cannot be

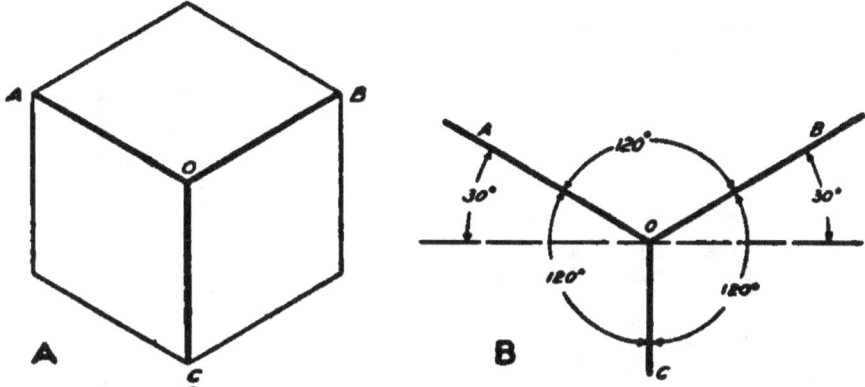

Fig. 344. Isometric Axes.

measured or laid off directly, but must be transferred from an orthographic projection. The method of doing this is shown in Figs. 345, 346, and 347, where both orthographic and isometric drawings are shown for several cases. Angles are drawn in isometric by transferring from the orthographic projection, as shown in Fig. 347, where B–C makes an angle with the other lines. It will be noticed that the effect of position in space would be lost

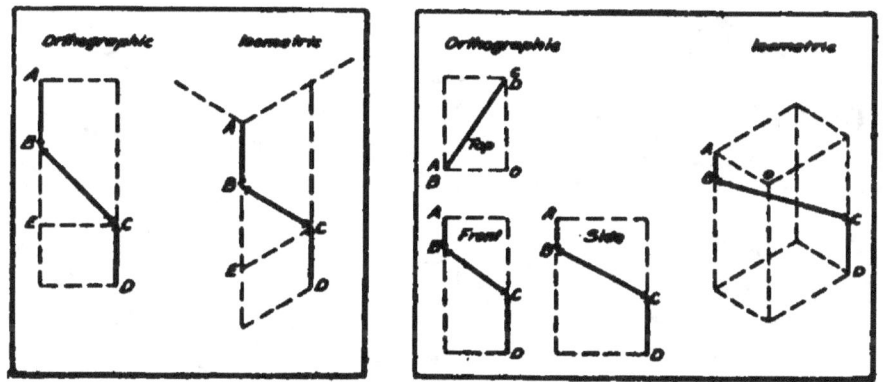

Figs. 345 and 346. Orthographic and Isometric Representations.

without the isometric lines in Fig. 346. Circles show as ellipses when drawn in isometric, but are generally drawn by approximate methods as shown on the three faces of the cube, Fig. 348,

322   A HANDBOOK ON PIPING

where two radii having centres at *A* and *B* are used. Circular arcs can be drawn by the same method.

In Fig. 349 the method of boxing in and laying out dimensions is shown for a plain ell. The orthographic projections of the ell

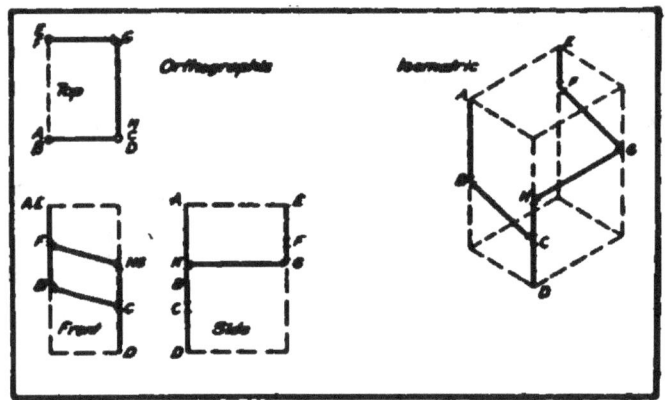

Fig. 347. Orthographic and Isometric Representations.

are shown at *A* and the points are numbered to correspond with the isometric views. The first step is to lay off the centre distances *2-3* and *3-4* as shown at *B*. The centre for the arc is found by the intersection of perpendiculars from *2* and *4*. The distances are indicated by dimension lines on Figs. *A* and *B*, and are the same length in both figures.

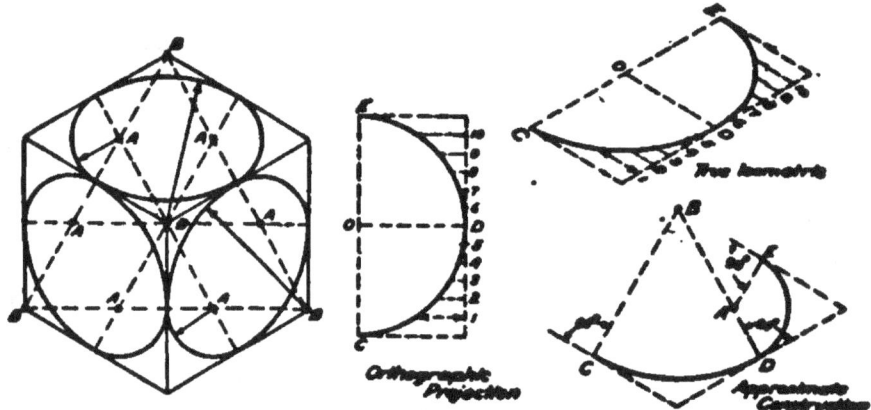

Fig. 348. Isometric Circles.

The next step is to lay out the diameters for the isometric circles, as shown at *C*. The centres for the arcs are shown at *D* and the completed ell at *E*.

PIPING DRAWINGS

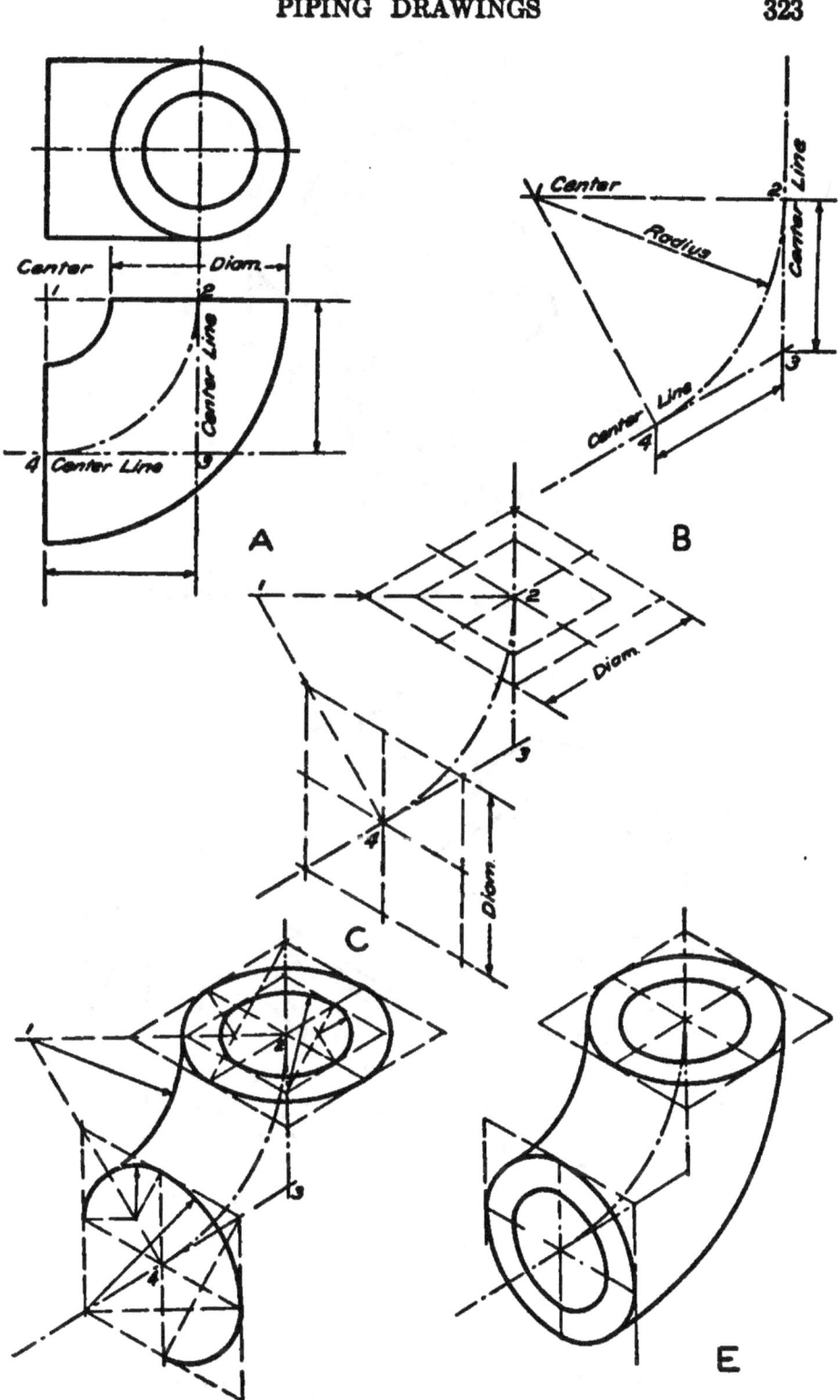

Fig. 349. Steps in Making Isometric Drawing of a Plain Elbow.

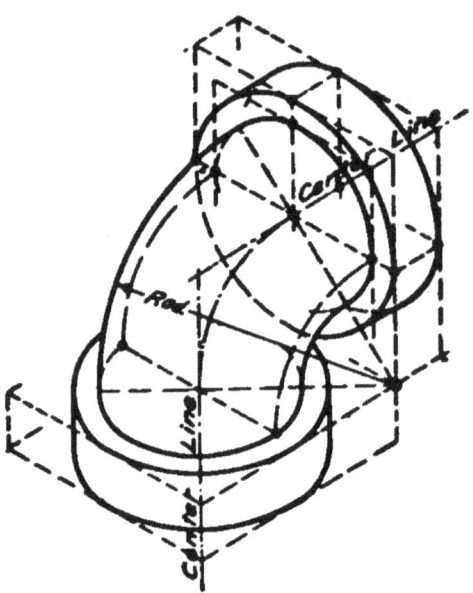

Fig. 350. Isometric Drawing of Screwed Elbow.

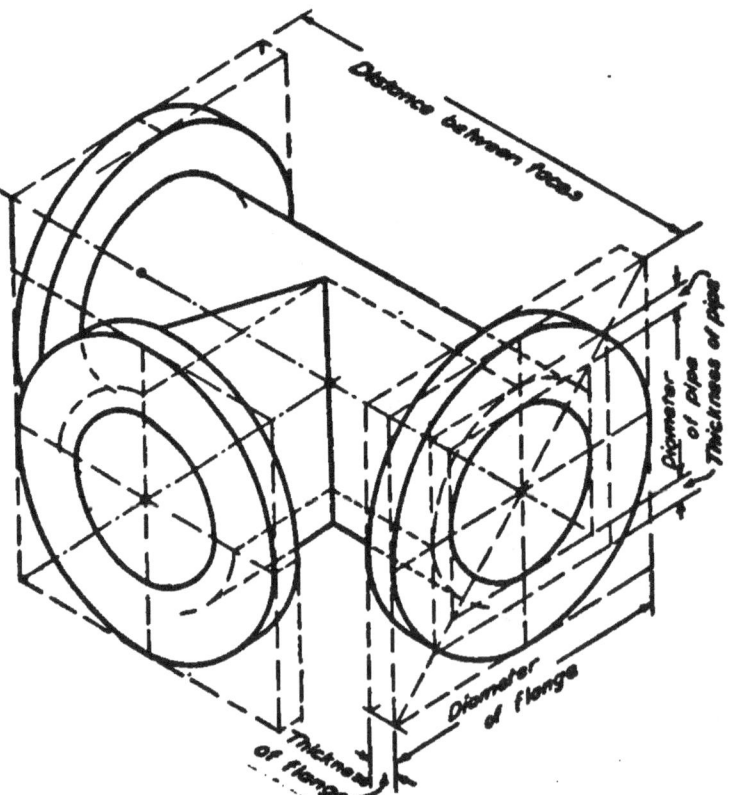

Fig. 351. Isometric Drawing of Flanged Tee.

PIPING DRAWINGS

The method of blocking in and drawing a screwed ell is indicated in Fig. 350. The construction for a flanged tee is indicated in Fig. 351, in which some of the dimensions are noted. The

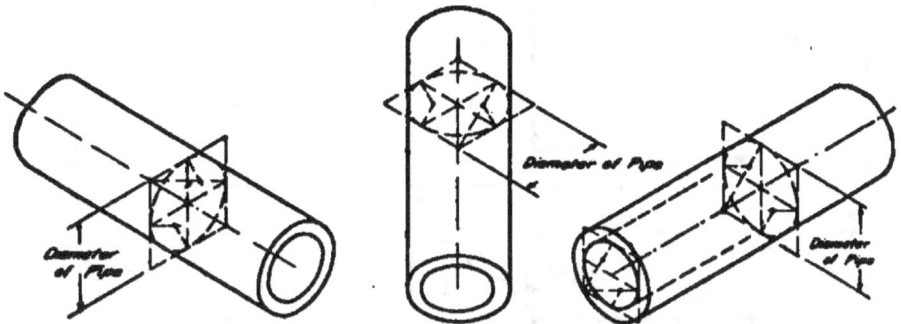

Fig. 352. Isometric Drawings of Pipe.

manner of obtaining the isometric diameter for piping is shown in Fig. 352, in which the measure of the actual diameter is marked. Some examples of piping as represented by isometric drawing are shown in Fig. 353 and other parts of the book.

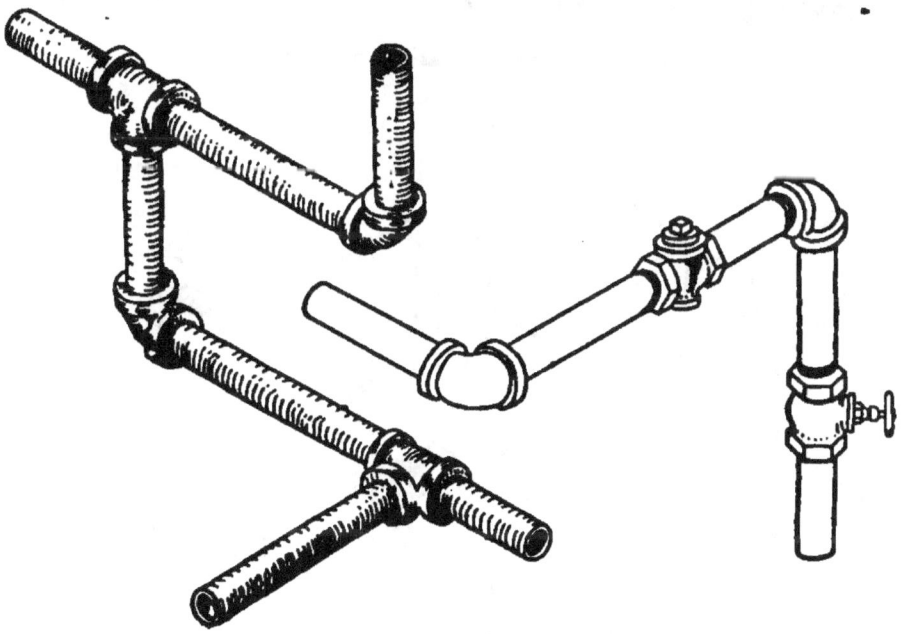

Fig. 353.

The method of laying out for a definite problem is shown in Figs. 354, 355 and 356. A sketch plan and elevation for an engine exhaust are shown in Fig. 354. The piping and engine room are

326 A HANDBOOK ON PIPING

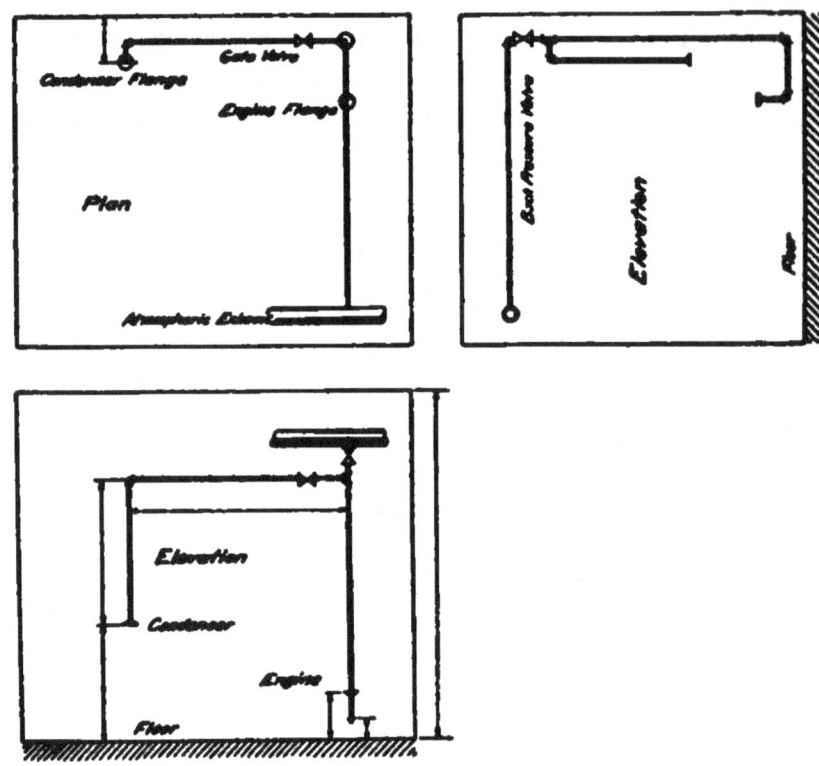

Fig. 354. Plan and Elevations of Piping.

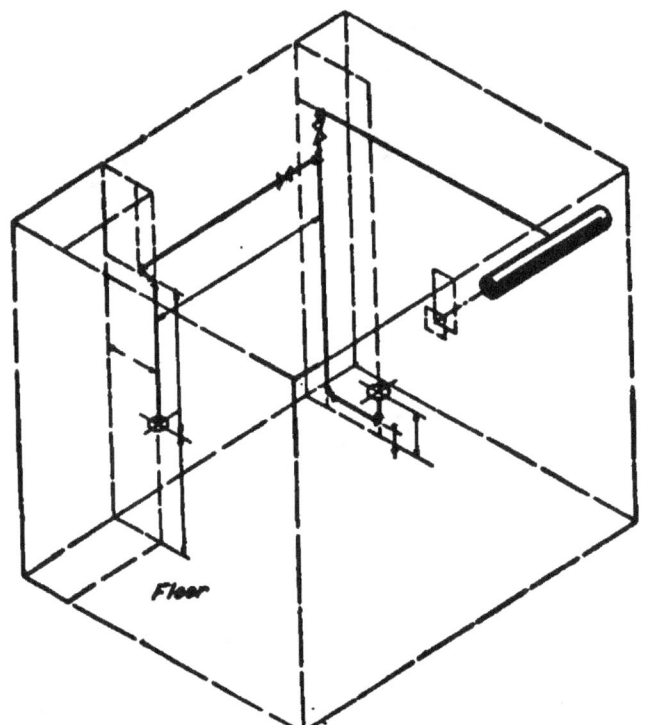

Fig. 355. Isometric Drawing.

boxed in, and the centres of pipe lines, valves, and fittings are measured off parallel to isometric lines as indicated in Fig. 355.

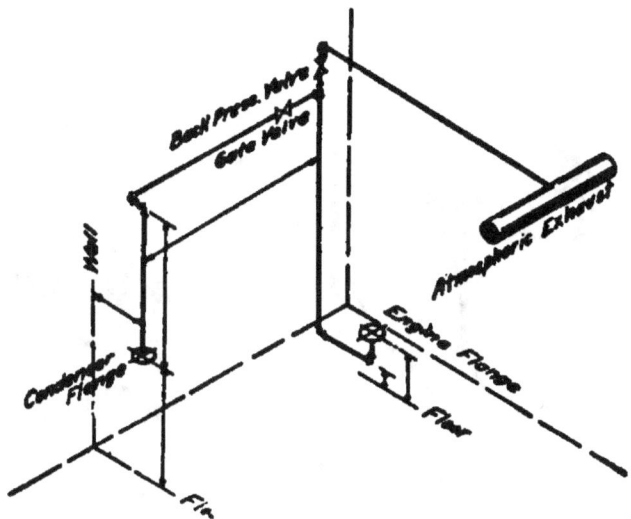

Fig. 356. Isometric Drawings.

The dimensions and notes are left off for the sake of clearness in showing the construction, but a few distances are indicated to show the manner of laying off measurements. Fig. 356 is the same as Fig. 355 except that the boxing has been left off.

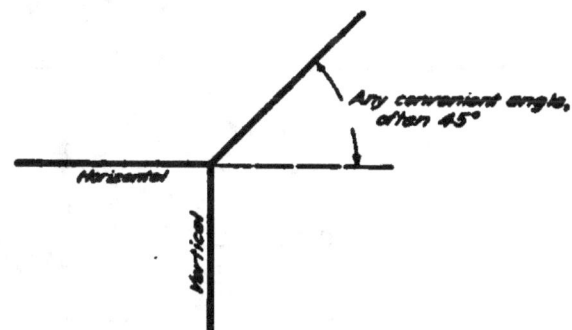

Fig. 357. Oblique Axes.

With a little practice it is possible to make free hand isometric drawings that are a great help in clearing up ideas and deciding locations.

Oblique Drawings. — Oblique drawings are made by the use of three axes located as shown in Fig. 357. Lines parallel to the plane of the front face of the cube show in their true length and angles in their true size. The drawing of circles is shown on the

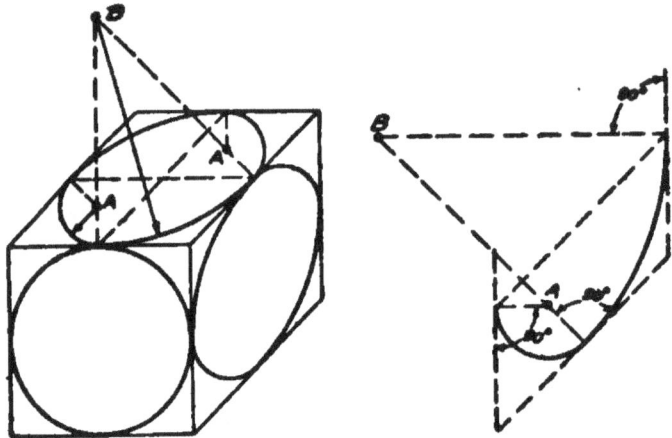

Fig. 358. Oblique Circles.

faces of the cube, Fig. 358. It should be noted that the centre for arcs is found by the intersection of perpendiculars erected at the points of tangency of the arcs. Except for the change in

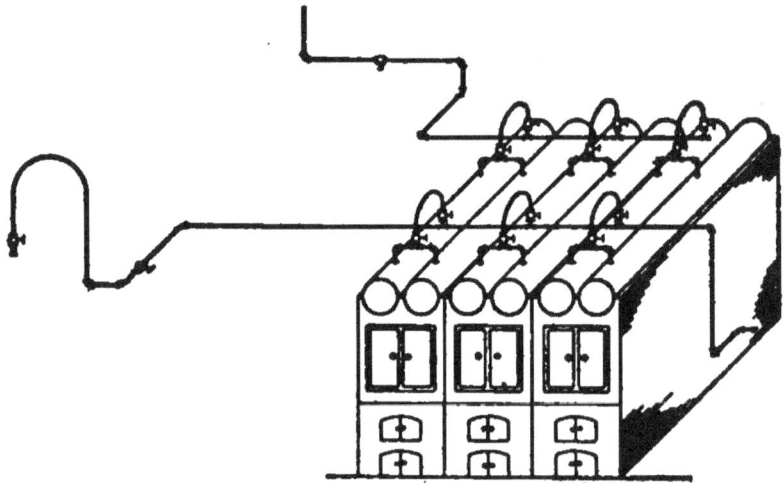

Fig. 359. Oblique Drawing.

angles this method is the same as for isometric. Fig. 359 shows an oblique drawing.

CHAPTER XVIII

SPECIFICATIONS

Specifications. — The specification of materials and piping apparatus for various purposes involves a knowledge of the conditions under which they are to be used. In the preceding chapters of this book an attempt has been made to describe piping materials, commercial sizes, and to indicate the uses for which they are adapted.

The possible consequences due to the failure of piping, often involving loss of life, are such that the best material, workmanship, and design should always be the end in view when preparing piping specifications.

Some fluids and the materials adapted for use with them are as follows:

For cold water — almost any material, but depending upon pressure and impurities.

For impure cold water — brass or similar composition.

For hot water — brass or similar composition, galvanized iron, cast iron.

For salt water or brine — brass or other composition.

For ammonia water — iron or steel.

For weak sulphuric acid — lead, lead lined iron or steel.

For strong sulphuric acid — wrought iron, wrought steel, cast iron.

For hydrochloric acid — lead, lead-lined pipe.

For fuel oil — steel tubing, extra heavy wrought iron or steel; galvanized pipe.

Specifications for piping can be very much simplified by the use of well made and accurate scale drawings showing the entire system with the sizes and makes of its various components. The specifications should cover whatever is not named on the drawings and should give the trade name, make or manufacturers' names, sizes and materials for all parts of the system which includes the following: kinds of pipe; method of support; provision for expansion; pipe bends; flanges; bolting and drilling; kinds

STANDARD PIPING SCHEDULE STONE & WEBSTER ENG'G. CORP'N

Service	Pipe	Fittings
Superheated steam Main and auxiliary (Not exceeding 200 lbs. pressure and 200° F. sup.)	2 and under E. S. steel 2½"–12" F. W. steel 14" and over O. D. steel ⅝" thick	2" and under E. H. scr. F. S. 2½"–12" E. H. flgd. C. S. 14" and over E. H. flgd. C. S. (special ports)
H. P. Saturated steam H. P. drips (Not exceeding 200 lbs. pressure)	2" and under E. S. steel 2½"–12" F. W. Steel 14" and over O. D. steel ⅜" thick	2" and under E. H. scr. C. I. 2½"–12" E. H. flgd. C. I. 14" and over E. H. flgd. C. I. and bronze
Feed discharge H. P. economiser (Not exceeding 275 lbs. pressure)	3" and under I. P. S. brass 3½ and over E. H. flgd. cast semi-steel	(special parts) 2" and under E. H. scr. C. I. 2½" and over E. H. flgd. cast semi-steel
Blow off Superheater blow off Economiser blow off (Not exceeding 275 lbs. pressure)	2" and under E. S. steel 2½" and over (inside sta.) E. S. steel 2½" and over (outside sta.) E. H. flgd. Cast semi-steel	2" and under E. H. scr. C. I. 2½" and over E. H. flgd. cast semi-steel

SPECIFICATIONS

Feed suction Low service and city water Hot well suct. and discharge Circulating water Misc. water	2″ and under E. S. gal. steel 2½″–3″ std. gal. steel 4″ and over std. flgd. C. I.	3½″ and under sta. scr. C. I. 4″ and over std. flgd. C. I.
Exhaust main and aux. Condenser air Compressed air L. P. drips Misc. L. P. steam	2″ and under E. S. steel 2½″–12″ std. steel 14″–22″ O. D. steel ⁵⁄₁₆″ thick 24″ and over std. C. I.	3½″ and under std. scr. C. I. 4″ and over std. flgd. C. I.
Free exhaust	12″ and under std. steel 14″–22″ O. D. steel ⁵⁄₁₆″ thick 24″ and over std. C. I. Vertical sp. riv. gal. steel	Std. flgd. C. I.
Oil	2″ and under E. S. steel 2½″ and over std. steel	3½″ and under std. scr. C. I. 4″ and over std. flgd. C. I.

NOTE. — By "Compressed asbestos" is meant either Durabla, Garlock 900, Permanite, Klingerct, Riverlate, Tauril, etc. Unless unavoidable, ⅜, ½, 4½, 7 and 9″ pipe shall not be used.

E. S. = Extra strong
E. H. = Extra heavy
C. I. = Cast iron
C. S. = Cast steel
F. S. = Forged steel
I. P. S. = Iron pipe size

STANDARD PIPE SCHEDULE — STONE & WEBSTER ENG'G CORP'N

Valves	Gaskets	Flanges	
1½" and under E. H. scr. F. S. and monel 2" and under E. H. scr. C. S. and monel 2½" and over E. H. flgd. C. S. and monel (O. S. and Y. pattern)	Compressed Asbestos 1/16" thick	2½"–5" E. H. F. S. Companion 6" and over E. H. F. S. Van Stone	Special specifications for higher pressure
2" and under E. H. scr., all bronze 2½" and over E. H. flgd. C. I. and bronze (O. S. and Y. pattern)	Compressed Asbestos 1/16" thick	2½"–5" E. H. F. S. Companion 6" and over E. H. F. S. Van Stone	Special specifications for higher pressure
2" and under E. H. scr., all bronze 2½" and over E. H. semi-steel and bronze (O. S. and Y. pattern)	Compressed Asbestos 1/16" thick	Cast integral	Special specifications for higher pressure
2" and under E. H. scr., all bronze 2½" and over E. H. flgd. C. I. and bronze (O. S. and Y. pattern)	Compressed Asbestos 1/16" thick	Cast integral	Special specifications for higher pressure
2" and under std. scr. composition (Inside screw pattern) 2½"–3½" std. scr. C. I. and bronze 4" and over std. flgd. C. I. and bronze (O. S. and Y. pattern)	Rainbow 1/16" thick	Cast integral	

SPECIFICATIONS

2" and under std. scr. composition (Inside screw pattern) 2½"–3½" std. scr. C. I. and bronze 4" and over std. flgd. C. I. and bronze (O. S. and Y. pattern)	Rainbow 1/16" thick	4"–12" std. C. I. Companion 14"–22" std. C. I. Van Stone 24" and over cast integral
Relief valve C. I. and bronze water sealed 18" and over with hydraulic cylinder and floor stand	Rainbow 1/16" thick	12" and under std. C. I. 14" and 22" std. Van Stone 24" and over cast integral Vert. spiral riv. std. C. I.
2½" and under std. scr. composition 4" and over std. flgd. C. I. and brass (Inside screw pattern)	Compressed Asbestos 1/16" thick	4" and over std. C. I. Companion

Unions to be Mark or Tuxedo
All Semi-Steel material to be tested to 600 lbs.

F. W. = Full weight.

of packing; fittings; steam valves; water valves; air valves; reducing valves; back pressure valves; blow-off valves; safety valves; non-return valves; relief valves; foot valves; separators; steam traps; injectors; meters, etc.

Standard Piping Schedule. — The standards for pipe and fittings of Stone & Webster Engineering Corporation are given in the accompanying tabulation. The different materials as used for power plant work and their variation to meet the needs of each particular service are made especially clear by this presentation.

Standard Specifications (Stone & Webster). — Local conditions are certain to vary any sample specifications that might be given, but the basis of the specification for high-class work should be very much the same. For this reason the author is pleased to be able to include the following standard piping specification which was kindly supplied by the Stone & Webster Engineering Corporation. It is used by them as a basis for detailed specifications on each particular job. It represents good modern practice and should prove of much value as a guide in the selection of proper materials, and in calling attention to the important factors involved in a piping installation.

STANDARD SPECIFICATION FOR PIPE AND FITTINGS

Stone & Webster Engineering Corporation

IN GENERAL

This specification covers the furnishing and installation of a complete piping system in the power station of the _____.

MATERIAL

All pipe steel, forged steel, cast steel, wrought iron, cast iron, and composition used in the various fittings, flanges, pipe, etc., shall have the following physical characteristics.

Pipe Steel

Tensile strength not less than 50,000 lbs. per sq. in.
Elastic limit " " " 30,000 " " " "
Elongation in 8 in., not less than 18%
Reduction of area, not less than 50%

Forged Steel

Tensile strength not less than 70,000 lbs. per sq. in.
Elastic limit " " " 40,000 " " " "
Elongation in 8 in., not less than 20%
Reduction of area, not less than 40%

SPECIFICATIONS

CAST STEEL

Tensile strength not less than 60,000 lbs. per sq. in.
Elastic limit   "   "   "   30,000 "   "   "   "
Elongation in 2 in., not less than 20%
Reduction of area, not less than 30%
The percentage each of phosporous and sulphur shall not exceed five one hundredths (0.05).
All castings shall be annealed and sample pieces shall satisfactorily stand bending cold around 1" radius and through 120°. Two test pieces from each melt shall be prepared to standard size for testing and shall be furnished free of charge.

WROUGHT IRON

Tensile strength not less than 50,000 lbs. per sq. in.
Elastic limit   "   "   "   26,000 "   "   "   "
Elongation in 8 in., not less than 18%
Reduction of area, not less than 50%

CAST IRON

All castings shall be of tough gray iron, free from all defects affecting either strength or tightness under pressure, true to pattern and of workman-like finish.
Sample pieces 1" square cast from the same heat of metal in sand molds, shall be capable of sustaining on a clear span of 4'-8" a central load of 500 lbs. when tested in the rough bar.
Turned test pieces shall show an ultimate tensile strength of not less than 24,000 lbs. per sq. in. One test piece from each melt for each of the above tests shall be prepared for testing and furnished free of charge to the Engineers.

COMPOSITION

All composition shall be a dense strong mixture especially selected for the particular service in which it is to be used and shall not suffer a serious loss of strength due to the temperature to which it is regularly subjected.
All wrought iron and steam pipe shall be made by the Youngstown Sheet and Tube Company.

HIGH PRESSURE STEAM PIPING

All fittings $2\frac{1}{2}"$ and above, for use with super-heated steam shall be of extra heavy flanged pattern, designed for 250 pounds per square inch working pressure and made of cast steel of a quality as previously specified. The section of all fittings shall increase gradually by a long taper at flanges.

The thickness of metal shall be not less than that given for corresponding sizes in the following table:

Size	15"	14"	12"	10"	8"	6"	5"	$4\frac{1}{2}"$	4"	$3\frac{1}{2}"$	3"	$2\frac{1}{2}"$
Thickness	$1\frac{3}{8}"$	$1\frac{1}{4}"$	$1\frac{1}{8}"$	$1\frac{1}{16}"$	1"	$\frac{7}{8}"$	$\frac{7}{8}"$	$\frac{7}{8}"$	$\frac{13}{16}"$	$\frac{13}{16}"$	$\frac{7}{8}"$	$\frac{7}{8}"$

All high pressure steam fittings 2½″ and above, for use with saturated steam shall be of the extra heavy flanged pattern designed for 250 pounds per square inch working pressure and made of cast iron.

The section of all fittings shall increase gradually by a long taper at flanges.

All high pressure steam fittings below 2½″, both for superheated and for saturated steam shall be extra heavy screw end pattern, made of cast iron and designed for a working steam pressure of 250 pounds per square inch.

All pipe 2½″ and above for high pressure steam piping, both for superheated and saturated steam shall be what is commercially known as full weight selected lap welded pipe made from the best quality of steel, as previously specified.

All steel bends must be bent to the radius designated and must be free from wrinkles, buckles, creases, etc., and flanges shall be faced at right angles to the centre line of the pipe.

All steel pipe and bends 6″ in diameter and above, for use with superheated steam, shall have extra heavy rolled or forged steel flanges of the Van Stone type.

All steel pipe and bends 6″ in diameter and above, for use with saturated steam, shall have extra heavy cast iron flanges of the Van Stone type.

All steel pipe and bends from 2½″ to 5″ inclusive shall have flanges screwed on and refaced in lathe. These flanges shall be of rolled or forged steel for superheated steam piping, and of cast iron for saturated steam piping.

All steel pipe and bends under 2½″ shall be extra strong and threaded for screw end fittings.

At the dead ends of all pipes blank flanges of approved design shall be furnished and shall be of cast steel for superheated steam piping and of cast iron for saturated steam piping.

All unions for high pressure steam piping under 2½″ shall be extra heavy bronze for 250 pounds per square inch working steam pressure and shall be of the ground joint type. They shall be of the Tuxedo or Economic make.

Main steam headers for use with superheated steam shall be made up by one of the following methods:

(1) With extra heavy cast steel fittings and full weight steel pipe with extra heavy rolled or forged Van Stone flanges.

(2) Full weight steel pipe with nozzles of full weight steel pipe welded on and with extra heavy rolled or forged steel Van Stone flanges made on. Fillets where nozzles are welded on to be long radius.

All flanges for high pressure steam work on pipe fittings and bends shall be faced off on the back or spot faced so as to provide a smooth even bearing for bolt heads and nuts. They shall be of dimensions and drilling as shown on attached sheet and shall be provided with a raised face inside bolt holes $\frac{1}{16}$″ in thickness.

HIGH PRESSURE WATER PIPING

All fittings 2½″ and above, for high pressure water piping, including feed water piping, shall be of extra heavy flanged pattern made of cast iron.

The section of all fittings shall increase gradually by a long taper at flanges.

SPECIFICATIONS

All fittings below $2\frac{1}{2}''$ shall be extra heavy cast iron, screw end pattern.

All pipe $4''$ in diameter and above, both for hot and cold water shall be extra heavy flanged cast iron.

Fillets at flanges shall be of long radius or tapered the same as specified above for high pressure water fittings.

All fittings and pipe shall be designed for a working pressure of 250 pounds per square inch.

All pipe for high pressure cold water from $2\frac{1}{2}''$ to $3\frac{1}{2}''$ inclusive shall be of full weight steel and shall have extra heavy cast iron flanges screwed on and refaced in lathe.

All pipe for high pressure cold water below $2\frac{1}{2}''$ shall be full weight lap welded steel and shall be threaded for screw end fittings.

All pipe for high pressure hot water from $2\frac{1}{2}''$ to $3\frac{1}{2}''$ inclusive shall be of iron pipe size brass pipe equal in every respect to that manufactured by the American Tube Company, and shall have extra heavy cast iron flanges screwed on and refaced in lathe.

All pipe for high pressure hot water under $2\frac{1}{2}''$ shall be of iron pipe size brass pipe threaded for screw end fittings.

All unions for high pressure water piping below $2\frac{1}{2}''$ shall be of the extra heavy bronze ground joint type and of Economic or Tuxedo make.

All flanges on above pipe and fittings shall be spot faced on the back to provide a smooth even bearing for bolt heads and nuts, and shall have a raised face inside bolt holes $\frac{1}{16}''$ in thickness. They shall conform to dimensions and drilling as shown on attached sheet.

BLOW-OFF PIPING

All fittings for blow-off piping $2\frac{1}{2}''$ and above shall be extra heavy flanged pattern made of cast iron.

The section of all fittings shall increase gradually by a long taper at flanges.

All fittings below $2\frac{1}{2}''$ shall be extra heavy screw end cast iron.

All pipe inside buildings $2\frac{1}{2}''$ and above shall be full weight lap welded steel with extra heavy cast iron flanges screwed on and refaced in lathe.

All pipe below $2\frac{1}{2}''$ in diameter shall be extra strong steel, threaded for screw end fittings.

All pipe $4''$ and above for blow-off outside of buildings shall be extra heavy flanged cast iron designed for a working pressure of 250 pounds per square inch.

Where blow-off piping outside of building is sufficient distance from boilers, extra heavy bell and spigot cast iron water pipe can be used in place of flanged cast iron pipe.

All blow-off piping outside of building shall be laid in wooden box and this box shall be filled with magnesia or asbestos or other suitable non-conducting material, thoroughly packed around the pipe.

Blow-off piping shall have free discharge above maximum water level wherever possible.

Blow-off piping outside buildings, where the free end is sealed by water, shall have $3''$ vent pipe installed in every 60 foot of length.

All flanges on both pipe and fittings shall be spot faced on the back to provide a smooth even bearing for bolt heads and nuts and shall have raised face inside bolt holes $1/16"$ in thickness.

All flanges shall be of dimensions and drilling as shown on attached sheet.

LOW PRESSURE EXHAUST PIPING

All fittings for low pressure exhaust piping 4" in diameter and above shall be standard weight flanged pattern designed for 100 pounds per square inch working pressure, and made of cast iron.

All fittings below 4" in diameter shall be standard weight cast iron, screw end pattern.

All pipe for low pressure exhaust from 4" to 12" inclusive, excepting vertical outboard exhaust pipe, shall be of standard weight steel of a quality as previously specified and shall have standard weight cast iron flanges made on.

All pipe under 4" in diameter shall be standard weight steel pipe threaded for screw end fittings.

All pipe from 14" to 22" inclusive, unless otherwise specified, shall be lap welded steel pipe $1/4"$ in thickness, with cast iron flanges riveted on.

Unless otherwise specified the sizes of lap welded steel exhaust pipe, from 14" to 22" inclusive, shall be taken as the inside diameter of the pipe.

All pipe 24" in diameter and above shall be standard weight flanged cast iron pipe designed for working pressure of 100 pounds per square inch.

Flanges on all pipe and fittings shall be plain faced and shall conform to dimensions and drilling shown on attached sheet.

Vertical outboard exhaust pipe beyond exhaust relief valve and back pressure valve shall be flanged galvanized spiral riveted pipe.

Exhaust heads shall be flanged galvanized of ample area, with inside parts of copper as made by the Wright Manufacturing Company.

All unions below $2\frac{1}{2}"$ in diameter shall be of ground joint type and made of brass.

All unions from $2\frac{1}{2}$ to $3\frac{1}{2}"$ inclusive, shall be flanged standard weight cast iron.

LOW PRESSURE WATER PIPING

All fittings for low pressure water piping 4" in diameter and above shall be standard weight flanged pattern and made of cast iron.

All fittings below 4" shall be standard weight cast iron screw end pattern.

All pipe for low pressure water 4" in diameter and above, for use both with cold and hot water, shall be standard weight flanged cast iron designed for a working pressure of 100 pounds per square inch.

All pipe for low pressure water below 4" shall be standard weight galvanized steel pipe and shall be threaded for screw end fittings.

Flanges on all pipe and fittings shall be plain faced and shall conform to dimensions and drilling shown on attached sheet.

All unions below $2\frac{1}{2}"$ shall be standard weight brass with ground joints.

All unions from $2\frac{1}{2}$ to $3\frac{1}{2}"$ inclusive shall be standard weight flanged cast iron.

SPECIFICATIONS

DRIP PIPING

(1) High Pressure

All fittings for high pressure drip piping below $2\frac{1}{2}''$ shall be extra heavy cast iron screw end pattern.

All fittings $2\frac{1}{2}''$ and above shall be extra heavy flanged cast iron.

All pipe for high pressure drips under $2\frac{1}{2}''$ shall be extra strong lap welded steel threaded for screw end fittings.

All pipe $2\frac{1}{2}''$ and above shall be full weight steel pipe with extra heavy cast iron flanges screwed on and refaced in lathe.

All unions below $2\frac{1}{2}''$ shall be extra heavy brass ground joint pattern of Economic or Tuxedo make.

Flanges on all pipe and fittings shall be spot faced on the back to provide smooth even bearing for bolt heads and nuts and shall conform to dimensions and drilling shown on attached sheet.

(2) Low Pressures

All fittings for low pressure drips shall be standard weight cast iron, screw end pattern.

All pipe shall be standard weight steel pipe threaded for screw end fittings.

All unions below $2\frac{1}{2}''$ shall be standard weight brass unions with ground joints.

All unions $2\frac{1}{2}''$ and above shall be standard weight flanged cast iron.

In all cases where possible, water seals made with pipe and fittings shall be used in place of traps.

DRY AIR PIPING

All fittings for dry air piping $4''$ and above, shall be standard weight flanged cast iron, designed for 100 pounds per square inch working pressure.

All fittings under $4''$ shall be standard weight cast iron screw end pattern.

All pipe shall be standard weight steel pipe and shall be provided for screw end fittings on sizes under $4''$.

Pipe $4''$ in diameter and above shall have standard weight cast iron flanges made on.

All flanges on both pipe and fittings shall be plain faced and shall conform to dimensions and drilling shown on attached sheet.

All unions below $2\frac{1}{2}''$ shall be standard weight brass unions with ground joints and all unions $2\frac{1}{2}''$ and above shall be standard weight flanged cast iron.

OIL PIPING

All fittings for oil piping shall be cast iron pattern brass fittings with screw ends.

All pipe shall be iron size brass pipe threaded for screw end fittings.

All unions below $2\frac{1}{2}''$ shall be standard weight brass ground joint pattern of Economic or Tuxedo make.

All unions $2\frac{1}{2}''$ and above shall be standard weight flanged cast iron.

AIR PIPING

All fittings for air piping shall be standard weight cast iron with screw ends.

All pipe shall be standard weight lap welded steel pipe threaded for screw end fittings.

All unions below $2^1/_2$" shall be standard weight brass with ground joints.

All unions $2^1/_2$" and above shall be standard weight flanged cast iron.

STEP BEARING PIPING

All fittings for step bearing piping to vertical turbines shall be of cast steel hydraulic pattern, designed for from 2000 to 3000 pounds per square inch working pressure with Economic ground joints made by the Edwards Steam Specialty Company.

All flanges and unions shall be of cast steel with Economic ground joints designed for same pressure as above fittings and made by the Edwards Steam Specialty Company.

All pipe shall be double extra strong lap welded steel threaded for flanges and unions specified above.

JOINTS

Joints for high pressure steam piping shall be made with Durabla gaskets $^1/_{16}$" in thickness, or of some other equally good packing as approved by Engineers.

Joints for high pressure water piping shall be made with Durabla gaskets $^1/_{16}$" in thickness, or with corrugated copper gaskets coated on both sides with Dixon's graphite or Callahan's cement.

Joints in low pressure steam and water piping shall be made with Rainbow gaskets $^1/_{16}$" in thickness, or of some other equally good rubber packing.

Where flanges are screwed on pipe they should be made on as tight as it is safe and the pipe shall be made entirely through the flange until it is flush with the face of the flange.

Joints made with screw end fittings shall have pipe threads thoroughly slushed with Dixon's graphite or Callahan's cement and made into the fittings as tight as it is safe to screw them.

All gaskets on both high and low pressure piping shall extend out to the inside edge of the bolt holes of flanges, except on low pressure piping above 14" in diameter, where they shall extend to the outside edge of flanges.

All bolts for both high and low pressure joints shall be made of bolt steel and shall have clean cut U. S. threads with upset square heads and semi-finished hexagonal cold pressed nuts.

SUPPORTS

All piping and apparatus shall be supported in a thorough and substantial manner.

Main steam headers, unless otherwise specified, shall be supported on a heavy cast iron adjustable pipe chair with concave rollers.

All other piping shall be supported by adjustable wrought iron hangers or by brackets.

All hangers and supports shall be installed so that they will not interfere in any way with the expansion and contraction of the piping.

SPECIFICATIONS

Exhaust risers to atmosphere shall be braced by means of stays fastened to walls or steel work in a thorough and substantial manner.

Wherever necessary, clamps, braces and anchors shall be installed in order to remove all vibration which is injurious or excessive. These clamps and braces shall not be fastened to pipe in such a manner as to interfere with their proper expansion and contraction.

ERECTING AND TESTING

All piping shall be erected so as to bring all the joints true and fair in order that flanges can be carefully faced and properly bolted.

Piping must not be subjected to unnecessary or excessive strain in making up.

All steel castings must be tested at the shop before shipment and shall be tight under an hydraulic pressure of 500 pounds per square inch.

All high pressure steam piping shall be tested after erection to 300 pounds per square inch hydraulic pressure.

All extra heavy cast iron feed pipe shall be tested under an hydraulic pressure of 500 lbs. per square inch before leaving the factory. The entire high pressure feed line shall be tested under an hydraulic pressure of 300 pounds per square inch after erection.

All piping under vacuum shall be tested and made tight against air leaks.

Pet-cocks shall be placed on any portion of the piping system where air is liable to collect.

All piping put together with screw end fittings shall have sufficient unions to allow of ease for removal or repairs.

After piping system has been erected it shall be blown clear of dirt and chips before connections are made to any apparatus.

All high pressure steam piping shall be blown with live steam.

Model Specifications (Walworth). — A valuable set of model piping specifications for three classes of power houses has been prepared by Mr. H. W. Evans, formerly manager of the power piping department of the Walworth Manufacturing Company, outlining standard practice. The following is condensed from the above.

SPECIFICATION OF MATERIALS FOR STEAM PLANTS OPERATING WITH SATURATED STEAM — PRESSURES UP TO 125 POUNDS PER SQUARE INCH

STEAM LINES

High pressure steam and drip pipe to be wrought steel, lap-welded. Sizes 12" and smaller to be full card weight. Sizes 14" and larger $3/8$" thick or heavier.

PIPE FOR BENDS to be same weight as straight lengths unless of short radius, when heavy pipe must be used. Bends to be finished accurately to dimensions to avoid forcing into position, except expansion bends, which should be cut shorter than dimensions and drawn into place which will allow the bend to expand into place and fit properly when the line heats.

FLANGES FOR PIPE and bends for sizes 3½" and smaller to be standard weight cast iron threaded type; 4" and larger to be Walmanco type.

FITTINGS 2½" and larger to be standard weight cast iron, flanged: 2" and smaller, standard cast iron, threaded.

VALVES 2" and larger, except stop and checks and other specialties, to be iron body, flanged, gate or angle valves, standard weight, outside screw and yoke; larger sizes fitted with by-pass. The seating faces of discs and the seat rings to be renewable bronze. Bonnet to be arranged for back seating when the valve is open for packing under pressure. Valves 1½" and smaller to be all bronze.

FLANGES, except Walmanco type, on pipe, valves, and fittings, to be faced straight across, rough finish.

BOILER FEED LINES

The feed water pipe from pumps to boilers to be full weight lap-welded wrought steel or iron. Use brass pipe if quality of water demands it.

FITTINGS 2½" and larger, except checks and feed valves (globes) to be iron body, flanged, gate or angle valves, standard weight, outside screw and yoke, with bronze stems. Valves 2" and smaller to be all bronze.

FLANGES on pipe valves and fittings to be faced straight across, rough finish.

Corrugated lead gaskets about 1/16" thick, cut in rings to fit inside the bolt holes.

EXHAUST LINES

Pipe for exhaust lines except cast iron to be lap-welded wrought steel, sizes 12" and smaller standard weight; 14" to 20" outside diameter, ¼" thick. Sizes 22" and over not less than ¼".

FOR BENDS, see specification for steam lines.

CAST IRON PIPE may be used for the exhaust to the condenser or for other lines if cheaper than wrought; weight, etc., to conform to specification for flanged fittings.

FLANGES FOR PIPE and bends for sizes 12" and smaller to be standard weight, cast iron, threaded type; for pipe 14" and larger to be standard weight cast iron attached by Walmanco method.

FITTINGS 3" and larger to be cast iron, flanged; 2½" and smaller, cast iron, threaded. Sizes 14" and smaller standard weight; 16" and larger may be low pressure.

VALVES for sizes 2½" and larger, except relief, back pressure, and other specialties to be iron body, flanged, gate or angle valves, preferably outside screw and yoke. Inside screw valves with brass stem; outside screw and yoke may have steel stem. Sizes 10" and smaller standard weight; 12" and larger may be low pressure, in which case they are to have standard weight flanges. The seating faces of discs and seat rings are to be renewable bronze; bonnet to be arranged for back seating when the valve is open for packing under pressure. Valves 2" and smaller to be all brass.

FLANGES on pipe valves and fittings to be faced straight across, rough finish.

SPECIFICATIONS

Garlock or Rainbow gaskets $1/16"$ thick cut in rings to fit inside the bolt holes.

WATER PIPING

Suction or discharge pipe (except cast iron) to be lap-welded wrought steel or iron. Sizes 12" and smaller standard weight; 14" and larger not less than $1/4"$ thick. Bends made as for steam piping.

CAST IRON pipe when used should conform to specifications for flanged fittings.

FLANGES for pipe and bends for sizes 12" and smaller to be standard weight cast iron, threaded type; for pipe 14" and larger to be standard weight cast iron attached by Walmanco method.

FITTINGS for sizes 3" and larger to be cast iron, flanged; 2" and smaller cast iron, threaded. Sizes 14" and smaller standard weight; 16" and larger either standard or low pressure as demanded by the service. Elbows long radius.

STOP VALVES $2\frac{1}{2}"$ and larger to be standard weight, iron body brass mounted, flanged, gate or angle valves. Preferably outside screw and yoke with brass stems. Valves 2" and smaller to be all brass.

FLANGES on pipe valves and fittings except Walmanco type, to be faced straight across, rough finish.

Cloth Inserted Rubber or Rainbow gaskets $1/16"$ thick, cut in rings to fit inside the bolt holes; for pipe in the ground use heavy canvas, full face, dipped in red lead.

BLOW-OFF LINES

PIPE AND BENDS to be full weight lap-welded steel. In all particulars same as for steam lines.

FLANGES for pipe and bends to be standard weight, cast iron, threaded, screwed on and refaced. (Same as for steam lines.)

FITTINGS to be standard weight cast iron, flanged. Elbows, long radius; use extra heavy malleable screwed ells if within the fire walls. Header fittings to be laterals or single sweep tees.

CAST IRON PIPE may be desirable for a header buried in the ground, then use heavy weight flanged pipe.

BLOW-OFF LINES from boilers to be double valved; use one heavy asbestos packed cock, and one Walworth angle pattern blow-off valve, flanged ends.

FLANGES on pipe, valves and fittings to be faced straight across, rough finish.

GARLOCK OR LEAD GASKETS $1/16"$ thick, cut in rings to fit inside the bolt holes.

SPECIFICATION OF MATERIALS FOR STEAM PLANTS OPERATING WITH SATURATED STEAM — PRESSURES UP TO 250 POUNDS PER SQUARE INCH

STEAM LINES

High pressure steam and drip pipe to be wrought steel, lap-welded. For pressures up to 200 pounds per square inch, sizes 7" and smaller to be full card weight; 8"–28 pounds per foot; 9"–34 pounds per foot; 10"–40 pounds per foot; 12″–50 pounds per foot. Sizes 14″ and larger ⅜″ thick or heavier. For pressures 200 pounds per square inch and over, 12″ and smaller to be extra strong; 14″ and larger ½″ thick.

Pipe for Bends to be same weight as straight lengths unless of short radius, when heavy pipe must be used. Bends to be finished accurately to dimensions to avoid forcing into position, except expansion bends, which should be cut shorter than dimensions and drawn into place which will allow the bend to expand into place and fit properly when the line heats.

Flanges for pipe and bends for sizes 3½″ and smaller to be extra heavy weight malleable iron or steel, threaded type, screwed on and refaced. For sizes 4″ and larger malleable iron or steel flanges (low hub section) attached by Walmanco method should be used.

Fittings 2″ and smaller to be extra heavy cast iron, threaded; sizes 2½″ and larger to be extra heavy weight cast iron or semi-steel, flanged.

Valves 2″ and larger, except stop and checks and other specialties to be iron body, flanged, gate or angle valves, extra heavy weight, outside screw and yoke. (For pressures up to 175 pounds medium weight valves may be used.) Sizes 8″ and larger to be fitted with one-piece by-pass valve. The seating faces of discs and the seat rings to be renewable hard bronze; bonnet to be arranged for back seating when the valve is opened for packing under pressure. Valves 1½″ and smaller to be all bronze.

Flanges, except Walmanco type, on pipe, valves and fittings to be faced with 1/16″ raised projection inside the bolt holes; bearing surface for bolt head and nut to be finished, i.e. spot faced.

BOILER FEED LINES

The feed water pipe from pumps to boilers to be extra strong lap-welded wrought steel or iron. Use brass pipe if the quality of water demands it. Flanges for the pipe and bends to be extra heavy weight malleable iron or steel (low hub section). Sizes 2½″ and smaller, threaded type; 3″ and larger, Walmanco method. Fittings 2½″ and larger to be extra heavy weight cast iron or semi-steel, flanged. Sizes 2″ and smaller to be extra heavy cast or malleable iron, threaded. Elbows, long radius.

Valves 2½″ and larger, except checks and feed valves (globes) to be iron body, flanged, gate or angle valves, extra heavy weight, outside screw and yoke, with bronze stem. (Medium weight valves may be used for pressures up to 175 pounds.) Valves 2″ and smaller to be all bronze. Flanges on pipe, valves and fittings to be faced with 1/16″ raised projection inside the bolt holes; bearing surface for bolt head and nut to be finished, i.e. spot faced.

Corrugated lead gaskets about 1/16″ thick cut in rings to fit the raised faced.

EXHAUST LINES

See exhaust lines under specifications for plant operating with 125 pounds steam pressure.

WATER PIPING

See water piping under specifications for plant operating with 125 pounds steam pressure.

SPECIFICATIONS

BLOW-OFF LINES

Pipe and bends to be extra strong lap-welded steel. In all particulars same as for steam lines.

FLANGES to be extra heavy malleable iron or steel (low hub section). Sizes $3^1/_2"$ and smaller, threaded type; $4"$ and larger Walmanco method. Semi-steel flanges may be used for pressures up to 150 pounds.

FITTINGS to be extra heavy weight cast iron, flanged. Elbows, long radius. Header fittings to be laterals or single sweep tees. Cast iron pipe, valves, facing, and gaskets same as for 125 pound plant.

SPECIFICATION OF MATERIALS FOR STEAM PLANTS OPERATING WITH SUPER-HEATED STEAM — PRESSURES UP TO 250 POUNDS PER SQUARE INCH

STEAM LINES

High pressure steam and drip pipe to be wrought steel, lap-welded. For pressures up to 175 pounds per square inch, sizes $7"$ and smaller to be full card weight; $8"$–28 pounds per foot; $9"$–34 pounds per foot; $10"$–40 pounds per foot; $12"$–50 pounds per foot. Sizes $14"$ and larger $^1/_2"$ thick or heavier. For pressure 175 pounds per square inch and over, $12"$ and smaller to be extra strong; $14"$ and larger $^1/_2"$ thick.

PIPE FOR BENDS to be same weight as straight lengths unless of short radius, when heavy pipe must be used. Bends to be finished accurately to dimensions to avoid forcing into position, except expansion bends, which should be cut shorter than dimensions and drawn into place which will allow the bend to expand into place and fit properly when the line heats.

FLANGES for pipe and bends for sizes $3^1/_2"$ and smaller to be extra heavy weight steel, threaded type, screwed on and refaced. For pipe $4"$ and larger to be steel (low hub section) attached by the Walmanco method.

FITTINGS $2"$ and larger to be extra heavy open hearth steel castings, having sweep outlets and large fillets back of the flanges. Sizes $1^1/_2"$ and smaller to be extra heavy malleable iron or cast steel, threaded.

VALVES $1^1/_2"$ and larger, except stop and checks and other specialties, to be extra heavy weight, flanged, gate or angle valves, outside screw and yoke; bonnet packed with Durabla gasket. Sizes $7"$ and larger to be fitted with one-piece by-pass valve. Body, Bonnet and Discs or Wedge to be open hearth steel castings — yoke may be cast iron. When temperature does not exceed 500° stem may be cold rolled steel; for higher temperatures use Monel metal stems. Valves $1^1/_4"$ and smaller to be all bronze, or of suitable composition to withstand high temperatures; fitted with renewable seat and disc.

FLANGES, except Walmanco type, on pipe valves and fittings, to be faced with $^1/_{16}"$ raised projection inside the bolt holes; bearing surface for bolt head and nut to be finished, i.e. spot faced.

BOILER FEED LINES

The feed water pipe from pumps to boilers to be extra strong lap-welded wrought steel or iron. Use brass if the quality of water demands it.

FLANGES for pipe and bends to be extra heavy weight malleable iron or steel (low hub section). Sizes $3^1/_2"$ and smaller to be threaded type; size $4"$

and larger to be attached by Walmanco method. Semi-steel flanges may be used for small sizes for pressures up to 150 pounds.

FITTINGS $2\frac{1}{2}''$ and larger to be extra heavy weight cast iron or semi-steel flanged. Sizes $2''$ and smaller to be extra heavy, cast or malleable iron, threaded. Elbows, long radius.

VALVES $2\frac{1}{2}''$ and larger, except checks and feed valves (globes) to be iron body, flanged, gate or angle valves, extra heavy weight, outside screw and yoke, with bronze stem. (For pressures up to 175 pounds medium weight valves may be used.) Valves $2''$ and smaller to be all bronze.

FLANGES except Walmanco type on pipe, valves and fittings, to be faced with $\frac{1}{16}''$ raised projection inside the bolt holes; bearing surface for bolt head and nut to be finished, i.e. spot faced.

EXHAUST LINES

See exhaust lines under specifications for plant operating with 125 pounds steam pressure.

WATER PIPING

See water piping under specifications for plant operating with 125 pounds steam pressure.

BLOW-OFF LINES

See blow-off lines under specifications for plant operating with 250 pounds steam pressure (saturated steam).

Corrugated lead gaskets about $\frac{1}{16}''$ thick, cut in rings to fit the raised face.

NOTES — (Common to all Piping)

Drilling. — Templates to be the "American Standard of 1915," for flanges, fittings and valves.

Supports. — Not more than 12 foot centres, designed to provide for movement in all directions; use substantial anchors where necessary.

Drainage. — Provide adequate drainage arrangements wherever necessary on all steam lines.

Unions. — Provide suitable unions on small threaded lines wherever necessary to insure quick repairs and at all valve connections.

Valves. — The seating faces of discs and the seat rings to be of renewable bronze (or suitable metal); bonnet to be arranged for back seating when the valve is open for packing under pressure.

CHAPTER XIX

LIST OF BOOKS AND REFERENCES

The following sources of information are included as a means of increasing the value of the book, which is necessarily limited in its treatment of the various phases of piping and allied subjects. It is not intended to be a complete list of books and articles, but is suggestive, and may be amplified by the reader.

ADAMS, A. I. — Wood Stave Pipe. Am. Soc. C. E. Transactions, Vol. 41, p. 27.

ALLEN, J. K. — Sizes of Flow and Return Steam Mains. 104 pp. ill. Pub. by Domestic Engineering, Chicago, 1907.

AMERICAN DISTRICT STEAM COMPANY. — Bulletins Nos. 103 to 143 covering subject of district heating. North Tonawanda, N. Y.

AMERICAN GAS INSTITUTE. — Standard Specifications for Cast Iron Pipe and Special Fittings. 55 pp. (Adopted Oct. 1911 and Oct. 1913.) The Chemical Publishing Co., Easton Pa., 1914.

THE AMERICAN STANDARD PIPE FLANGES, FITTINGS AND THEIR BOLTING. — Report of Committee of Am. Soc. M. E. Revised to Mar. 7 and 20, 1914. N. Y.

ARMSTRONG CORK AND INSULATION COMPANY. — Nonpareil High Pressure Covering. 80 pp., 1916. Nonpareil Cork Covering for Cold Pipes. 60 pp., 1916. Pittsburgh, Pa.

BATCHELLER, B. C. — The Rapieff Joint is described in the American Machinist, April 23, 1908.

BJORLING, PHILLIP R. — Pipe and Tubes. 344 pp. ill. Whittaker and Co., London, 1902.

BOOTH, WM. H. — Steam Pipes. 187 pp. ill. A. Constable & Co., Ltd., London, 1905.

BROWNING, WILLIAM D. — Dimensions of Pipe, Fittings and Valves. 88 pp. ill. 3rd ed., 1910. For sale by National Book Co., Collinwood, Ohio.

CHANDLER, S. M. — Bursting Strength of Cast-iron Elbows and Tees. Tests at Case School of Applied Science. American Machinist, Mar., 1906.

COLLINS, HUBERT E. — Pipes and Piping. 140 pp. ill. $1.00. McGraw-Hill Book Co., N. Y. 1908.

CONDENSED CATALOGUES OF MECHANICAL EQUIPMENT. — Gives names and addresses of manufacturers of piping and equipment engineers, etc. 6th Vol., Oct., 1916. Am. Soc. M. E., N. Y.

CRANE COMPANY. — The Effect of High Temperatures on the Physical Properties of Some Metals and Alloys, by I. M. Bregowsky and L. W. Spring. Power Plant Piping Specifications. Chicago.

CRANE, R. T. — Early History of Gas Pipes. Engineering Record, July 8, 1893.

DUDLEY, ARTHUR W. — Experiments with Wood Pipe in New Hampshire Journal of the New England Waterworks Association. Sept., 1916.

DURAND, W. L. — Flow of Steam in Pipes (A Chart). Mechanical World, May 26, 1916.

ELLIS, GEORGE A. — Tables Relating to the Flow of Water in Cast Iron Pipes. 53 pp. Press of Springfield Printing Co., Springfield, Mass. 1883.

ENGINEERING STANDARDS COMMITTEE. — Leslie S. Robertson, M. INST. C. E. Sec'y. Published for the Committee by C. Lockwood & Son, London.

    Report No. 10, 1904. British Standard Tables for Pipe Flanges.

    Report No. 21, 1905. British Standard Pipe Threads for Iron or Steel Pipes.

    Report No. 40, 1908. British Standard Specifications for Cast Iron Spigot and Socket Low Pressure Heating Pipes.

    Report No. 44, 1909. British Standard Specification for Cast Iron Pipes for Hydraulic Power.

    Report No. 58, 1912. British Standard Specification for Cast Iron Spigot and Socket Soil Pipes.

    Report No. 59, 1912. British Standard Specification for Cast Iron Spigot and Socket Waste and Ventilating Pipes, for other than Soil Purposes.

EVANS, W. H. — Model Piping Specifications. Walworth Mfg. Co., 1915, Boston, Mass.

FORSTALL, WALTON. — The Installation of Cast Iron Street Mains. 121 pp. The Chemical Publishing Co., Easton, Pa. 1913.

FOSTER, E. H. — Flow of Superheated Steam in Pipes. Am. Soc. M. E. Transactions, Vol. 29, p. 247.

FRIEND, J. NEWTON. — The Corrosion of Iron and Steel. Longmans, Green Co., N. Y. 1911.

GARRETT, JESSE. — Making Cast Iron Pipe. Journal of N. E. Waterworks Association, Sept., 1896.

GERHARD, W. P. — Gas Piping and Gas Lighting. 306 pp. $3.00. McGraw-Hill Pub. Co., N. Y. 1908.

GIBSON, A. H. — Water Hammer in Hydraulic Pipe Lines. 60 pp. ill. D. Van Nostrand Co., N. Y. 1909.

GUILLAUME, M. — Table, Determination of Pressure Fall in Steam Piping. Journal Am. Soc. M. E., 1914, p. 0129.

HARRISON SAFETY BOILER WORKS. — Philadelphia, Pa. "The Exhaust Steam Heating Encyclopedia," Bulletins and Catalogs, Cochrane Heaters, Separators, Multiport Valves, etc.

HAWLEY, W. C. — Wooden Stave Pipe. 18 pp. ill. Engineers' Society of Western Pennsylvania, Pittsburgh, Pa. Mar. 21, 1905.

HERSCHEL, CLEMENS. — 115 Experiments on the Carrying Capacity of Large, Riveted, Metal Conduits. 122 to 130 pp. J. Wiley & Sons, N. Y. 1897.

HILLS, H. F. — Gas and Gas Fittings. 243 pp. ill. Whittaker & Co., N. Y. 1902.

HOLE, WALTER. — The Distribution of Gas. 837 pp. ill. $7.50. J. Allen & Co., London, 1912.

HOLLIS, I. N. — Cast Iron Fittings for Superheated Steam. Am. Soc. M. E. Transactions, Vol. 31, p. 989.

HOWE, H. M. — The Relative Corrosion of Steel and Wrought Iron Tubing. Am. Soc. for Testing Materials. Vol. 8.

HUBBARD, CHAS. I. — Heating and Ventilation. 213 pp. American Technical Society. Chicago, Ill.

HUTTON, WILLIAM. — Hot Water Supply and Kitchen Boiler Connections, etc. 211 pp. ill. $1.50. David Williams Co., N. Y. 1913.

JAYNE, STEPHEN O. — Wood Pipe for Conveying Water for Irrigation. 40 pp. U. S. Dept. of Agriculture Bulletin No. 155. Government Printing Office, Washington, D. C. 1914.

KELLOG, M. W. — Pipe, Fittings, Valves, Joints, Gaskets for Superheated Steam. Am. Soc. M. E. Transactions, Vol. 29, p. 355.

KENT, WM. — The Mechanical Engineer's Pocket-Book. $5.00. John Wiley & Sons, N. Y.

LEWIS, W. K. — The Flow of Viscous Liquids Through Pipes. The Journal of Industrial and Engineering Chemistry, July, 1916.

LOVEKIN, S. D. — Joints for High Pressure Superheated Steam or Hydraulic Work are described in the American Machinist, June 8, 1905.

MACHINERY DATA SHEET BOOK No. 12. — Pipe and Pipe Fittings. 44 pp. ill. $0.25. The Industrial Press, N. Y. 1910.

MACHINERY REFERENCE SERIES No. 72. — Pumps and Condensers, Steam and Water Piping. 48 pp. ill. $0.25. The Industrial Press, N. Y. 1911.

MANN, A. S. — Cast Iron Valves and Fittings for Superheated Steam. Am. Soc. M. E. Transactions, Vol. 31, p. 1003.

MARKS, LIONEL S. — Mechanical Engineers' Handbook. 1836 pp. $5.00. McGraw-Hill Book Co., N. Y. 1916.

McMILLAN, L. B. — The Heat Insulating Properties of Commercial Steam Pipe Coverings. Journal of Am. Soc. M. E., Jan. 1916.

METER CONNECTIONS. — Report of Committee of American Gas Institute, N. Y. 1916.

MILLER, E. F. — The Effect of Superheated Steam on the Strength of Cast Iron, Gun Iron, and Steel. Am. Soc. M. E. Transactions, Vol. 31, p. 998.
The Flow of Superheated Ammonia Gas in Pipes. Am. Soc. Refrig. Eng'rs Journal, Sept. 1916.

MORRIS, WILLIAM L. — Steam Power Plant Piping. 490 pp. ill. $5.00. McGraw-Hill Book Co., N. Y. 1909.

"NATIONAL" BULLETINS NOS. 1 TO 24. — National Tube Co., Pittsburgh, Pa.

NATIONAL TUBE COMPANY, BOOK OF STANDARDS. — 559 pp. $2.00. National Tube Co., Pittsburgh, Pa.

PEABODY, ERNEST H. — Oil Fuel. Paper No. 214, Trans. International Engineering Congress, 1915. The Neal Pub. Co., San Francisco, Cal.

PIPING. — Practical Engineer, Jan. 1, 1917.

PIPING FOR STEAM GENERATING PLANTS FROM A SAFETY POINT OF VIEW. — The Travelers' Standard, Vol. IV, No. 8.

PRESTON, ARTHUR C. — Experiments on the Flow of Oil in Pipes. Journal of Engineering of the University of Colorado, Dec. 1915.

PLUMBING & GAS FITTINGS. — Prepared for students of the International Correspondence Schools. The Colliery Engineer Co., Scranton, Pa. 1897.

SANG, A. — The Corrosion of Iron and Steel. McGraw-Hill Book Co., N. Y. 1910.

Specifications for Cast Iron Soil Pipe and Fittings. 31 pp. Hitzelberger, Tietenberg & Co., N. Y. 1915.

SCOBEY, FRED C. — The Flow of Water in Wood-Stave Pipe. 96 pp. U. S. Dept. of Agriculture Bulletin No. 376. Government Printing Office, 1916. Washington, D. C.

SNOW, WILLIAM, G. — Pipe Fitting Charts. 285 pp. ill. $1.50. David Williams Co., N. Y. 1912.

STANDARD PIPE AND PIPE THREADS. — Report of Committee. Am. Soc. M. E. Transactions. Vol. 7, pp. 20, 414; Vol. 8, p. 29.

STANDARD SPECIFICATIONS. — Am. Soc. for Testing Materials. Edgar Warburg, Sec'y Treas., Philadelphia, Pa.

A 53-15. For Welded Steel and Wrought Pipe.
A 44-04. For Cast Iron Pipe and Special Fittings.

STANDARDIZATION OF SPECIAL THREADS FOR FIXTURES AND FITTINGS (Straight Threads). — Report of Committee of Am. Soc. M. E. Trans. Vol. 37, p. 1263.

STANLEY, W. E. — Loss of Head in Pipes, Bends, Valves and Other Fittings. The Purdue Engineering Review, May, 1916.

STEWART, R. T. — Strength of Steel Tubes, Pipes and Cylinders under Internal Fluid Pressure. Am. Soc. M. E. Transactions, Vol. 34.

WALKER, W. H. — "The Relative Corrosion of Iron and Steel Water Pipes." N. E. Water Works Association, Boston, Dec. 1911.

WEHRLE, GEORGE. — Instructions for Gas Company Fitters. The Gas Age. An Extensive Series of Articles beginning Sept., 1916.

WESTON, E. B. — Tables Showing the Loss of Head Due to Friction of Water in Pipes. 170 pp. D. Van Nostrand Co., N. Y. 1896.

Among the technical magazines which contain much information on piping the following may be mentioned.

Compressed Air Magazine.
Engineering News, N. Y.
The Gas Age, N. Y.
Journal of the A. S. M. E.
Journal of the N. E. Waterworks Association.
Power, N. Y.
Practical Engineer, Chicago.
The Valve World, Chicago.

APPENDIX

The drawings shown on Plates 1 to 8 inclusive are re-drawn for reproduction from piping drawings prepared by Stone & Webster Engineering Corporation for a steam power plant (Cannon Street Station) which they are constructing for the New Bedford Gas & Edison Light Company, New Bedford, Massachusetts. A brief description of the plant is contained in The Walworth Log for December, 1916, which says that it is, perhaps, the last word in every detail as regards efficiency and low cost of operation, and continues: "The coal is brought to the company's wharf in barges, transferred by an electric unloading tower through the coal crusher into storage, only crushed coal being stored. It is transferred from storage by locomotive crane and dump cars into hoppers at the east end of the station; from here by skip chutes to bunker storage at end of firing aisle.

"From bunker storage to automatic stokers the coal is transferred by a traveling coal weigher, same having two compartments, one for north and the other for south boilers. By the use of bunkers and traveling ash cars the ashes are removed and disposed of in a correspondingly modern way. By the use of force draft and Babcock & Wilcox boilers they are able to meet peak loads with a liberal boiler overload. The steam leads and mains are figured to provide enough steam to meet any emergency which may arise." All of the high pressure piping, and most of the low pressure work in this station was furnished by the Walworth Manufacturing Company.

On the original drawings all figures and lettering are made large and very distinct. The large reduction necessary for reproduction has of course caused a loss in the matter of clearness. A great deal of valuable information in connection with the preparation of piping drawings can be obtained by a careful study of these plates. The completeness of the notes, descriptions of valves and special fittings, old and new material, location of centre lines for present and future apparatus, together with the location of building features should be noted. The grade lines specified on the elevations and the location of the north point on the different plans make comparisons easy.

These drawings are considered typical for modern plants operating at about 200 pounds pressure.

Plates 1 and 2 show the main steam pipe lines in plan and elevation. Expansion is cared for by bends and loops. Connections from the boilers to the 12 inch header are made by 6 inch bends. The location of connections for indicating pressure gauge and recording temperature and pressure gauges is indicated on Plate 1.

Plates 3 and 4 give the plan and elevation of the auxiliary exhaust lines.

Plates 5 and 6 show the boiler feed lines in plan and elevation. Note the enlarged detail for the connections at the Bailey Meter.

Plate 7 gives the plan and elevation for the boiler blow-off lines. Note the location of the valves.

Plate 8 shows the plan and elevation for the heater suction and city water lines.

For the use of these valuable drawings the author is indebted to the Stone & Webster Engineering Corporation, who were kind enough to supply them for this purpose.

INDEX

Abendroth & Root, spiral riveted pipe, 22

Air, equivalent volumes of free, 244; piping, 237, 339, 340; weight of, 239

Air lift pumping system, 244; well pipe sizes, 246

Aluminum Co. of America, 284

Aluminum piping, 284; sizes and weights, 287

American District Steam Co., 216, 219, 220, 223

American Gas Institute, 251

American Metal Hose Co., 284

American pipe threads, 35

American Radiator Co., 203, 208

Am. Soc. Mech. Engrs., 13, 18, 37, 134, 138, 140, 289, 305, 314

Am. Soc. Test Materials, 13

American Spiral Pipe Works, 25

American standard flanged fittings, 58-70

Ammonia fittings, 71

Apparatus, conventional representation, 311

Armstrong Cork & Insulation Co., 295, 299

Asphalted riveted pipe, 22

Atwood line weld, 76

Auld Co., 125, 128

Automatic valves, Crane-Erwood, 121; Fisher exhaust relief, 132; Foster, 117

Babcock's formula, 143

Back pressure valves, 130

Baldwin, Wm. J., 43

Bamboo tubes, 1

Barlow's formula, 20

Barometric condenser, 183; piping for, 184

Bell and spigot joint, 89

Bending pipe, 281; machine, 282

Benjamin, C. H., 13

Bends, pipe, 275-281; dimensions of, 275

Blake & Knowles Pump Works, 180, 185

Blow-off piping, 169, 337, 343; tanks, 170, 171.

Blow-off valves, 114; arrangement of, 169

Boiler feed piping, 232; specifications, 342, 344, 345

Boiler stop valves, 117

Boiler tubes, 287

Bolt circles and drilling, 79

Bolted socket joint, 89

Books and references, 347

Brackets, 273; dimensions of, 275

Brass, pipe, 29; fittings, 54; uses of, 8

Brass tubing, 285

Bregowsky, I. M., 146

Bridgeport Brass Co., 228

Briggs, Robert, 35

Briggs standard, 2

British pipe threads, 42

British standard flanges and fittings, 72

Bronze, gun, 10

Bull head tees, 60

Burhorn, Edwin, 27

Bursting pressures of, cylinders, 13; flanged fittings, 57; wrought pipe, 18, 21

Bushings, 47

Butterfly valves, 114

Butt weld pipe, 6
By-pass valves, 103

Caps, 47
Casing, wood, 297
Casting alloys, U. S. Navy, Bureau of Steam Engineering, 9
Cast iron, bosses, 312; cylinder tests, 13
Cast iron pipe, Am. std., 64, 65; dimensions of hub and spigot, 7, 13, 15; fittings, 49; flange ends, 7; formulae for, 12; joints, 96; plain, 16; uses of, 7; weights of hub and spigot, 14; weight of plain, 16
Cast steel fittings, 71
Cast steel screwed fittings, 57
Central station heating, 217; condensation meter, 225; interior piping, 224
Chadwick-Boston Co., 30
Chasers, number of, 40
Check valves, 111; hydraulic, 235
Clark, Walter R., 228
Clearance, 40
Closed heater piping, 190
Cochrane steam-stack and cut-out valve, 193
Coils, 316; drawings of, 317
Cold pipes, coverings for, 296
Color system, 288
Compressed air piping, 237
Compressed air transmission tables, 238, 240–243
Condensers, 176
Conductivity chart for gas pipes, 249
Conduit, split tile, 299
Connections, boiler to header, 152; exhaust main, 174; gas engine, 256; gas meter, 252–254; hot water radiator, 208; lubricator, 267; special, 88; steam radiator, 203
Converse joints, 90
Copper pipe, 8, 29; flanges for, 93; method of manufacture, 8; uses of, 8
Copper tubing, 285
Corrosion of pipe, 2

Couplings, 44, 45
Coverings, pipe, 289; forms of, 296; tests on, 289; thicknesses of, 295
Crane Co., 53, 54, 57, 78, 82, 103, 104, 121, 146
Crimped end, 94
Crosby Steam Gage & Valve Co., 100
Crosses, 46
Cylinder tests, 13

Detail drawing, 308
Dimensioning drawings, 312
Dimensions of, Am. std. flanged fittings, 61–70; boiler tubes, 287; brass fittings, 55; British pipe threads, 42; British std. flanged fittings, 72–75; cast iron bosses, 312; cast iron screwed fittings, 50–54; Converse lock joint pipe, 91; expansion joints, 278
Dimensions of flanges, standard weight Walmanco, 85; extra heavy Walmanco, 86; extra heavy Cranelap, 84; extra heavy shrunk and peened, 86; extra heavy tongued and grooved, 87; extra heavy male and female, 88
Dimensions of globe and gate valves, 104–111; hub and spigot pipe, 15; lead pipe, 32; malleable iron fittings, 56, 57; Matheson joint pipe, 92; pipe, 11; pipe bends, 275, 281; pipe brackets, 275; pipe saddles, 283; riveted pipe flanges, 95; screwed unions, 78; spiral riveted pipe, 22–26; straight riveted pipe, 27; Universal C. I. pipe, 97; Whitworth pipe threads, 43
Dopes, pipe, 270
Double extra strong wrought pipe, 3
Drainage, 161
Drainage fittings, 167
Draining exhaust pipe, 173
Drawings, conventional representation, 307; dimensioning, 312; erection, 306; flanged, 315; gas piping, 260; isometric, 319–327; oblique, 328; oil piping, 266; pictorial, 319–

INDEX

328; piping, 306; single plane, 318, 320; sketching, 316; steam piping, 309; steam power plant, 351
Drilling for bolt circles, 79
Drip and blow-off piping, 161
Drip piping, 339
Drip pockets, 163
Drips from steam cylinders, 167

Eductor condenser, 185; piping for, 186
Efficiency of pipe coverings, 290
Elbows, 46, 59
Emergency stop valves, 118, 121
Engineering Standards Committee, 73
Engines, steam lines for, 154; exhaust from, 173
English pipe, 22; formula for, 22
Equalization of pipes, formula for, 144; tables, standard wrought pipe, 147; extra strong, 148; double extra, 149
Equivalent lengths of pipe, 90° elbow, 145; elbow, tee, etc., 230
Erecting, specifications, 341
Erection drawings, 306; pipe, 269
Evans, H. W., 341
Exhaust heads, 174
Exhaust piping, 172; method of draining, 173; specifications, 338, 342
Exhaust relief valves, 132
Expansion, 274
Expansion bends, 275, 276; radii for, 280; thickness of pipe, 281; values, 280
Expansion chart, 279
Expansion joints, 96, 277; exhaust pipe, 176
Extra heavy Am. std. C. I. pipe, 65; flanged fittings, 63
Extra strong wrought pipe, 3; dimensions of, 18; weight of, 18

Farnsworth Mfg. Co., 64
Feed piping, 232
Feed water heaters, 188

Feed water purifier, live steam, 157
Field riveted joint, 89
Filling-in piece, 315
Fisher Governor Co., 129, 131, 132
Fisher reducing valve, 126
Fittings, flanged, Am. std. C. I., 58–70; ammonia, 71; British std., 72–75; conventional representations, 310; distance pipe enters, 313; drainage, 167; form for listing, 307; gas, 247; hydraulic, 233; oil pipe, 264; riveted steel plate, 172; screwed, 44–57; sizes of water supply, 233
Flanged fittings, strength of, 57
Flanged unions, 79
Flanges, Am. std., 65–67; British std., 72, 73; dimensions of, 85; drilling, 315; for copper pipe, 93; facing, 80; male and female, 81; raised face, 80; riveted, 89; straight face, 80; tongued and grooved, 81; with follower rings, 89
Flow of water in pipes, 227; chart, 229
Foreign pipe threads, 43
Formula, Barlow's, 20
Formula for, air lift pumping system, 245; cast iron pipe, 12; compressed air transmission, 238; copper pipe, 29; English pipe, 22; flow of water, 227, 228; gas pipes, 248; lead pipe, 30; safety valves, 136; spiral riveted pipe, 22; steam pipes, 143, 144; strength of pipe, 11; wooden stave pipe, 34
Forstall, Walton, 255
Foster Engineering Co., 117, 131
Fuel piping, oil, 267; U. S. Navy, 268

Gages, pipe thread, 37; steam, 160
Gas engine connections, 256
Gas fitting, 246
Gaskets, 271; ammonia, 71
Gas meters, 250; connecting, 252; sizes of, 251
Gas pipe, sizes of, 247, 257; testing, 250

INDEX

Gas piping, arms, 261; drawings, 260; location of, 247; obstructions and joining, 255; outlets, 256; pressure tests, 255; schedule, 257; slope of, 255; specifications, 255; stems, 261

Gate valves, 99–103; standard pressures and dimensions, 104–111; strength of, 104

Giesecke, F. E., 228

Globe valves, 99; standard pressures and dimensions, 104–111

Governors, pump, 128

Gravity pipe lines, 226

Gun bronze, 10

Handling pipe, 269

Header, live steam, 152

Heads and pressures of water, 227

Heaters, feed water, 188; piping for, 199

Heating systems, piping for, 201

High temperature, effect of, 146, 150

Hirshfield, C. F., 138

Homestead Valve Mfg. Co., 116

Hoppes Mfg. Co., 157, 163, 175, 197

Hose, metal, 284

Hot water heating, 206; down feed system, 208; forced circulation system, 208; mains and risers, 210; open tank system, 207; pipe sizes, 209

Hot water suction pipe, 232

Hub and spigot pipe, 13; weights of, 14; dimensions of, 15

Hydraulic pipe and fittings, 233

Hydraulic stop valves, 236

Ingersoll-Rand Co., 238, 244

Injector piping, 156

Insulation, 289; for water stand pipe, 304

Interlock welded necks, 76

Interior water piping, 233

Int'l Ass'n for Test. Mat'ls, 146

Isometric drawing, 319–327

Jayne, S. O., 34

Jenkins Bros., 100

Jet condensers, 180; piping for, 181

Joints, expansion, 277; flanged for steel pipe, 81; pipe, 76; specifications, 340; welded, 76

Kewanee flanged union, 79

Lap weld furnace, 4

Lap welding rolls, 5

Lap weld process, 3

Laterals, 59

Lead pipe, formula for, 30; history, 1; joints, 93; manufacture of, 30; uses of, 8

Lip angle, 39

Live steam header, 152

Location of valves, 113

Long bends, British std., 75

Long, H. E., 217

Long radius fittings, 59

Lubricator connections, 267

Lunkenheimer Co., 54, 101

Main header, pipe lines from, 154

Malleable iron fittings, 55

Mason Regulator Co., 123

Materials for valves, 99; specifications, 334; strength of, 9; symbols for, 314

Matheson joints, 90

McMillan, L. B., 289

Metal hose, 284

Meter cock, 247

Meters, gas, 250; steam condensation, 225

Mill tests of wrought pipe, 20

National Pipe Bending Co., 192

National Tube Co., 20, 38, 51, 56, 78, 102

New Bedford Gas and Edison Light Co., 351

Nipples, 47, 48

Nozzles, 282

Nut, pipe, 47

Oblique drawing, 328
Oil fuel piping, 267; U. S. Navy, 268
Oil piping, 339; drawing, 266; fittings, 264; for lubrication, 263; Phenix system, 264; Richardson system, 263
Open heater piping, 198
Operation of valves, 112
Outlets, gas, 256
Out-of-doors piping, 301
Outside diameter wrought pipe, 3; weight of, 19

Philadelphia Gas Works, 255
Pictorial drawing, 319, 328
Pilot valve, 120
Pipe coverings, forms of, 287, 296
Pipe joints, 76–83
Pipe nut, 47
Pipe saddles, dimensions of, 283
Pipe sizes. See Dimensions.
Pipe threading, 38; machine, 39
Pipe threads, 35; foreign, 43; symbols, plan and section, 316; table of standard, 36; Whitworth, 41
Pipe tools, 38–41
Piping drawings, 306
Piping for various liquids, 329
Piping schedule, service, pipe, fittings, valves, gaskets, flanges, 330
Pittsburgh Valve, Foundry and Construction Co., 76
Plain cast iron pipe, 16
Plan of gas piping, 260
Plugs, 47
Plug valves, 116
Pohle, E. S., 244
Pope, Henry G., 299
Pop safety valves, 132; installation of, 134
Pottery tubes, 1
Power plant piping, 330
Power plant piping drawings, 351
Preference heater, 199
Pressures, bursting, 18
Pump, and receiver, 168; and surface condenser, 179; discharge piping, 231; governors, 128; suction piping, 228; well, 231

Pumping system, air lift, 244; well pipe sizes, 246
Pumps, exhaust from, 173; gas proving, 250; steam lines for, 154
Purifier, feed water, 157; method of piping, 158
Radiator connections, hot water, 208; steam, 203
Radiators, pipe sizes, 204, 209, 212
Reducing elbows, 58
Reducing fittings, 54; Am. std., 67–69
Reducing valves, 122; sizes of, 127
Reference books, 347
Relief valves, 132, 232
Representation, conventional, 307; fittings, 310; apparatus, 311
Return trap, 163; setting for, 167
Richardson-Phenix Co., 263
Riveted pipe, joints, 94; spiral, 22; straight, 27
Roller support, 301
Russell, James, 1

Saddles, 283
Safety valves, 132; hydraulic, 235; requirements, 134
Schedule, standard piping, 330; gas piping, 257
Schutte & Koerting Co., 185, 235
Scott, J. B., 140
Screwed fittings, 44; cast iron, 50–54; malleable, 56, 57; reducing, 54; X cast steel, 57
Screwed unions, 77, 78
Sections, conventional, 314
Separators, 161
Service cock, 247
Short bends and tees, British std., 74
Side outlet elbows, 60
Side outlet tees, 60
Sizes of, gas engine pipes, 256; gas pipes, 247, 257
Sizes of pipes. See Dimensions.
Sizes of, safety valves, 135; steam pipes, 143; tile conduit, 300; water supply fittings, 233
Sketching, 316
Slip joint, 94

Special connections, 88
Special valves, 114
Specifications, 329
Spiral riveted pipe, 22-26
Spring, L. W., 146
Standard pipe threads, 35, 36
Standard valves, 98
Standard wrought pipe, dimensions of, 17; weight of, 17
Steam cylinders, drips from, 167
Steam gages, piping of, 160
Steam heating, atmospheric system, 216, 223; central station, 217; down flow system, 203; exhaust, 211; mains and risers, 206; one pipe circuit system, 203; pipe sizes, 203; two pipe system, 204; Webster vacuum system, 211-215
Steam line, out door, 301
Steam loop, 156
Steam mains, underground, 218
Steam pipe casing, 218
Steam piping, 137; drawing, 309; specifications, 335, 341, 343, 345
Steam power piping, direct system, 138; duplicate system, 142; header system, 137; ring system, 140
Steam power plant piping drawings, 351
Steam, superheated, 146
Steam traps, 163
Steam turbine, and eductor condenser, 187; and jet condenser, 182; and surface condenser, 179
Steam velocity, 142
Steel pipe, manufacture of, 3; uses of, 2
Step bearing piping, 340
Stewart, Reid T., 18
Stone & Webster Engineering Corporation, 352; standard specifications, 334
Stove cock, 247
Straight seam riveted pipe, 27
Strength of, gate valves, 104; pipe, 11; piping materials, 9
Suction piping, 228; arrangement of, 231
Superheated steam, 146

Supports, pipe, 273; for insulated pipe, 296; roller, 301; specifications, 340; thin pipe, 284
Surface condensers, 176; piping for, 177

Tanks, blow-off, 170, 171
Tap drills diameter of, 36
Taper of pipe threads, 35
Taylor's spiral riveted pipe, 25, 26
Teague, W. E., 305
Tees, 46; specification of, 47
Testing gas pipes, 250
Testing, specifications, 341
Tests, cylinder, 13; pipe covering, 290
Thermometer well, 159; placing of, 160
Thickness of, Am. std. C. I. pipe, 64, 65; wrought pipe, 17
Thin pipe, supporting, 284
Thoroughfare heater, 199
Tile conduit, 299; sizes of, 300
Transmission, compressed air, 238
Tubes, bamboo, 1; brass and copper, 285; boiler, 287; wrought iron, history of, 1
Tubing, aluminum, 284; sizes, 286;
Twin elbows, 60

Underdrainage, 220
Underground piping, insulation of, 297
Underground steam mains, 218; installation of, 221
Unions, flanged, 79; hydraulic flanged, 234; screwed, 77, 78
United Gas Improvement Co., 254
U. S. Navy, Bureau of Steam Engineering, casting alloys, 9
Universal C. I. joint, 96

Vacuum exhaust pipes, 175
Valve seats, 99
Valves (see type wanted)
Valves, care of, 272; hydraulic, 235; location of, 113; operation of, 112; standard, 98; special, 114
Vibration, 273

INDEX

Walmanco flanges, 85
Walworth Mfg. Co., 50, 54, 57, 71, 82, 100, 115, 341, 351
Warren Webster & Co., 212
Water, equivalent pressures and heads, 227; flow in pipes, 227; chart, 229
Water column piping, 159
Water piping, 226; coverings for, 295; interior, 233; specifications, 336, 338, 343
Water stand pipe, insulation for, 304
Watson-Stillman Co., 146, 234
Webster heaters, piping for, 195
Wehrle, George, 248
Weight of, brass pipe, 29; copper pipe, 29; hub and spigot pipe, 16; lead pipe, 31; O. D. wrought pipe, 19; plain cast iron pipe, 16; spiral riveted pipe, 22-26; straight riveted pipe, 27
Welded joints, 76

Welding rolls, 5
Whitworth pipe threads, 41
Wolfang, W. H., 299
Wood, Albert C., 303
Wood casing, 297
Wooden stave pipe, 33
Wood pipe, 1
Wrought pipe, bursting pressures of, 18; dimensions of, 17; history, 1; lengths of, 18; tests of, 20; uses of, 2; weight of, 17; weight of O. D., 19; X, dimensions of, 18; X, weight of, 18; XX, dimensions of, 19; XX, weight of, 19
Wyckoff, A. & Son Co., 297

X wrought pipe. *See* Extra strong wrought pipe.
XX wrought pipe. *See* Double extra strong wrought pipe.

Yarnell-Waring Co., 115

www.ingramcontent.com/pod-product-compliance
Lightning Source LLC
Chambersburg PA
CBHW082321220526
45470CB00008B/2366